The OEE Primer

Understanding Overall Equipment Effectiveness, Reliability, and Maintainability

The OEE Primer

Understanding Overall Equipment Effectiveness, Reliability, and Maintainability

D.H. Stamatis

CRC Press
Taylor & Francis Group
Boca Raton London New York

CRC Press is an imprint of the
Taylor & Francis Group, an **informa** business

A PRODUCTIVITY PRESS BOOK

Productivity Press
Taylor & Francis Group
270 Madison Avenue
New York, NY 10016

© 2010 by Taylor and Francis Group, LLC
Productivity Press is an imprint of Taylor & Francis Group, an Informa business

No claim to original U.S. Government works

Printed in the United States of America on acid-free paper
10 9 8 7 6 5 4 3 2 1

International Standard Book Number: 978-1-4398-1406-2 (Paperback)

This book contains information obtained from authentic and highly regarded sources. Reasonable efforts have been made to publish reliable data and information, but the author and publisher cannot assume responsibility for the validity of all materials or the consequences of their use. The authors and publishers have attempted to trace the copyright holders of all material reproduced in this publication and apologize to copyright holders if permission to publish in this form has not been obtained. If any copyright material has not been acknowledged please write and let us know so we may rectify in any future reprint.

Except as permitted under U.S. Copyright Law, no part of this book may be reprinted, reproduced, transmitted, or utilized in any form by any electronic, mechanical, or other means, now known or hereafter invented, including photocopying, microfilming, and recording, or in any information storage or retrieval system, without written permission from the publishers.

For permission to photocopy or use material electronically from this work, please access www.copyright.com (http://www.copyright.com/) or contact the Copyright Clearance Center, Inc. (CCC), 222 Rosewood Drive, Danvers, MA 01923, 978-750-8400. CCC is a not-for-profit organization that provides licenses and registration for a variety of users. For organizations that have been granted a photocopy license by the CCC, a separate system of payment has been arranged.

Trademark Notice: Product or corporate names may be trademarks or registered trademarks, and are used only for identification and explanation without intent to infringe.

Library of Congress Cataloging-in-Publication Data

Stamatis, D. H., 1947-
   The OEE primer : understanding overall equipment effectiveness, reliability, and maintainability / D.H. Stamatis.
      p. cm.
   Includes bibliographical references and index.
   ISBN 978-1-4398-1406-2 (pbk. : alk. paper)
   1. Reliability (Engineering) 2. Maintainability (Engineering) 3. Machinery--Maintenance and repair. 4. Machine parts--Failures. 5. Service life (Engineering) I. Title.

TS173.S695 2010
621.8'16--dc22
                                      2009049404

Visit the Taylor & Francis Web site at
http://www.taylorandfrancis.com

and the Productivity Press Web site at
http://www.productivitypress.com

Dedicated to my friend A. A. A. ("Tasso")

Contents

Acknowledgments .... **xxiii**

Introduction .... **xxv**

1 Total Preventive Maintenance 1
    A Brief History of TPM 2
    The Goals and Activities of Total Preventive Maintenance 3
    An Overview of the Concept of Lean Manufacturing 5
        How to Make Your Organization Lean 6
        Map Your Processes 7
        Constructing a Value Stream Map 8
        How to Map the Value Stream of a Process 10
            Step 1: Create a List of Products and Group Them
            in Families 10
            Step 2: Determine Which Product, Machine, or Service Is
            Considered Primary 10
            Step 3: Document the Steps of the Process 11
        Develop Both Current-State and Future-State Maps 11
        Detailed Process Mapping 13
        Problems with Lean 15
    The "5S" Methodology of Organizing the Work Area 15
        What 5S Is 16
        How to Implement 5S in Your Organization 17
        Benefits of 5S 17
    Developing a Visual Factory to Make Decisions Quickly 18
        The Visual Factory Requires Standardization 19
    Summary 20

2 OEE: Understanding Loss of Effectiveness 21
OEE Comprises Six Key Metrics 22
  Metric #1: OEE (Overall Equipment Effectiveness) 24
  Metric #2: TEEP (Total Effective Equipment Performance) 24
  Metric #3: The Loading Portion of the TEEP Metric 25
  Metric #4: Availability 25
  Metric #5: Performance 26
  Metric #6: Quality 26
Observe and Monitor Time and Other Considerations 27
  Observe and Monitor Equipment Uptime 27
  Observe and Monitor Equipment Downtime 27
  Determining the Ideal Cycle Time 28
  Quantifying Total Defects and the Cost of Quality 28
  Process Line Considerations 29
  Process Line Quality Issues 29
  What to Do with All This Information 29
  Measure Equipment Availability 29
  Measure Equipment Setup Time 30
  Identify Other Production Losses 30
  Measure Equipment Performance 31
OEE and Capacity 32
Using OEE in the Automotive Industry 38
Selecting the Appropriate Process for Demonstrated OEE 42
  How to Improve Demonstrated OEE 43
Reducing Required OEE 44
Summary 45

3 Measuring OEE: A Traditional Simple Approach 47
Collecting Data about OEE 47
  Clarify Your Purpose for Collecting Data 48
Data Collection by Sampling 49
  What Is Sampling? 50
  When Is Sampling Necessary? 50
  Why Collect Data? 50
Basic Tools for Evaluating Data 51
  Brainstorming to Gather Ideas 51
    Procedures for Brainstorming 52
    Rules for Brainstorming 52

Using Check Sheets to Collect Data 53
    Designing a Check Sheet.. 53
    Check Sheet for Measurable (Continuous) Data 55
    Check Sheet for Attribute Counting 56
Using Histograms to Chart Data 57
    Why Histograms Are Useful 58
    How to Construct a Histogram 58
    What a Histogram Shows 59
Using Pareto Diagrams to Chart Data 59
    How to Use Pareto Diagrams 60
    How to Create a Pareto Diagram 61
Using Cause-and-Effect Diagrams to Display Data 62
    Types of CE Relationships 63
    Common Mistakes in Analyzing the Cause 64
    Constructing a CE Diagram 65
    Types of CE Diagrams 66
Using Scatter Diagrams to Show Data 67
    Reading Scatter Diagrams 69
    Scatter Diagrams Showing Positive Correlation 69
    Scatter Diagrams Showing Possible Positive Correlation 69
    Scatter Diagrams Showing No Correlation 70
    Scatter Diagrams Showing Possible Negative
    Correlation 70
    Scatter Diagrams Showing Negative Correlation 71
Using Run Charts to Display Data 71
Using Is-Is Not Analysis to Compare Data 72
Using 5-Why Analysis to Identify the Root Cause of a Problem 75
Using Process Flow Diagrams to Visualize Processes 78
Using Kanban to Collect and Analyze Equipment Repair Data 79
Using Kaizen to Improve OEE 79
Using Control Charts to Show Variations in Processes 81
    Variable Data 85
    Attribute Data 85
    Steps for Control Charting a Production Process 86
    Interpreting Control Charts 87
Summary 88

4 Measuring OEE: The Reliability Approach 89

Calculating Reliability Point Measurement 90
Measuring MTBE (Mean Time between Events) 91
Measuring MTBF (Mean Time between Failures) 92
Data Collection to Monitor Equipment Performance 93
    Method #1: Manual Recording of Data 95
        R&M Continuous Improvement Activities 95
        Structure of a Manual Recording 96
        Failure Report Form 96
    Method #2: Direct Machine Monitoring (Electronic) 96
        Direct Machine Monitoring 96
        Maintenance System Data 98
        Field History/Service Reports 99
        Component Supplier Failure Data 99
FRACAS: Failure Reporting, Analysis, and Corrective Action System 100
    Step 1: Problem Investigation Responsibilities of Machinery & Equipment OEMs during Manufacturing and Runoff 100
    Step 2: Failure Investigation Responsibility of Component Manufacturer or Supplier 101
    Step 3: Problem Responsibility of Customer's Manufacturing R&M Team 102
Summary 103

5 Improving OEE on Existing Machinery 105

Using Pareto Analysis to Identify What Needs Improvement Most 106
Using the 8-Discipline (8D) Process to Resolve Machinery Failures or Problems 106
Implementing a Quick Changeover to Improve Process Flexibility 110
    Benefits of Quick Changeover 119
    Roles and Responsibilities (R&R) 120
        Manufacturing Tech 120
        Coordinator 120
        Advisor 121
        Maintenance Supervisor 121
        Manufacturing Planning Specialist (MPS) 121
        Production Superintendent 122

> Maintenance Superintendent 122
> Business Unit Manager 122
> Plant Manager 122
> Examples of Changeover 122
Implementing Mistake Proofing to Prevent or Mitigate Errors 124
> Six Mistake-Proofing Principles 125
> When to Conduct Mistake Proofing 126
> How to Minimize Human Errors 126
> How to Prioritize Mistake-Proofing Opportunities 127
Using P-M Analysis to Analyze and Eliminate Chronic
Machinery Problems 127
Using Finite Element Analysis (FEA) to Improve
Product Refinement 128
> Types of Engineering Analysis Using FEA 130
Using Failure Mode and Effect Analysis (FMEA) to Predict and
Prevent Machine Failures 131
Using Fault Tree Analysis to Show the Causes of Machine Failures 131
Using Equipment Failure Mode and Effect Analysis (EFMEA) to
Identify Potential Machine Failures and Causes 134
> Components of the Form 137
>> MFMEA Header Information 137
>> System/Subsystem Name Function 138
>> Failure Mode 138
>> Potential Effects 138
>> Severity Ratings 139
>> Classification 140
>> Potential Causes 140
>> Occurrence Ratings 141
>> Current Controls 141
>> Detection Rating 142
>> Risk Priority Number (RPN) 142
>> Recommended Actions 142
>> Date, Responsible Party 143
>> Actions Taken/Revised RPN 143
>> Revised RPN 143
Summary 143

6 Improving OEE on New Machinery: An Overview of Mechanical Reliability 145

Understanding Equipment Design Variables.. 146
    Developing Design Input Requirements.................................... 147
        Equipment under Design 147
        Ideal Function of Equipment 148
        Control Factors.. 148
        Failure States of Equipment.................................... 148
    Implementing Factory Requirements.. 149
        Jobs per Hour (JPH).. 149
        Required Quality Levels 149
        Duty Cycle Operating Patterns................................ 149
        Operator Attention.. 149
        Maintenance Required.. 150
        Management.. 150
    Conducting an Environmental Analysis before Installing New Equipment.. 150
    Understanding Possible Environmental Failures to Prevent Them from Occurring 151
    The "Top 10 List" of Equipment Failures and Causes.................. 151
    Developing an Equipment Performance Profile.......................... 151
        Design Concept 154
        Design Simplicity 155
        Commonality of Design 156
    Improving Mechanical Reliability by Understanding Mechanical Failures.. 156
        Designing Equipment to Prevent and Maintain against Fatigue or Fracture.. 157
        Preventing and Minimizing Equipment Wear......................... 158
        Methods to Reduce Wear 158
        Preventing and Maintaining against Equipment Corrosion ... 160
        Developing Equipment Safety Margins................................... 160
        Selecting the Proper Materials and Components 169
        Designing for Equipment Maintainability............................... 170
        Visual Factory 172
        Designing Equipment Components to Ensure Accessibility for Maintenance 172
        Design with Maintenance Requirements in Mind 174

Minimize Maintenance Handling 174
Design with Maintenance Tools and Equipment in Mind 175
Design for Removal and Replacement of Components
Requiring Repair 175
Design for Interchangeability and Standardization of Parts ... 175
Review the Working Environment 176
Use R&M Validation/Verification Techniques 176
Conduct a Component Application Review 176
Use Control Point Analysis to Measure Failed
Component Reliability 177
Use the "Bucket of Parts" System to Compile Failure
Reports 177
Collect Failure Data by Tracking Spare Part Utilization 178
Summary 178

7 Improving OEE on New Machinery: An Overview of Electrical Reliability 179

Consider the Thermal Properties of Electrical Equipment 179
Understanding the Benefits of Thermal Analysis 181
Conducting a Thermal Analysis 183
    Electrical Design/Safety Margins 185
Derating to Improve Reliability and Availability of
Electrical Components 185
    An Example of Electrical Stress 186
Preventing Problems with Electrical Power Quality 188
Preventing Electrical Failures 188
Selecting the Appropriate Electrical Components
for the Environment 190
Summary 191

8 Improving OEE on New Machinery: Selected Methodologies 193

Using the TRIZ Method—The Theory of Inventive Problem
Solving to Improve Machine Design 194
    Striving for a Higher Level of Innovation 195
    A Step-by-Step Guide to Using the Core Tools of the
    TRIZ Methodology 195
    Using TRIZ in a Design Situation 196

Using Pugh Diagrams to Improve Machine Design 199
Using Geometric Dimensioning and Tolerancing (GD&T) to
Improve Machine Design 200
    An Overview of the Key Concepts of GD&T 201
Using Short-Run Statistical Process Control (SPC) to Improve
Machine Design 202
    Clarifying Misunderstandings of What SPC Is 203
    Understanding the Problems of Using SPC for Short-Run
    Production Cycles 204
Using Measurement System Analysis (MSA) to Improve
Machine Design 204
Using Capability Analysis to Improve Machine Design 206
Using Design of Experiments (DOE) in Reliability Applications to
Improve Machine Design 209
Using Hazard Analysis and Critical Control Points (HACCP) to
Improve Machine Design 210
    Using HALT and HASS Test Processes to Improve
    Machine Design 210
        FAST: Ford's Accelerated Stress Test 211
    PASS: Ford's Production Accelerated Stress Screen 213
Using ROCOF Analysis to Improve Machine Design 215
    Benefits of Using ROCOF Analysis 218
    Cautions for Using ROCOF Plots with Warranty Data 218
Summary 221

9 Reliability Growth 223
Calculating Reliability Growth 224
    Step 1: Collect Data on the Machine and Calculate the
    Cumulative Mean Time between Failures (MTBF) Value
    for the Machine 224
    Step 2: Take the Data and Plot the Data on
    Log–Log Paper 225
    Step 3: Calculate the Slope 225
How the Equipment Supplier Can Help Improve
Machine Reliability 226
The Customer's Responsibility in Improving Machine Reliability 227
How to Implement a Reliability Growth Program 227
    Initiate an R&M Feedback Process 227
    Create a Failure Data Feedback Process Flowchart 228

    Implement a Performance Data Feedback Plan 228
    R&M Information Systems 231
    Preparing Reports on Things Gone Right/Things Gone
    Wrong (TGR/TGW) 231
Reliability 235
    Using a Reliability Allocation Model to Evaluate
    Equipment Design 235
    Using a Series Model to Evaluate Equipment Design 237
    Using a Parallel Model to Evaluate Equipment Design 238
Reasons for Performing Reliability Tests 239
    When to Conduct Reliability Testing 240
    Setting Your Objectives for Reliability Testing 241
Classical Reliability Terms 242
    Calculating MTBF 242
    Determining MTBF Point Estimates and Confidence Intervals 242
    Calculating Equipment Failure Rates 243
        Calculating Failure Rates vis-à-vis the Total Population of
        Equipment Components 244
    Measuring Equipment Maintainability 244
Five Tests for Defining Failures 246
    1. Sudden-Death Testing 246
    2. Accelerated Testing 248
    Constant-Stress Testing 250
    Step-Stress Testing 251
    Progressive-Stress Testing 252
    Using the Inverse Power Law Model 253
    Using the Arrhenius Model 255
Characteristics of a Reliability Demonstration Test 257
    Types of Reliability Demonstration Tests 258
        Attributes Tests 259
        Variables Tests 259
        Fixed-Sample Tests 259
        Sequential Tests 259
    Test Methods 260
    When to Use Each Type of Attributes Tests 260
        Small Populations—Fixed-Sample Test Using the
        Hypergeometric Distribution 260
        Large Population—Fixed-Sample Test Using the
        Binomial Distribution 263

Large Population—Fixed-Sample Test Using the
Poisson Distribution 265
Success Testing 266
Sequential Test Plan for the Binomial Distribution 267
Failure-Truncated Test Plans—Fixed-Sample Test
Using the Exponential Distribution 270
Time-Truncated Test Plans #7—Fixed-Sample Test
Using the Exponential Distribution 271
Weibull and Normal Distributions 273
Sequential Test Plans 273
Exponential Distribution Sequential Test Plans 273
Interference (Tail) Testing 278
Summary 278

10 Maintenance Issues and Concerns 279
Benefits of Improved Equipment Maintainability 279
Key Concepts Pertaining to Maintainability 280
Calculating the Availability of Machinery/Equipment 280
Addressing Failures of Equipment or Machinery 282
Failures during the Infant Mortality Period 282
Failures during the Useful Life Period 283
Failures during the Wearout Period 284
Determining Failure Rates 285
Ensuring Safety in Using Equipment 286
Keeping an Eye on Life-Cycle Costs (LCC) 286
Successful Implementation of Equipment R&M 286
Phase 1: Concept—Establishing System Requirements for
Equipment R&M 287
Phase 2: Developing and Designing for Equipment R&M 288
Phase 3: Building and Installing for Equipment R&M 289
Phase 4: Operation and Support of Equipment R&M 290
Phase 5: Converting and/or Decommissioning Equipment 290
Option #1: Decommission and Replace with a
New Machine 291
Option #2: Rebuild the Equipment to a "Good-as-New"
Condition 291
Option #3: Modernize the Equipment 291
Summary 293

11 Requirements of Phase 1 of Implementing Equipment R&M: The Concept 295
Overview of Phase 1 Functions: Establishing Specs for Reliability and Maintainability (R&M) 295
Developing R&M Equipment Specifications 299
    Step 1: Determining Factory Requirements 299
    Step 2: Develop Current Equipment Baseline 300
    Step 3: Utilize the Information from Steps 1 and 2 to Determine Fit and Areas of Leverage 300
    Step 4: Document Performance Requirements 300
    Step 5: Develop Environmental Specifications 300
    Step 6: Analyze Performance 301
Determine Equipment R&M Goals 303
    Determining Goals for Acceptable Failure Requirements for the Equipment 304
    Determining Goals for Equipment Usage 304
    Determining Maintainability Requirements 304
    Setting Goals for Documenting Failure Definition 304
    Setting Goals for Environmental Considerations 304
Improving Equipment Performance 305
    Evaluating ISO-Capacity 305
    Calculating Available Production Time 305
    Calculating Mean Time between Failures Parameters 306
    Calculating Mean Time to Repair or Replace 307
    Calculating Scheduled Downtime 307
    Calculating the % Starved Time 309
    Calculating the Number of Assists 309
    Evaluating Equipment Conversions 311
    Calculating the 1st Run % 311
Summary 313

12 Requirements of Phase 2 of Implementing Equipment R&M: Development and Design 315
Techniques for Designing Equipment Maintainability 315
    Minimize Maintenance Requirements 316
    Minimize Maintenance Handling Requirements 316
    Design for Interfacing Parts Reliability 316
    Fault Tolerance Design 316
    Maintenance Tools and Equipment 317

- Remove and Replace 317
- Strive for Interchangeability and Standardization of Equipment 317
- Keep in Mind Working Environment Considerations 318
- Customer's Equipment Design Responsibilities 318
- Supplier Equipment Design Responsibilities 320
- Conducting Design Reviews 320
- Designing Equipment Reliability 321
- Creating Parallel Reliability Models 321
- Creating Series Reliability Models 322
  - Allocation of Reliability Goals 323
  - Apportioning the Reliability Model 324
  - Identifying the Subsystem Tree Model 324
- Conducting Equipment Failure Mode and Effect Analysis (EFMEA) 324
  - Who Prepares an EFMEA? 325
  - Who Updates an EFMEA? 326
  - When Is an EFMEA Started? 326
  - When Is an EFMEA Updated? 326
  - When Is an EFMEA Completed? 326
  - When Can an EFMEA Be Discarded? 327
  - Using an EFMEA Form to Analyze Equipment Failures 327
    - Line (1) FMEA Number 327
    - Line (2) Equipment Name 327
    - Line (3) Design Responsibility 327
    - Line (4) Prepared By 327
    - Line (5) Model Line 329
    - Line (6) Review Date 329
    - Line (7) EFMEA Date 329
    - Line (8) Core Team 329
    - Line (9A) Subsystem Name 329
    - Line (9B) Subsystem Function Performance Requirements 330
    - Line (10) Failure Modes 331
    - Line (11) Potential Effects 331
    - Line (12) Severity 332
    - Line (13) Class 332
    - Line (14) Potential Causes 332
    - Line (15) Occurrence 333
    - Line (16) Current Design Controls 334
    - Line (17) Detection 334

Line (18) Risk Priority Number...335
Line (19) Recommended Actions...335
Line (20) Area/Individual Responsible and Completion
Date 338
Line (21) Actions Taken..338
Line (22) Revised RPN 339
Creating Fault Tree Analysis (FTA) Diagrams 339
The Failure/Causal Event Symbol 340
The OR Gate Symbol 340
The AND Gate Symbol 340
The Diamond Symbol 340
The Circle Symbol 341
Hints on Performing Fault Tree Analysis..342
Summary 343

13 Requirements of Phase 3 of Implementing Equipment R&M: Build and Install 345

Selecting Machinery Parts...346
Completing Tolerance Studies 346
Performing Stress Analysis..346
Conducting Reliability Qualification Testing (RQT)..............................346
Collecting Reliability Data at the Supplier's Facility...............................347
Collecting Reliability Data at the Customer's Plant................................347
Performing Root Cause/Failure Analysis of Equipment 347
Eliminating Testing Roadblocks..347
Do Not Test Everything...348
Reduce the Test Time..348
Establish Confidence through Design 348
Perform Simultaneous Engineering 348
Conduct Reliability Testing...348
Common Testing Programs 349
Conduct a 24-Hour Dry Cycle Run..349
Conduct a Vibration Measurements Test 349
Perform a Dimensional Prove-Out Test...350
Conduct a Preliminary Process Capability Study 350
Test for Overall Equipment Effectiveness (OEE).............................350
Perform Reliability Qualification Testing 351
Establishing Existing Equipment's MTBF Value 351
Validating MTBF Parameters 353

RQT Assumptions 354
RQT Test Plan 354
Using RAMGT Software 355
Establishing System Requirements 355
Select Area for Test Focusing 356
Look for Existing Data 357
Look for Increased Sample Size 357
Strive to Simulate Real-World Conditions 358
Harsh Environment 358
Processing Parts 358
Imperfect Operators 359
Accelerated Testing Considerations 359
Automating the Test 359
Acceleration Testing for Subsystems 360
Integrated System Testing 360
Testing Guidelines 361
Subsystem Testing 361
Test Progress Sheets 361
Reliability Growth 362
Equipment Supplier 362
The Customer's Responsibility during Building and
Installation of Equipment 363
How to Implement a Reliability Growth Program 363
Summary 363

14 Requirements for Implementing Equipment R&M in Phase 4 (Operations and Support) and Phase 5 (Conversion and/or Decommission) 365

Introduction to Life-Cycle Costs (LCC) 365
Calculating Nonrecurring Costs for Acquiring Equipment 367
Calculating Support Costs for Continued Operation of
Equipment 368
Calculating Equipment Maintenance Costs (M_C) 369
Calculating Scheduled Equipment Maintenance (S_M) Costs 369
Calculating Unscheduled Equipment Maintenance (U_M) Costs 369
The Impact of Life-Cycle Costs (LCC) on Concept and Design 370
Improving LCC 372
Documenting the LCC Process 372

Overall Items of Concern for New Machinery 373
    Develop a Life-Cycle Costing (LCC) Model 374
    Conduct Periodic R&M Reporting 374
    Continuously Review Field History Data 374
    Conduct a Failure Mode and Effects Analysis (FMEA) 374
    Conduct Machinery FMEA 375
    Conduct Design FMEA 375
    Perform Electrical Stress Analysis and Derating 375
    Perform Mechanical Stress Analysis to Check Design Margins 375
    Perform Thermal Analysis 376
    Applications Engineering 376
    R&M Design Review Guidelines 377
    Conducting R&M Testing and Assessment of Results 377
    Provide an Equipment Maintainability Matrix 377
    Performing Total Productive Maintenance (TPM) 378
    Documenting Things Gone Right/Things Gone Wrong
    (TGR/TGW) 378
    Developing an R&M Matrix 378
    Developing an R&M Checklist 380
    Performing Design Reviews 382
    Developing Runoff Assessment Techniques 383
    Conducting a 50/20 Dry-Cycle Run 383
Summary 385

15 Weibull Distribution 387
Analyzing Life Data 388
Calculating Statistical Distributions for Equipment Lifetimes 388
Estimating the Parameters of Equipment Lifetime Distribution to
Analyze Failures of Machinery or Equipment 390
    Estimating the Shape Parameter 391
    Estimating the Scale Parameter 391
    Estimating the Location Parameter 393
    Calculating Results and Plots 393
    Use Confidence Bounds to Quantify Uncertainty in
    Analyzing Data 394
Introduction to Failure Data Modeling 395
    Use of a Probability Density Function Model 395
    Calculating Rate of Failure and Hazard Functions 397

Weibull Probability Density Function 401
　　　　　Three-Parameter Weibull 401
　　　　　Two-Parameter Weibull 402
　　　　　Characteristic Life 403
　　　　　Graphical Estimation of Weibull Parameters 404
　　　　　Weibull Slope 405
　　　　　Median Rank Tables 407
　　　　　Confidence Limits for Graphical Analysis 408
　　　Constructing Confidence Limits 409
　　　　　Suspended Item Analysis 409
　　　　　New Increment Method 410
　　　　　Sudden Death Testing 411
　　Summary 424

Glossary **425**

Index **447**

About the Author **465**

Acknowledgments

Writing a book such as this is the endeavor of many individuals contributing to the ideas that finally appear on its pages. To be sure, the final work is the author's responsibility, but that does not diminish the contribution of others.

Friends and colleagues have contributed to this work either directly or indirectly in many ways, and it is unfortunate that I cannot thank them all individually. There are some, however, who need special recognition and my gratitude for their help.

First and foremost I want to thank Prof. Leonard Lemberson from Western Michigan University for the countless and thought-provoking discussions on the topics of reliability and Weibull distribution applications. Thanks, Leonard, for your insight and guidance.

A thank-you also goes to Dr. R. Munro and Dr. E. Rice for their continuing encouragement at the low points of my writing. They stood by me offering suggestions and positive reinforcement in many ways. I really appreciated their effort to keep me on target.

Special thanks also to Stephen Stamatis (my son), a Ph.D. candidate at Purdue University, for generating the statistical and Weibull tables in the appendices.

Also special thanks go to the editors of the book for their suggestions for making it even better.

Finally, as always, my greatest "thank you" belongs to my number one critic, editor, supporter, motivator, and partner in life, my wife, Carla. Thanks, Carla Jeanne, for all you have done and continue to do. You are indeed the GREATEST!

Introduction

The essence of overall equipment effectiveness (OEE), reliability, and maintainability is to establish system effectiveness. That means that a machine individually or as part of a subsystem or as a system must be operating as designed. If it happens however, to have an unscheduled downtime, this downtime must be at the very minimum. This is very important because as the unscheduled overtime increases, production decreases, as shown in Figure I.1. Notice that as the individual downtimes increase, so does the overall unscheduled time. Therefore, it is imperative for anyone who evaluates OEE to be cognizant of the Mean Time to Repair (MTTR).

System effectiveness is often expressed as one or more figures of merit representing the extent to which the system is able to perform the intended function. The figures of merit used may vary considerably depending on the type of system and its mission requirements; however, the following should be considered:

1. *System performance parameters,* as defined by the customer and agreed by the supplier.
2. *Availability,* which is the measure of the degree to which a system is in the operable and committable state at the start of a mission when called for at an unknown random point in time. This is often called "operational readiness." Availability is a function of operating time (reliability) and downtime (maintainability and/or supportability).
3. *Dependability,* which is the measure of the system operating conditions at one or more points during the mission, given the system condition at the start of the mission (i.e., availability). Dependability also is a function of operating time (reliability) and downtime (maintainability and/or supportability).

Reliability and Maintainability (R&M), contrary to general perception, are not tools to be used in specific tasks. Rather, R&M is a *discipline.* It is founded

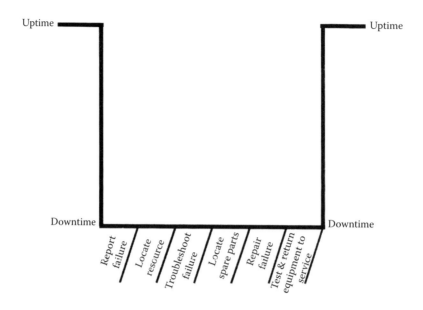

Figure I.1 Relationship of uptime and downtime.

on several techniques that are meant to direct both machine suppliers and users beyond the question of, "Will it work?" to a quantifiable analysis of "How long it will work without failure?"

To understand R&M, you must understand its components. First, *reliability* is the probability that machinery/equipment can perform continuously, without failure, for a specified interval of time when operating under stated conditions. Second, *maintainability* is a characteristic of design, installation, and operation usually expressed as the probability that a machine can be retained in, or restored to, a specified operable condition within a specified interval of time when maintenance is performed in accordance with prescribed procedures.

R&M is the vital characteristics of manufacturing machinery and equipment that enable its users to be "world-class" competitors. After all, efficient production planning depends on a process that yields high-grade parts at a specific rate without interruption. What makes R&M worth pursuing is the fact that it allows the manufacturer of a specific equipment or tool to be able to predict a specified quality level. This predictability is the key ingredient in maintaining production efficiency and the effective deployment of just-in-time principles.

As important as R&M is to any organization, in order for it to work, there must be a cooperative effort between the supplier and the user (i.e., the customer) of manufacturing machinery and equipment. Both must understand which equipment performance data are needed to ensure continued improvement in equipment operation and design, and they must exchange this information on a regular basis.

R&M, as it is used in most of the manufacturing, tooling, and equipment industry, is organized in five phases:

1. Concept
2. Development/design
3. Building and installation
4. Operation and support
5. Documentation, conversion, and transition

These phases are known as the life-cycle costs phases. Most companies, through their R&M specification documentation, have embraced the concept of life-cycle cost (LCC) instead of Fixed and Test (F&T) costs in the equipment selection process. The LCC is typically performed to estimate the overall cost of ownership of the equipment over the life cycle. The LCC analysis should be completed to compare the cost characteristics of one machine against another. LCC is a powerful tool to use to perform a cost/benefit analysis when considering different designs, architecture, or potential equipment improvement activities.

When developing an LCC analysis, the cost of ownership must be evaluated over the equipment's life cycle (i.e., the five phases). For example, Table I.1 illustrates the effectiveness of the LCC analysis. Notice, however, that this example represents only the *general costs* associated with LCC. It does not show the *itemized costs* associated with each of the general costs.

Observe that Machine A has a greater acquisition cost than Machine B, but the LCC of Machine A is less than the LCC of Machine B. This example shows that cost effectiveness should not be based on a few selected costs.

Table I.1 Example of Effectiveness of the LCC Analysis

General Costs	Machine A	Machine B
Acquisition costs (A)	$2,000,000.00	$1,520,000.00
Operating costs (0)	$9,360,000.00	$10,870,000.00
Maintenance costs (M)	$7,656,000.00	$9,942,500.00
Conversion/decommission costs (C)		
Life-cycle costs (LCC)	$19,016,000.00	$22,332,500.00

Instead, you should analyze *all* of the costs associated with the purchase of machinery and/or equipment.

During the first three life cycle phases, more than 95% of the reliability dollars are allocated for R&M improvements, leaving only 5% of the R&M dollars for improvements after the build and installation phase. The moral of the story, then, is to act early for improvement, thereby spending less money over the life of a machine.

The first three phases of the machinery/equipment's life cycle are typically identified as *nonrecurring* costs. The remaining two phases are associated with the equipment's *support* costs.

Finally, R&M as it relates to Overall Equipment Effectiveness (OEE) focuses primarily on three items:

1. Availability—features and repairs, which are set-up time and other losses
2. Performance—speed of machinery, which focuses on reduced operating speed and minor stoppages
3. Quality—defect losses, which focuses on scrap and rework, as well as start-up and rework

These three items collectively produce a calculated number that measures the effectiveness of the machines in the work environment. They depend on accurate and timely data and, above all, on an understanding of when and how to do the R&M. This understanding is illustrated in a flowchart shown in Figure I.2.

In addition to the three characteristics, you also must be cognizant about testing, its planning, and the results of these tests. Traditionally, testing is based on three test levels:

1. Test to bogey
2. Test to failure
3. Degradation testing

These three levels are shown in Figures I.3–I.5.

The first level of testing, *test to bogey,* is the process of conducting a test to a specified time, mileage, or cycle, then stopping and recording the number of items failed during that time. Bogey testing requires a clear definition

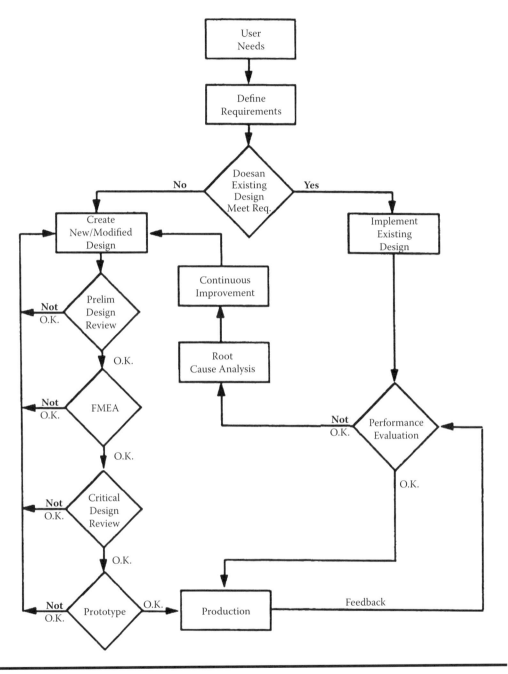

Figure I.2 Applying the machine R&M process.

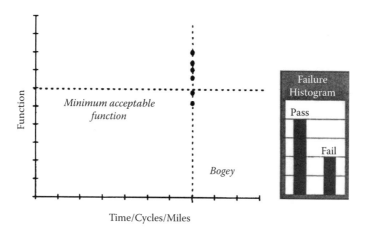

Figure I.3   Test to Bogey.

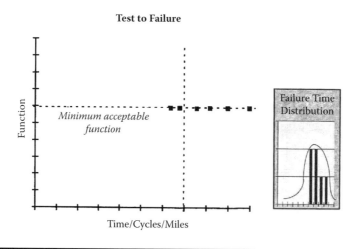

Figure I.4   Test to failure.

of loading cycles, the number of cycles equal to "life" (exposure time), and a clear definition of failure. Bogey testing:

- Has bivariate (pass/fail) pass criteria. The test is passed if there are no failures by end of test.
- Focuses on "Has it broken yet?"; thus, if there are no failures, the testing yields no information on how the system breaks.
- Will estimate the proportion of failures at a particular number of cycles.

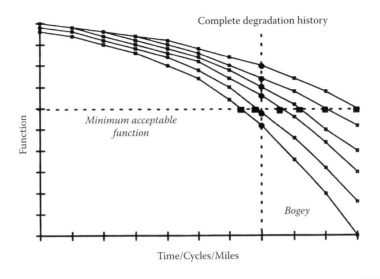

Figure I.5 Degradation testing.

- Does not yield information on what would happen in the next instant of time.
- At best, the data (results) can be summarized in a pass/fail histogram.

The second level of testing, *test to failure,* is the process of conducting a test until some or all items fail and then recording the failure time(s). Testing to failure requires a clear definition of both loading cycles and failure before commencing the test. Testing to failure:

- Provides all of the information provided by bogey testing.
- Allows lifetime prediction based on life data.
- Allows examination of hardware to understand failure mode.
- Indicates kind of design changes that might be necessary.
- Does not yield information on the gradual loss of customer satisfaction due to deterioration of performance prior to failure.
- Can summarize test data in a failure time histogram or distribution.
- Enables the prediction of failure and the estimate of mean time to failure (MTTF). Furthermore, if plotted using the Weibull technique, the failure generic type can be inferred—that is, infant mortality, useful life, or wearout.

The third level of testing, *degradation testing,* is a measure of performance at regular intervals throughout testing and recording the deterioration of function over time. Degradation testing yields an order of magnitude more information per test or prototype than testing-to-failure. Degradation data are more customer-focused and specifically:

- Yield a complete picture of the deterioration of ideal function over time. For example, the development of "play" in the steering system of an automobile: customer satisfaction decreases as the play increases; there is no point that a hard failure has occurred.
- Enable robustness analysis by increasing understanding of the pattern of widening variation over time.
- Allow failure analysis for varying levels of minimum acceptable function.
- Can provide an analytical model of the failure mechanism, which allows extrapolation of life data to different stress levels.
- Provide basis for preventive maintenance recommendations.
- Can summarize test data by a collection of degradation curves, which provides failure distributions and deterioration rates over time.

Figure I.6 shows functional deterioration over time for three different designs. Design A performed best in the beginning, but worse over time. Items that perform best at first do not always perform best over time. Degradation trends for each design could be predicted early on, based on a combination of deterioration rates from the data and engineering knowledge of the degradation.

Based on an understanding of the degradation pattern, an engineer may be able to predict long-term performance using shorter test periods (reduce testing to failure or life-testing) and fewer testing resources.

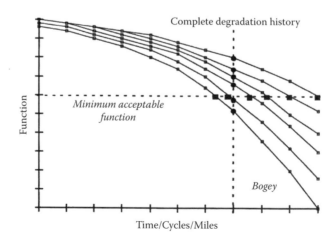

Figure I.6 Functional degradation of several design alternatives.

It must be noted here that when the degradation pattern is unknown, degradation testing is preferable to testing-to-failure or bogey testing. However, although the quality of information received from degradation testing is high, the cost may also be high. Because of this high cost, you must know your testing objectives clearly in advance to make decisions that will provide the greatest value from testing resources in the testing process. That is why you should always concentrate on limited testing resources on high-impact items.

Also, because cost is an issue in testing, it is paramount to recognize that accelerating testing is a way of testing designs in such a way that the integrity of the test will give you the results applicable to the particular objective. This is especially correct when you have less time to design. In other words, accelerated testing can reduce the total test time required. In fact, you should use accelerated testing to:

- Accumulate product stress history in a short period of time
- Generate failures that can be examined in hardware—especially in components that have long life under normal conditions.

Accelerated testing may be applied during the following phases:

- Define Requirements phase, for determining requirements through benchmarking
- Design for Robustness phase, for robust design study and prototype testing, and
- Verify Design phase, for Key Life Testing or other Design Verification (D) tests.

Accelerated testing is accomplished by reducing the cycle time, such as by compressing the cycle time by increasing stress cycles, eliminating non-damaging cycles, or reducing dormant time in the normal operating cycle. Another way to reduce the cycle time in order to perform accelerated testing is to increase stress levels, or intensify environmental exposure.

Always keep in mind your specific application and objective. For example, automotive parts are subjected to multiple stresses and combinations of stresses. Identify these stresses and combinations early in the design phase. When performing accelerated tests, make sure all critical stresses are represented in the test environment. Further, the stresses will need to be accelerated to the maximum. Often, the maximum of all stresses is difficult to ascertain. However, strive to simulate the real-world environment as closely as possible. This is where Key Life Testing comes into play.

To be sure, accelerated testing is a very good way to test, however, correlation to real-world use and failure modes is just as critical, and you should be very careful to recognize it as well as to evaluate it. If there is little correlation to "real" use, such as aging, thermal cycling, corrosion, etc., then it will be difficult to determine how accelerated testing affects these types of failure modes. If you compare the results from the accelerated stress levels to those at the design stress or normal operating conditions, you can discover discrepancies. If you remove rest periods to create accelerated tests, be careful it is not important to a system: sometimes, materials recover during these rest periods. Be aware that sometimes it is not the high stresses that cause so much damage because they occur so infrequently; instead, small stresses with higher frequency can be worse.

Make sure no unrepresentative failure modes are introduced when accelerating a test by increasing stress. More important, recognize that not all failure modes can be accelerated. Understanding the system's physics and stress-life model is essential to avoid questionable results. If in doubt, consult with a reliability implementation engineer and system hardware specialist to construct accelerated tests.

The discipline of obtaining the real-world usage and incorporating key noises into a customer-correlated accelerated test that emphasizes product is embodied in Key Life Testing (KLT). KLTs should correlate to a critical percentile (CP) of real-world usage to measure and assess product function and performance over time. (The percentage depends on the industry's expectations: for example, in the automotive industry, the CP is generally set at the 90th percentile; in the commercial airline business, the percentile is set at about 45%. That means that 90% and 45% of the customers are satisfied). This may be accomplished by applying noises in the tail of the frequency distribution. Of course, more than one test may be required to cover the multiplicity of functions/failures involved.

During the design process, system solutions to customer requirements are evolved through an iterative process. Therefore, robustness experiments are conducted to discover interactions and to understand the best design and manufacturing settings for the total system (given the noises are present). Each iteration refines knowledge of the system and its parts, and provides test data evidence on which to base further optimization. Once you identify all the interactions and noises, you can analyze the effects of the noise and mimic these states as part of your Key Life Tests. Knowledge gained should be fed back to core engineering to create more simplified test procedures that better relate to customer usage profiles.

The idea is that components and subsystems should be tested to ensure:

- Ideal output.
- Input to the next subsystem is tuned to suit the next system's needs.
- Systems should be tested to ensure that customer requirements are met in the presence of noise.

As already mentioned, testing and the data resulting from such testing are important. However, it all starts with appropriate and applicable planning. Without this appropriate and applicable "test planning," the analyst and/or the engineer is more likely to have a misguided understanding of the machinery's failure. Why? Because reliability of a product in the field is often estimated based on "established" tests from a sample of the population, or in some situations through surrogate data. Therefore, the accuracy of such estimates depends on both the integrity of the data acquisition process and the correlation to real-world usage representing the wide range of demographics.

There are three key ingredients to "test planning":

1. Define your goals: Define the object of the experiment or test, and the scope and vision of what constitutes success. In addition, agree on a work plan that is focused on measuring specific characteristics based on the equipment's ideal function.
2. Create a test plan: There are three items to be concerned with:
   a. Decide on the proper test level: bogey, test-to-failure, or degradation.
   b. Clearly define how to measure function and what constitutes failure.
   c. Develop a statistically valid test plan by applying the fundamentals of experimental design.

3. Establish engineering confidence: Include critical, real-world noise in the test.

A Note About the CD

Throughout the text, there are references to material appearing in various appendices. All of these appendices are included on the CD supplied in the back of the book. These appendices feature statistical tables, outlines, case studies, guidelines, and standards.

Chapter 1

Total Preventive Maintenance

In our modern world of always being conscious of productivity and efficiency, we have become very cognizant of the importance of measurement. This chapter introduces some of the basic concepts in measuring and improving both productivity and efficiency. The metrics we have identified are total preventive maintenance (TPM), overall equipment effectiveness (OEE), lean, 5S, and the virtual factory.

The traditional name for this topic is total productive maintenance. However, here I am using it as total preventive maintenance (TPM) because I feel it is more appropriate in the actual usage of the term: *preventive* maintenance is a new way of looking at maintenance, or conversely, a reversion to old ways but on a mass scale. In TPM, machine operators perform much, and sometimes all, of the routine maintenance tasks themselves. This automaintenance ensures that appropriate and effective efforts are expended because the machine is wholly the domain of one person or team.

TPM is a critical and necessary adjunct to lean manufacturing. If machine uptime (i.e., availability) is not predictable and if process capability is not sustained, the process must keep extra stocks to buffer against this uncertainty, and flow through the process will be interrupted. One way to think of TPM is "deterioration prevention" and "maintenance reduction," not fixing machines. For this reason, many people refer to TPM as "Total Productive Manufacturing" or "Total Process Management." TPM is at the very minimum a proactive approach that essentially aims to prevent any kind of slack before occurrence. Its motto is "zero error, zero work-related accident, and zero loss."

A Brief History of TPM

TPM is originally a Ford idea (traced back to Henry Ford in the early 1900s), but it was borrowed and fine-tuned by the Japanese in the 1950s when preventive maintenance was introduced into Japan from the United States. Nippondenso, part of Toyota, was the first company in Japan to introduce plantwide preventive maintenance in 1960. In preventive maintenance, operators produced goods using machines, and the maintenance group was dedicated to the work of maintaining those machines. However, with the high level of automation of Nippondenso, maintenance became a problem because so many more maintenance personnel were now required. So, the management decided that the routine maintenance of equipment would now be carried out by the operators themselves. (This is autonomous maintenance, one of the features of TPM.) The maintenance group then focused only on "maintenance" works for upgrades.

The maintenance group performed equipment modification that would improve its reliability. These modifications were then made or incorporated into new equipment. The work of the maintenance group is then to make changes that lead to maintenance prevention. Thus, *preventive maintenance* along with *maintenance prevention* and *maintainability improvement* were grouped as *productive maintenance*. The aim of productive maintenance was to maximize plant and equipment effectiveness to achieve the optimum life-cycle cost of production equipment.

Nippondenso already had quality circles, which involved the employees in changes. Therefore, all employees took part in implementing Productive Maintenance. Based on these developments, the Japanese Institute of Plant Engineers (JIPE) awarded Nippondenso the distinguished plant prize for developing and implementing TPM. Thus, Nippondenso of the Toyota group became the first company to obtain the TPM certifications.

The Goals and Activities of Total Preventive Maintenance

TPM has five goals:

1. To maximize equipment effectiveness
2. To develop a system of productive maintenance for the life of the equipment
3. To involve all departments that plan, design, use, or maintain equipment in implementing TPM
4. To actively involve all employees
5. To promote TPM through motivational management

TPM identifies the seven types of waste (*Muda*), and then works systematically to eliminate them by making improvements, primarily through the incremental approach of Kaizen.

TPM also has eight pillars of activity, each being set to achieve a "zero" target. These pillars are

1. **Focused improvement** (*Kobetsu-Kaizen*): It is aimed at eliminating waste. The basic wastes are
   - *Unnecessary transport of materials:* In moving products between factories, work operations, desks, and machines, all that is added is lead time—in other words, no value is created.
   - *Inventories beyond the absolute minimum:* Caused by overproduction, inventories take up floor space—something that is always at a premium in factories and offices. There is always a tendency to use inventories to mask other problems. Remember, if you have got plenty of spares, there is no incentive to fix problems with quality!
   - *Motions of employees:* When looking for parts, bending or reaching for materials, searching for tools, etc.
   - *Waiting for the next process step:* While waiting, the product is just soaking up "overheads"—the last thing that the customer actually wants to pay for!
   - *Overproduction ahead of demand*: This exposes the organization to risks in changing demands from customers, and is a disincentive to the firm to reduce the other wastes, because there is always plenty of extra material to use in case of problems.
   - *Overprocessing of parts*: Running parts on machines that are too fast or too slow, or even too accurate to achieve the customer's

definition of value. What is the problem with doing *too good* of a job? Generally, it means it is really *too expensive* a job for the market's expectations.

- *Production of defective parts:* If processes produce defects, then extra staffs are needed to inspect, and extra materials are needed to take account of potential losses. Worse than this, Inspection does not work. Eventually, you will miss a problem, and then someone will send a defective product to a customer. And that customer will notice at which point the customer is dissatisfied. Manual inspection is only 79% effective. In some cases, however, it is the only control you have. Therefore, it is used, but with reservations.

2. **Autonomous maintenance** (*Jishu-Hozen*): In autonomous maintenance, the operator is the key player. This involves daily maintenance activities carried out by the operators themselves that prevent the deterioration of the equipment. The steps for this autonomous maintenance are
   - Conduct initial inspection and cleaning.
   - Fix all sources of contamination.
   - Fix all areas of inaccessibility.
   - Develop and test all procedures for cleaning, inspection, and lubrication for possible standards.
   - Based on the previous task, conduct and develop inspection procedures.
   - Conduct inspections autonomously.
   - Apply the standardization of the inspection procedures done previously, and apply visual management wherever possible in the proximity of the machine.
   - Continue to conduct the autonomous maintenance for continual improvement.

3. **Planned maintenance:** for achieving zero breakdowns.
4. **Education and training:** for increasing productivity.
5. **Early equipment/product management:** to reduce waste occurring during the implementation of a new machine or the production of a new product.
6. **Quality maintenance** (*Hinshitsu-Hozen*): This is actually "maintenance for quality." It includes the most effective quality tool of TPM: *Poka-yoke* (which means mistake proofing or error proofing), which aims to achieve zero loss by taking necessary measures to prevent

loss, due to human intervention in design or manufacturing or even both.
7. **Safety, hygiene, and environment:** for achieving zero work-related accidents and for protecting the environment.
8. **Office TPM:** for involvement of all parties in TPM because office processes can be improved in a similar manner as well.

In the final analysis, TPM is "Success Measurement." This means that it is a set of performance metrics that is considered to fit well in the overall equipment effectiveness (OEE) methodology for improvement.

An Overview of the Concept of Lean Manufacturing

Lean manufacturing addresses the growing need for all types of organizations that drive process change and performance improvements in their organization environment and supports the evolution toward demand-driven supply networks. As customer service demands continue to challenge supplier capabilities, companies are forced to incur more costs just to remain competitive. The only way for organizations to effectively manage the cost side is

- *To change the product flow from push to pull*—to become a truly lean enterprise, beginning at the operational level and extending outward,
- *To decrease cycle time* of value-added operations—that is, to increase the speed of delivering the product or service to the customer.

Becoming lean and sustaining the value due to that transformation over time requires a platform from which managers and supply chain executives can propel the business forward. The philosophy of lean manufacturing provides all the best practice tools, processes, and controls needed to define, run, and continually improve operations within and beyond the organizational unit—from a simple process mapping to value stream and factory design to lean production scheduling and sequencing; from simple Kanban and supplier collaboration to distribution inventory management.

It is very important for an organization dealing with lean concepts to be very cognizant of the term *value*. Yes, all organizations without exception want value creation because value dictates that sales and profitability will follow. However, all organizations must recognize that value is not abstract and cannot always be 100%. Organizations quite often find themselves with non-value-added, but *necessary* activities, as well as non-value-added, *not necessary* activities. In lean, and especially with value stream mapping, you should focus on these three categories of value (value, non-value-added/necessary, and non-value-added/not necessary), and you should try to optimize them for the particular organization.

By enabling major gains in demand response and operating performance simultaneously, including dramatically reduced cycle times and inventories, lean manufacturing helps solve organizational issues of customer service versus profitability challenge.

Is "Lean" the silver bullet for efficiency and productivity increases? No. Is lean the answer to consistent profitability? No. Lean is a methodology that can help any organization to achieve these goals by focusing on fundamental changes in the organization, for value-added activities. However, these changes are not limited to imitating the Toyotas and Hondas of the world in reducing waste and variation; instead, in addition to identifying the best practices, they have to change the critical paradigms of their operations. From my own experience, you need to

1. Recognize that knowledge at any level cannot be bought. History has demonstrated time and again that mergers generally result in a net loss of knowledge.
2. Recognize that all employees at all levels can create knowledge for the organization—if given the opportunity and appropriate recognition.
3. Recognize that with "things learned," you can reduce costs (labor, material, capital, etc.), improve efficiency, and improve cycle time.
4. Recognize that unless things learned are implemented, there will be no improvement.
5. Recognize that Things Gone Wrong (TGW) have to be fixed without having to be fixed again and again, and Things Gone Right (TGR) have to be implemented, so that the success can be repeated again and again.

How to Make Your Organization Lean

So, what are the main ingredients that will make an organization Lean? Fundamentally, they are the following:

- Recognizing that change is inevitable. As a consequence, you need to plan accordingly.
- Allowing for dissent and open communication within the organization.

- Demanding that the leadership (i.e., top management) of your organization be both *committed* to improvement and involved in *communicating* that commitment to the entire organization.
- Recognizing that before improvement, the organization must find out where it is currently at and where it wants to go. Process mapping will help in identifying the "current state" as well as the "should- and could-be state."
- Recognizing that the change from the current state to either should- or could-be state is a matter of developing a system and following it.
- Recognizing that the ultimate change will occur if and only if the leadership of the organization can articulate the need as well as the benefits of the change to the entire organization. This means that the leadership must constantly reinforce the need for the change, as well as their confidence of its success.
- Recognize that the change to lean will be a long-term commitment.

I believe these ingredients are the starting recipe for a successful implementation of lean. In essence, lean is about doing more with less: less time, less inventory, less space, less people, and less money. Lean is about speed and getting it right the first time.

Map Your Processes

To optimize the process with all these reductions, first and foremost, you must know the current process and then project the ideal. The first employs the traditional *Process Flow Chart,* and the second uses *Process Stream Value Mapping.* For those of you who are just starting your improvement journey and are wondering which process map to use, it is important to first understand the differences between the two. After all, the selection process—generally—depends on resource availability and deadlines, as well as project experience. However, in general terms, here is the difference:

- *Value stream mapping* identifies waste within and between processes.
- In contrast, *detailed process mapping* identifies both the voice of the customer and the process outputs, and it identifies and classifies process inputs.

Once you understand this difference, the selection is very straightforward. Let us take a closer look at each.

Value stream mapping. Value stream mapping takes a high-level look at a company's flow of goods or services from customer to customer. It usually contains no more than 10 steps, quite often about 7. Practitioners can drill down to find the true bottleneck in a company's processes. Key metrics captured are *cycle times, defect rates, wait times, headcount, inventory levels, changeover times,* etc.

Detailed process mapping. In comparison, detailed process mapping provides a more detailed look, with a much deeper dive into a process. One captures the inputs and outputs of every step in a process and classifies each as *critical, noise, standard operating procedure,* or *controllable.* The key to using this tool is controlling inputs and monitoring outputs. Detailed process mapping also helps document decision points within a process.

Constructing a Value Stream Map

Although most people are familiar with the concepts of detailed and high-level process maps, many need clarification on value stream maps. Value stream mapping helps companies avoid randomly making improvements by allowing them to identify and prioritize areas of improvement up front as well as to set measurable goals for improvement activities. This is accomplished in three stages:

- *Stage 1: Create a current-state map,* showing how the company serves its customers today.
- *Stage 2: Create a future-state map,* showing the reduction of waste and the effects of the changes.
- *Stage 3: Develop and implement a plan* to reach the future state.

An interesting definition is given by the iSixSigma dictionary:

> A value stream is all the steps (both value-added and non-value-added) in a process that the customer is willing to pay for in order to bring a product or service through the main flows essential to producing that product or service.

One of the key elements of value stream mapping is that it can provide a baseline of defined processes. The critical phrase in this definition is "that the customer is willing to pay for": in other words, it has some inherent value—hence, the name *value stream.* If a company's customer walked through its process, how would that customer react? That customer will be very happy when value is added to their product, but very unhappy when they see processes that not only do not add value but also take away value—such as scrap, rework, inventory, inspection, delays, and so on. In other words, the customer sees value and waste in the production of their product. To be sure, although no one can eliminate all waste, using value stream mapping to identify waste helps determine a plan for eliminating it.

However, before a company can identify its value stream, it needs to determine:

- The value in the process that the customer is willing to pay for.
- The steps required to deliver the product or service to the customer.
- What is significant in each?

To be sure, there are steps in any process that create value and those that do not. However, it is very important to recognize that some non-value-added steps (perhaps because of regulations, policies, and current technologies) cannot be eliminated—or at least, they cannot be eliminated immediately. However, they can be minimized. So, the effort of stream value is to eliminate or at least minimize waste in the process that benefits a company's bottom line.

But what is value and waste? *Value* is an activity that transforms or shapes raw materials or information to meet customer needs. Another definition of value is the willingness of the customer to pay for the product and its functions. On the other hand, *waste* is any activity that consumes time, resources, and/or space, but does not contribute to satisfying customer needs. Examples of waste include

- Overproduction
- Inventory
- Transportation
- Waiting
- Motion
- Overprocessing
- Correction
- Not utilizing the talent and knowledge of human resources

In addition to these examples, here are some *causes* of waste:

- Layout (distance)
- Long setup time

- Incapable processes
- Poor maintenance
- Poor work methods
- Inadequate training
- Product design
- Performance measures
- Ineffective production planning and scheduling
- Equipment design and selection
- Poor workplace organization
- Supplier quality/reliability, and more

How to Map the Value Stream of a Process

So, how should one proceed to value stream a process? The answer is a simple three-step approach—assuming that the team has a very good understanding of the value stream.

Step 1: Create a List of Products and Group Them in Families

Some companies offer varied products and services or have different equipment or machinery. For example, an investment company offers different investment opportunities such as mutual funds, 401Ks, stocks, etc. A finance company offers different types of loans, including first mortgages, home equity, car loans, and small-business loans. A manufacturing plant has different machinery. It is relatively easy to group products into families by constructing a simple table. The goal is not only to identify all product families but also to identify what process steps each product utilizes. This will be a living, breathing table, so a project team should be prepared to make further revisions as it dives deeper into its analyses.

Step 2: Determine Which Product, Machine, or Service Is Considered Primary

Although a product or service may utilize different processes, a company needs to concentrate on one process at a time, focusing on processes critical to company goals. In the case of equipment or machine, it is also important to focus on the most critical in the stream. In many instances, a company's improvement plans may be filled with process improvement projects with no clear link to its overall goals or vision. With limited resources available, efforts need to be concentrated only on those projects that really need to be done. Selecting which product family to analyze will depend on the individual business situation. Examples of products or services to analyze include those that

- Stem from company goals/vision
- Utilize the most process steps
- Are known to have high defect rates
- Represent the voice of the customer and offer the highest customer rate of return
- Are high volumes in dollars and/or units

Step 3: Document the Steps of the Process

In this step, you should perform an initial walk-through. Use a SIPOC diagram (which maps Suppliers, Inputs, Process, Output, and Customers) to document the process steps. Begin with the customers, and work backward. A project team will gain more insight by working in reverse order. During the walk-through, think about the customer and ask these questions:

- How does the customer receive the product or service?
- What triggers the product or service to be delivered to the customer?
- What are the inputs?
- From where are these inputs supplied?

Once the walk-through is completed, there should be enough initial data to understand the value stream, and begin creating a current-state value stream map with a more detailed depiction of the value stream. Figure 1.1 is an example of a typical process flow diagram.

Develop Both Current-State and Future-State Maps

Value stream mapping requires both current- and future-state process maps (see Figures 1.2 and 1.3). However, future-state maps are often less well defined in services or administrative organizations. These organizations typically require a strategic perspective, such as what the new service delivery model looks like. Value stream mapping typically focuses on a single product family, but choosing only one product family may not be appropriate in a service organization—especially if the customer can choose between

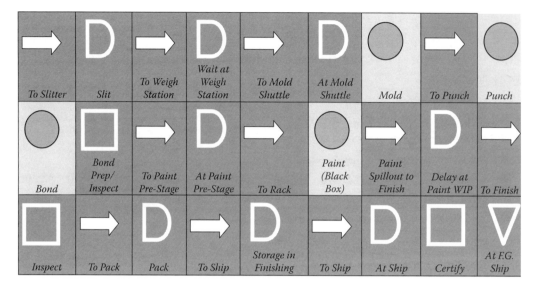

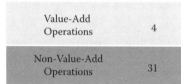

Figure 1.1 An example of a typical process flow diagram.

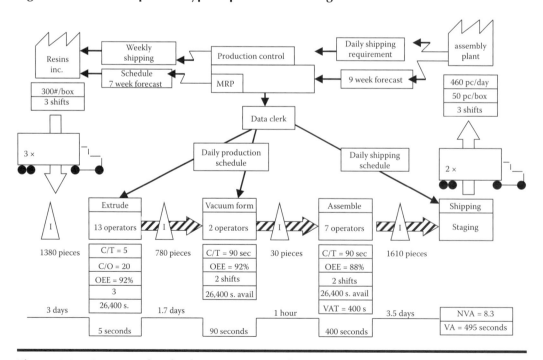

Figure 1.2 An example of value stream mapping—current state.

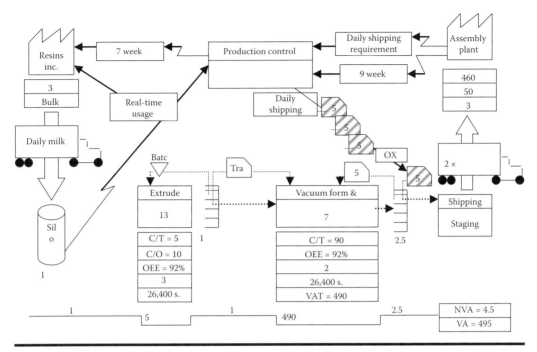

Figure 1.3 An example of Value stream mapping—future state.

different channels. For example, in banking, the customer may choose channels such as online, e-mail, or telephone banking. Focusing on a single product family may not provide the insight needed to identify all available improvement opportunities. In such cases, the value stream mapping methodology can be combined with other tools such as a bottleneck analysis. For example, Figure 1.2 is an example of value stream mapping—current state.

Detailed Process Mapping

As mentioned earlier, in some cases, there is no clear way to select the traditional process flow diagram or the value stream. In such cases, you should take advantage of both models by using detailed process mapping and adding value stream mapping data into it. Although each type of map is used to identify different variables, there is more value in combining components of value stream with detailed process mapping. Detailed process mapping has all the process components the value stream map does, and it can be broken down in much greater detail. Due to the time involved in constructing detailed process maps, one could include detailed process mapping after value stream mapping has located the bottleneck. For example, Figure 1.3 is an example of value stream mapping—future state.

In the world of machinery and equipment, the focus is always on making sure that the work flows continuously with minimal inventory by making sure that:

- Production is synchronized to shipping schedules, not based on machine utilization.
- Defects are prevented.
- Organizations are team-based with multiskilled operators.
- Measurable matrices are used to solve problems.
- Operators are empowered to make decisions and improve operations with few indirect staff.
- Top management and workers are actively involved together in troubleshooting and problem solving to improve quality and eliminate waste.
- Value stream is closely integrated from raw material through finished goods with the support of suppliers and customers.

Also, it must be remembered that Order-to-Delivery (OTD) lead time is the time required to deliver a product to a customer from the time that the product was ordered. The OTD process is a set of business practices that reduce this lead time but is affected by the OEE of the equipment. It is important, therefore, to identify as many sources of waste as possible—if not all. However, one always remember that

- Not all waste can be eliminated immediately.
- Identification of waste makes the opportunity for improvement visible.
- Process creates results—use results to define areas for improvement.

Never focus on average performance, because the average leads to complacency:

Average companies say "we're better than the status quo,"
and lose their motivation to get better.
Great companies say "How can we be the best in the world?"
and strive to be the best across all industries

By using the value stream map, make decision at the value-added task, or:

- Develop systems to encourage change to occur where the value is added.
  - Develop systems to encourage to occur where the value is added by not focusing on a suggestion system. Rather, we focus on developing an implementation system.

- Have a bias for action: A plan for improvement is worth nothing until it is implemented.
- Implement ideas: This is the true measure of lean deployment success.

Finally, value stream mapping is a powerful tool that helps identify the vital few Lean and Six Sigma projects that will yield the most value to the process tagged for improvement. And its approach of current- and future-state maps allows the practitioners to know where they are starting from, where they are going, and how they will get there. When a company looks back tomorrow, it will be much more rewarding if it knows the route it followed to arrive at its current state.

Problems with Lean

In the area of maintenance, there are some issues that one should be concerned with. For example:

- In equipment and machinery, a lean process requires reduced variability and uncertainty in the supply chain to reduce needs for raw materials and finished goods inventories.
- In order to reduce lot sizes and achieve reduced work-in-process inventory in an economical fashion, fixed costs of batching and ordering would have to be reduced first.
- Seasonal effects might give incentive to buffer the production process against known shifts in demand in order to reduce costly shifts in production levels.
- Without a reduction in fixed costs or uncertainty, Lean supply chains are subject to longer flow times and reduced throughput and potentially reduce economic performance.

All this means that the machinery and equipment must be always taken care of in terms of preventive maintenance, so that they can keep up with the demand of production and scheduling without unnecessary downtime.

The "5S" Methodology of Organizing the Work Area

The 5S methodology is a concept developed by Toyota: it is commonly identified as a method for organizing a workplace, especially a shared workplace (such as a shop floor or a maintenance area), and keeping it organized.

What 5S Is

It is sometimes referred to as a *housekeeping methodology*; however, this characterization can be misleading because organizing a workplace goes beyond housekeeping (see the discussion of "*Seiton*" in the following text).

The key targets of 5S are workplace morale and efficiency. The assertion of 5S is, by assigning everything a location, time is not wasted by looking for things. Additionally, it is quickly obvious when something is missing from its designated location. 5S advocates believe the benefits of this methodology come from deciding what should be kept, where it should be kept, and how it should be stored. This decision-making process usually comes from a dialog about standardization that builds a clear understanding, between employees, of how work should be done. It also instills ownership of the process in each employee.

In addition to the preceding text, another key distinction between 5S and "standardized cleanup" is *Seiton*. *Seiton* is often misunderstood, perhaps due to efforts to translate into an English word beginning with "S" (such as "sort" or "straighten"). The key concept here is to order items or activities in a manner that is conducive to work flow. For example, tools should be kept at the point of use; workers should not have to repetitively bend to access materials; flow paths can be altered to improve efficiency, etc. The 5S's are described as follows.

Seiri (Sorting). This function involves going through all the tools, materials, etc., in the plant and work area and keeping only essential items. Everything else is stored or discarded.

Seiton (Straighten Out or Set in Order). This function focuses on efficiency. When this is translated to "Straighten or Set in Order," it sounds like more sorting or sweeping, but the intent is to arrange the tools, equipment, and parts in a manner that promotes work flow. For example, tools and equipment should be kept where they will be used (i.e., straighten the flow path), and the process should be set in an order that maximizes efficiency.

Seiso (Sweeping or Systematic Cleaning). This function is the need to keep the workplace clean as well as neat. Daily activity at the end of each shift, the work area is cleaned up and everything is restored to its place, making it easy to know what goes where; knowing that everything is where it should be is essential here. The key point is that maintaining cleanliness should be part of the daily work—not an occasional activity initiated when things get too messy.

Seiketsu (Standardizing). This function consists of standardized work practices or operating in a consistent and standardized fashion. Everyone knows exactly what his or her responsibilities are to keep the above 3S's.

Shitsuke (Sustaining). Finally, this last function refers to maintaining and reviewing standards. Once the previous 4S's have been established, they become the new way to operate. Maintain the focus on this new way of operating, and do not allow a gradual decline back to the old ways of operating. However, when an issue arises such as a suggested improvement or a new way of working, or a new tool, or a new output requirement, then a review of the first 4S's is appropriate.

How to Implement 5S in Your Organization

The way to implement the 5S in any organization is by first identifying the areas of concern, and then proceed with fixing or improving. Specifically, you should use a tagging system, as follows:

- Red tag = Unneeded or rarely used items that need to be moved or thrown away
- Yellow tag = Items that need to be repaired or adjusted

5S is used with other lean concepts such as single minute exchange of die (SMED), TPM, and just-in-time (JIT). The 5S discipline requires clearing out things that are not needed in order to make it easier and faster to obtain the tools and parts that are needed. This is the foundation of SMED, which in turn enables JIT production. The first step in TPM is operator cleanup of machines, a mandate of 5S.

Benefits of 5S

5S methodology is definitely more than just housekeeping. It is the beginning of waste reduction, and it is built on the foundation of discipline. And you should do it because it helps to set a standard, develop discipline, eliminate waste, and more. It is a place for everything, at the right time to

- Separate—sort and eliminate unnecessary items.
- Straighten—put necessary items in order for easy access.
- Scrub—clean everything; tools and workplace.

And everything is in its place, all the time:

- Sustain/standardize—make cleaning and checking (previous three steps) routine.
- Systematize—propagate the previous four steps throughout the organization and continually improve.

Developing a Visual Factory to Make Decisions Quickly

Visual factory is an operational philosophy based on fast absorption of information to make effective decisions. The basic philosophy of the visual factory is based on:

- Organize, standardize, communicate.
- A picture is worth a thousand words.
- Similar to an aircraft controller (they use color and patterns to identify problems in the air traffic).

The tools for the visual factory are based on:

- Work cells
- Production control boards
- Shadow boards—tools, tables
- Color-keyed bins—scrap, rework
- Horns, whistles, and lights
- Inventory control cards

The effectiveness of the visual factory is based on the visual layout. This means that the layout of the machines or the operator can (or should)

- See everything from one point
- Highlight abnormalities
- Designed for visual control

Obviously, for this layout to work at optimum efficiency all the time, it must have some form of standardization (in process standards), and the people involved must be committed to continual improvement. (Remember, however, that improvement comes before innovation; it should be no-cost or low-cost; and *every improvement*, no matter how small, is worthwhile, because small improvements over time add up to a large total improvement.)

The Visual Factory Requires Standardization

Standardization in the visual factory depends on three elements of standardized work: *takt time, standard work-in-process inventory,* and *work sequence.*

Takt time is the maximum time allowed to produce a product in order to meet demand. It is derived from the German word *taktzeit,* which translates to clock cycle, pace, and rhythm. There is logic, therefore, to setting the pace of production flow to this takt time. The pace of product flow is expected to be less than or equal to the takt time. In a lean manufacturing environment, the pace time is set equal to the takt time. It is calculated as

$$\text{Takt time} = \frac{\text{Available time}}{\text{Customer requirements}}$$

Note: Available time: run time if everything was perfect.

The benefits of standardization are

- Balanced production with minimum inventory and labor
- Focus management
    - Maintenance and improvement
- Basis for training
- Basis for audit and diagnosis
- Control variability
- Ensure sharing of "best practices"

In addition, in terms of improvement, the following must be—at least—considered:

- (The improvement must) innovate
- Best, easiest, safest way you know today

- Only one standard at a time
- Documentation of know-how
- Objective, simple, and conspicuous
- Consistent with quality, cost, and delivery requirements
- Show relationship between cause and effect

Summary

This chapter introduced the concepts of TPM, OEE, lean, 5S, and the virtual factory, with some key examples of each concept. The next chapter introduces the concepts of loss and effectiveness.

Chapter 2

OEE: Understanding Loss of Effectiveness

Chapter 1 introduced some of the key concepts in maintenance, and this chapter deals with understanding loss and effectiveness. I begin with overall equipment effectiveness (OEE) as the key factor in measuring both productivity and efficiency. In addition, I will address the new movement in the automotive industry's use of OEE as a metric for capacity.

OEE is a hierarchy of metrics that focus on how effectively a manufacturing operation is utilized. The results are stated in a generic form that allows comparison between manufacturing units in differing departments, organizations, machines, and industries. In essence:

- OEE is a measure that identifies equipment potential.
- OEE identifies and tracks loss.
- OEE identifies windows of opportunity.

And its main objective is to

- Increase productivity
- Decrease cost
- Increase awareness of the need of machine productivity
- Increase equipment life

The results of these objectives are to

- Increase profits
- Attain (or maintain) a competitive edge

- Identify equipment ownership
- Reduce expenses

As a measure, OEE measures how effective capital equipment is used by identifying constraints, and how the constraints impact the OEE. The effectiveness is measured by multiplying availability and performance efficiency by the rate of quality product produced. The actual calculations are

Availability = ([available time − downtime]/[net available time]) × 100
(Remember that the potential availability is always 24 h or 1440 min)
Performance = ([ideal cycle time and total parts run]/[operating time]) × 100
Quality = ([total parts run − defects amount]/[total parts run]) × 100

OEE Comprises Six Key Metrics

The six big losses in equipment or machines generally are thought to be

1. Breakdowns
2. Setup/adjustments
3. Idle/stops
4. Reduced speed
5. Scrap
6. Start-up yield

A world-class OEE value is considered to be 85% or higher (see Figure 2.1, shown later in this chapter, for the derivation of the 85%).

OEE measurement is also commonly used as a key performance indicator (KPI) in conjunction with lean manufacturing efforts to provide an indicator of success. This can be best illustrated by a brief discussion of the six metrics that comprise the system. The hierarchy consists of two top-level measures and four underlying measures. OEE and total effective equipment performance (TEEP) are two closely related measurements that report the overall utilization of facilities, time, and material for manufacturing operations. These top-view metrics directly indicate the gap between actual and ideal performance.

OEE quantifies how well a manufacturing unit performs relative to its designed capacity, during the periods when it is scheduled to run. On the other hand, TEEP measures OEE effectiveness against calendar hours, that is, 24 hours per day, 365 days per year.

Overall Equipment Effectiveness (OEE)

Equipment ID: Date:
Part ID: Shift:

Equipment Availability:

A. Total available time   _____ minutes
B. Planned Downtime   _____ minutes
C. Net available time (A-B)   _____ minutes
D. Unplanned downtime (from downtime reports)
     # of breakdowns _____ Total Minutes _____ +
     # of setups & adjustments ____ Total Minutes _____ +
     # of minor breakdowns _____ Total minutes _____ = _____ minutes
E. Operating time (C-D)   _____ minutes
F. Equipment availability (E/C)x100   _____% [90%]

Performance Efficiency:

G. Total parts run (good + bad)   _____ parts
H. Ideal cycle time   _____ minutes/part
I. Performance efficiency (HxG)x100   _____% [95%]

Quality Rate:

J. Total defects (rework + scrap)   _____ parts
K. Quality rate [(G-J)/G]x 100   _____% [99%]

Overall Equipment Effectiveness:
                     (F x I x K) x 100 _____% [85%]

Figure 2.1   A typical OEE form.

In addition to the preceding measures, there are four underlying metrics that provide understanding as to why and where the OEE and TEEP performance gaps exist. These measurements are

1. *Loading:* The portion of the TEEP metric that represents the percentage of total calendar time that is actually scheduled for operation.
2. *Availability:* The portion of the OEE metric represents the percentage of scheduled time that the operation is available to operate. Often referred to as *Uptime*.
3. *Performance:* The portion of the OEE metric represents the speed at which the work area runs as a percentage of its designed speed.
4. *Quality:* The portion of the OEE metric represents the Good Units produced as a percentage of the Total Units Started. Commonly referred to as *First Pass Yield* or *First Time Throughput (FTT)*.

What follows in the next sections is a detailed presentation of each of the six OEE/TEEP metrics and examples of how to perform calculations. The calculations are not particularly complicated, but care must be taken as to standards that are used as the basis. Additionally, these calculations are valid at the work area or part number level but become more complicated if rolling up to aggregate levels.

Metric #1: OEE (Overall Equipment Effectiveness)

OEE breaks the performance of a manufacturing unit into three separate but measurable components: Availability, Performance, and Quality. Each component points to an aspect of the process that can be targeted for improvement. OEE may be applied to any individual work area, or rolled up to Department or Plant levels. This tool also allows for drilling down for very specific analysis such as a particular Part Number, Shift, or any of several other parameters. It is very unlikely that any manufacturing process can run at 100% OEE. Many manufacturers benchmark their industry to set a challenging target; 85% is not uncommon. The calculation formula for the OEE is

$$OEE = Availability \times Performance \times Quality$$

Example

A machine exhibits the following measurements: Availability of 86.7%; Performance is 93.0%; Quality is 95.0%. The OEE is calculated to be

$$OEE = 86.7\% \times 93.0\% \times 95.0\% = 76.6\%$$

Metric #2: TEEP (Total Effective Equipment Performance)

Although OEE measures effectiveness based on scheduled hours, TEEP measures effectiveness against calendar hours, that is, 24 hours per day, 365 days per year. The TEEP, therefore, reports the "bottom line" utilization of assets. The calculation formula for the TEEP is

$$TEEP = Loading \times OEE$$

Example

A machine exhibits the following measurements:

OEE of 76.67%; Loading is 71.4%. The TEEP is calculated to be
TEEP = 71.4% × 76.7% = 54.8%

Stated in another way, TEEP adds a fourth metric "Loading." Therefore:

$$TEEP = Loading \times Availability \times Performance \times Quality$$

Metric #3: The Loading Portion of the TEEP Metric

The Loading portion of the TEEP metric represents the percentage of time that an operation is scheduled to operate compared to the total Calendar Time that is available. The Loading Metric is a pure measurement of Schedule Effectiveness and is designed to exclude the effects of how well that operation may perform. The formula for this calculation is

$$Loading = Scheduled\ Time/Calendar\ Time$$

Example

An operation is scheduled to run 5 days per week, 24 hours per day. For a given week, the Total Calendar Time is 7 days at 24 hours. Therefore:

$$Loading = (5\ days \times 24\ hours)/(7\ days \times 24\ hours) = 71.4\%$$

Metric #4: Availability

The Availability portion of the OEE metric represents the percentage of scheduled time that the operation is available to operate. Another way of saying this is that the Availability is a percentage of time a machine is available to produce parts. The Availability metric is a pure measurement of Uptime that is designed to exclude the effects of Quality, Performance, and Scheduled Downtime Events. The formula for this calculation is

$$Availability = Available\ Time/Scheduled\ Time$$

Example

A machine is scheduled to run for an 8 h (480 min) shift. The normal shift includes a scheduled 30 min break when the machine is expected to be down. The machine undergoes 60 min of unscheduled downtime. Therefore, the availability is calculated in the following manner:

Scheduled Time = 480 min − 30 min break = 450 min; Available Time = 450 min scheduled − 60 min unscheduled downtime = 390 min
Availability = 390 available minutes/450 scheduled minutes = 86.7%

Metric #5: Performance

The Performance portion of the OEE metric represents the speed at which the machine runs as a percentage of its designed speed. In other words, it is the actual speed of the machine as related to the design speed of the machine. The Performance metric is a pure measurement of speed that is designed to exclude the effects of Quality and Availability. The formula for calculating Performance is

$$\text{Performance} = \text{Actual Rate}/\text{Standard Rate}$$

Example

A machine is scheduled to run for an 8 h (480 min) shift with a 30 min scheduled break. The Available Time is = 450 min scheduled − 60 min unscheduled downtime = 390 min; The Standard Rate for the part being produced is 40 Units/hour; and the machine produces 242 Total Units during the shift.

Note: The basis is Total Units, not Good Units.

The Performance metric does not penalize for Quality; Actual Rate = 242 Units/(390 Avail min/60 min) = 37.2 Units/hour.

Therefore, the Performance is calculated to be

$$\text{Performance} = 37.2 \text{ Units/hour}/40 \text{ Units/hour} = 93.0\%$$

Metric #6: Quality

The Quality portion of the OEE metric represents the Good Units produced as a percentage of the Total Units Started. In other words, it is the percentage of resulting parts that are within specifications, as defined by the customer. The Quality Metric is a pure measurement of Process Yield that is designed to exclude the effects of Availability and Performance. The formula for calculating Quality is

$$\text{Quality} = \text{Good Units}/\text{Units Started}$$

Example

A machine produces 230 Good Units during a shift. 242 Units were started in order to produce the 230 Good Units. Therefore, the Quality metric is

$$\text{Quality} = 230 \text{ Good Units}/242 \text{ Units Started} = 95.0\%$$

Observe and Monitor Time and Other Considerations

In dealing with both OEE and TEEP, one must be conscious of the time. Especially when calculating the OEE, the operator or the engineer of the machine must be familiar with the categories of time. These times should be observed and monitored on a continual basis. These time references are identified as uptime and downtime, described in more detail in the next two sections. In addition, the ideal cycle time is also important, as are defects, cost of quality, process line considerations, and process line quality.

Observe and Monitor Equipment Uptime

- Check Gross Operating Time
  - Net operating
  - Machine cycle
  - Scrap
- Check Nonoperating Time
  - Blocked
  - Starved
- Administration

Observe and Monitor Equipment Downtime

- Check Scheduled Downtime
  - Predictive and Preventive Maintenance
    - Access
    - Part removal
    - Part replacement
    - Startup
    - Test/calibration
    - Upgrade/modify
  - Delay Time
    - Setup/changeover
    - Startup
    - Meetings
    - Lunch breaks

- Check Unscheduled Downtime
  - Corrective Maintenance Time
  - Access
  - Diagnostic
  - Part removal
  - Part replacement
  - Startup
  - Test
- Check Delay Time
  - Obtain parts
  - Obtain tools
  - Obtain skilled resources
- Report failure

Determining the Ideal Cycle Time

The ideal cycle time is the fastest speed you can operate before quality is compromised. This is usually thought of as the theoretical cycle time. In real terms, cycle time is equated with total parts run during a specified period, regardless of quality. Therefore, when cycle time is considered, some of the following questions may arise:

- Do cycle times vary with each product?
- Should you use the process or individual unit cycle times?
- Should you use an average based on possible combinations?
- Do shift personalities affect cycle times?
- Do cycle times vary with the seasons?
- Do manual operations (load or unload time, fatigue factors) have an effect on cycle time?

Quantifying Total Defects and the Cost of Quality

Total defects are the total number of rejected, reworked, or scrapped parts produced during a period. Of course, the goal is to have zero defects.

The *cost of quality* is the cost of generating defects and inefficiencies of the process. To have a good system of cost of quality, the following guidelines may be helpful:

- Identify the sources of your data.
- Specify who collects the data.
- Specify who records the data.
- Identify how the data are reported.
- Specify who calculates and produces the report.
- Publish the results with dollars gained.

Process Line Considerations

In any maintenance environment, process considerations must be a critical part of the evaluation toward "zero defects." Typical considerations are

- Maintain reliability of all equipment, focusing on constraint machines.
- Manufacturing faster depends on the constraint machines.
- Provisions for accumulation of in-process material between machines.
- Process must accommodate stoppages or surges.
- Constraint machine is starved or blocked downstream.

Process Line Quality Issues

As you evaluate the machine or equipment, you must be also cognizant of the quality issues presented in the cycle of performance. Typical issues that you should look for and evaluate are

- Defects upstream are not serious unless they starve constraint.
- Defects at constraint or downstream points are more serious because they affect production output.
- Whether or not all quality defects should be recorded (e.g., upstream affects cost; downstream affects cost and production).

What to Do with All This Information

In order to figure out what the OEE is, you need to understand and measure its components, which are described in the following sections.

Measure Equipment Availability

Availability focuses on features and repairs. As already mentioned, availability is the measure of the degree to which machinery or equipment is in an operable and committable state at any point in time; specifically, the percentage of time that machinery or equipment will be operable when needed. Availability depends on reliability and maintainability (R&M), and it combines R&M into one measure.

Furthermore, it is impacted by the planned maintenance schedule for a particular machine or equipment. In fact, one of the main goals of a planned maintenance schedule is to maximize the availability of the machinery or equipment. It is essential to ensure that the machinery and equipment are designed at the highest possible level of R&M to maximize availability. The following areas have an impact on availability: breakdown loss, setup and adjustment loss, and others.

When calculating this metric, it is assumed that maintenance starts as soon as the failure is reported. So,

$$\text{Availability} = \frac{\text{MTBF}}{\text{MTBF} + \text{MTTR}}$$

Example

Calculate the Inherent Availability of a system that has a mean time between failures (MTBF) of 50 h and a mean time to repair (MTTR) of 1 h:

$$\text{Availability} = \frac{\text{MTBF}}{\text{MTBF} + \text{MTTR}} = \frac{50}{50 + 1} = 0.98 \text{ or } 98\%$$

Measure Equipment Setup Time

It is common knowledge that machines do not always perform exactly as designed the first time they operate. It takes them some time to come up to speed and turn out quality parts. The time it takes to reach this optimum production (stable production) is the set time. During this phase, companies experience scrap, rework, and delays.

Identify Other Production Losses

Quality problems happen any time the optimum design conditions do not exist at the moment when everything is set for production to begin. Any deviation from that plan is a quality issue—whether that part or service is used by the customer or not.

Measure Equipment Performance

Performance pertains to the speed of the machinery. In this area of investigation, performance reduction due to speed is of interest. There are two ways performance may be reduced because of loss of speed:

1. *Reduced operating speed:* These are losses that occur due to defects when the machine or equipment is operating at lower than recommended speeds. The reason for this lower speed is that operators are not sure of the quality level produced under normal speeds.
2. *Minor stoppages:* These are the minor interruptions that occur during production and usually are not recorded. The result is that you should include them as part of performance losses that reduce the product output. Unless you get a handle on minor stoppages, it will be very difficult to automate the machinery or equipment at hand, due to their irregular behavior.

A typical OEE form is shown in Figure 2.1.

The OEE measurable is meant to be used as a tool to track machine improvement progress. The key definitions are

- *Total available time* = The time that the equipment could run during a shift, given that there was no downtime either planned or unplanned.
- *Planned downtime* = The time that the equipment is down due to planned activities such as meetings, breaks, lunch, and so on.
- *Unplanned downtime* = The time that the equipment is down due to breakdowns, setups, adjustments, and so on.
- *Ideal cycle time* = Can be the best cycle time achieved, the design cycle time, or an estimation.

A typical first-time throughput form is shown in Figure 2.2.

First-time throughput = $C_1 \times C_2 \times C_3 \times C_4 \times C_5 \times C_6$

where $C_1, C_2, ..., C_6$ are the quality rates.

A typical dock-to-dock form is shown in Figure 2.3.

Part Number:		Part Name:	
Date:		Shift:	

Process # Description	A: # of parts entering process	B: (Scrap + on line rework + off line rework)	C: Quality rate: [(A-B)/A]
1			
2			
3			
4			
5			
6			

Figure 2.2   A typical first-time throughput form.

OEE and Capacity

To be sure, OEE is a shop floor tool. No doubt about it. However, it is certainly possible to use certain elements as useful reference information (e.g., it may be used as a benchmark for performance and capacity). To do this, it must be done in an appropriate manner.

In the past year or so, the application of OEE in the automotive industry is taking a new twist. It is looked upon as a metric to measure and evaluate capacity. It has the potential of helping in this process, if it is used properly.

As mentioned earlier, it is impossible to have 100% effectiveness in the long run because of maintenance, downtime, and setups. The guideline is that 85% is a realistic "world-class" value for "traditional" machines (see Figure 2.1). This implies that the installation, for example, produces 99% of the products "First Time Right within Specs"; that it operates at a speed of 95% of the theoretical maximum speed; and that it is actually running 90% of the operating time (99% quality × 95% speed × 90% running time = 85% effectiveness).

In a three-shift system, it means that the installation runs for 90% × 24 h = 21:36 h at 95% speed with 99% quality. Consequently, there will be 03:24 h available for maintenance, setups, and other possible waiting times. Incidentally, the 85% mentioned is a rather conservative figure; nowadays, automotive industry equipment typically runs over 90% in robotic applications. On the other hand, typical tooling machinery may be at 45%; plastic extrusion may be at 95%.

From my experience, the average assembly installation in a non-total preventive maintenance (TPM) environment runs at an effectiveness rate of 35%–45%. Of course, there are always cases that stick out: for example,

Part number: Part name:
Date: Shift:

Raw material inventory	Inventory @ Press line	WIP before weld	Inventory @ weld	WIP after weld	Repair area	Inventory @ paint line	Finished goods inventory	Total inventory (# of pieces)
Dock to Dock (Days)						(Total Inventory) (Daily Customer Requirements)		
	=					DAYS		

Figure 2.3 Dock-to-dock form (an example).

values in the pharmaceutical industry may lie considerably lower, and there are also cases that show considerably higher values.

So, is there really a steady or consistent standard for OEE? Well, let us examine first the 100% OEE level. The designation 100% OEE means the theoretical maximum capacity of the equipment. This is very important to keep in mind. The reason is that this 100% implies perfection: that everything is 100% all the time, every time. It is for this reason that it should be taken very seriously. If not, situations may (and do) occur where the shop floor is filled with scrap, the machine is suffering one breakdown after the other, and it still accounts for 80% or more OEE. How is that possible?

For example, if you have a 70% Availability, 80% Quality, 143% Performance = 80% OEE! However, the minute you have a performance over 100%, you are warned that the standard of choice—usually cycle time—is too low, and the OEE parameters are broken. The focus then is taken away from what it is all about: identifying and reducing losses.

In cases where the maximum speed has to be determined based upon a Best-Of-Best (BOB) analysis, it should be considered that this BOB is achieved under the former and current circumstances, including current losses. This is why in the automotive industry the OEE is gaining ground as a metric for capacity.

Because in the end, even standards are broken by product and equipment improvement; so the BOB should not be considered too easily as a maximum value. It only becomes a benchmark for improvement. As a rule of thumb, the BOB value should be raised with at least 10%–25% to serve as a standard. A well-chosen standard will only change when the product or the machine fundamentally changes.

So what is the relationship between efficiency, effectiveness, and productivity? And what is the path that can be followed to bring about "improvement" in capacity? Let us review again some of the fundamental definitions:

- *Efficiency is the allocation of resources within the organization.* That is determined by the amount of time, money, and energy (i.e., resources) that is necessary to obtain certain results. To meet your daily production quota, you commit a specific machine that uses up energy, you make operators and maintenance personnel available, and you provide raw materials. For example, if you are able to meet your daily production with less energy and fewer operators, you have operated more efficiently.

- *Effectiveness is the level of satisfying the customer's requirement via the set specifications.* That is determined by comparing what a process or installation *can* produce with what it *actually produces*; therefore, *effectiveness* does not tell anything about the *efficiency*—the amount of resources that have to be committed to obtain that output. However, if you are successful in manufacturing "more good product" in the same time period, effectiveness will increase.
- *Productivity is determined by looking at the production obtained (effectiveness) versus the invested effort in order to achieve the result (efficiency);* in other words, if you can achieve more with less effort, then productivity increases. Goldratt (1992) defines productivity as "the extent in which a company generates money." The goal of a production company is, therefore, not to reduce expenses but to generate as much money as possible!

A visual representation of the relationship between effectiveness, efficiency, and process performance is shown in Figure 2.4.

The most striking observation—when looking at the traditional approach to improvement—is that the focus is often exclusively on efficiency. That is interpreted as follows:

- How much room for improvement is still left at the input (efficiency) side? 10%? 20%?
- And does it still make sense to try to reduce another operator or engineer, or to put pressure on the buyers to negotiate even more competitive prices?

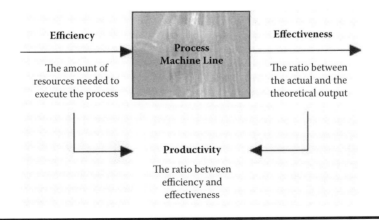

Figure 2.4 The relationship of effectiveness, efficiency, and process performance.

As is often the case, the previous question cannot directly be answered. If the supplier can give you a better price because you help him with managing his production process better or (as in the automotive industry) if you force the supplier to go through basic improvement processes (such as Six Sigma, SPC, lean manufacturing, TPM, and so on), not only will the price decrease but also the quality and the reliability in delivery will increase. That is good news for both parties.

However, if you manage solely by keeping the cost price down, you run a large risk of saving pennies per product but losing many dollars due to stoppages, quality losses, etc.; in other words, you become penny wise but pound foolish. There are many examples of this in many industries (see A. Kotch at www.oeetoolkit.com).

Strangely, the *output* side (i.e., the effectiveness) of the equipment is examined less often. Apparently, the output is more or less considered to be "as it is." However, every line manager knows that the installation will spontaneously start to run better simply by standing beside it and giving it attention. When you check the logbooks, there are days that the installation produced spectacular amounts of good output.

If you ask the team how that happened, you will hear a precise rundown of all elements that went right that day—for example, "the raw materials arrived on time and were of the correct quality; the installation kept on running and was set correctly; the right people were present; it was not too warm; etc., etc." This is often regarded as a fluke, however, so nobody is wondering how you could create a similar situation a second time. That is strange actually, for if it can happen once, why should it not be possible for it to happen again? And if it can happen a second time, why not always? Usually, a whole series of "Yes, but's" will follow.

Suppose you write down those "Yes, but's" and turn them into a list of action items. What would that give you? To be able to answer that question, we will have to dive a little deeper into the world of effectiveness. What determines the effectiveness of an installation? First of all, you must address the question of whether it does or does not run. Roughly, there can be three reasons why an installation is not running:

1. The installation stopped working; it broke down.
2. The installation could be running—technically speaking—but it is waiting for something, for example, materials, an operator, filling, to be set up, etc.
3. The installation could be running, but it is not planned to run because there is more capacity than demand.

Of course, the ideal machine would never break down and would never have to wait for anything; therefore, it would be running all the time as long as there is demand for the product. Subsequently, the effectiveness is determined by the speed at which the installation is running. This is always a tricky topic, for what is the maximum speed? The answer, of course, is the speed at which it is on the verge of breaking down. On the other hand, is it the speed at which the quality of the output reaches the bottom limit of its spec? The choice of definition will certainly guarantee a lively discussion.

It is useful to see that the maximum speed is often unknown, while the maximum speed that people come up with is usually based on various assumptions (which in turn could be turned into an interesting list of action points! At this point, by definition, the installation is *not* running at the maximum, but at its *optimum*).

An example of such an assumption is: "If I ran the machine any faster and the material got jammed, you would suffer major damage." Why does it get jammed? Is that always the case? When is it not the case? What must happen to prevent it from jamming again? Why does damage occur when it gets jammed? What do you have to do in order to …? etc.

If the actual speed is determined versus the theoretical speed, the next effectiveness-determining factor can be looked at: Does the realized output meet the set quality standards? It can be quite an eye-opener that if you ask 10 different people on the shop floor to indicate very clearly when a product does or does not meet the specs, you will receive 10 or even 11 different answers. It becomes even worse when it turns out that the one who produces the product (i.e., the operator) cannot determine (or cannot unequivocally determine) this.

Also, here lie many opportunities to solve all "Yes but's" and to ensure that the person who makes the product is also able to determine whether he is manufacturing a good product, so that he can keep the quality within preset specifications.

If it turns out that an installation has an effectiveness of 40%, while people always thought that there were limited options left for potential improvements, it is extremely good news: this means that twice as much good product can be manufactured (your effectiveness rate would be 80%!) at the present cost level. Or, you can manufacture the same product with one shift instead of two.

It is often assumed that achieving such improvements will necessitate an enormous increase in costs, for instance, for maintenance. That is sometimes partly true—for example, when it concerns overdue maintenance and you are then actually paying off a loan, because a fundamental design flaw has to be solved (and, therefore, you can also see this as paying off a postponed cost item). However, by activating the knowledge that is present on the shop floor in the right way, 80% of the improvements can often be implemented without any capital expenditures and at minimal costs.

It is not so hard to imagine that an installation that halts on a regular basis for various reasons, or whose process is not stable enough to operate at high speed without any losses in quality, automatically requires more resources at the *input* side as well! On the other hand, it may be that lowering the efficiency (for instance, by spending a little bit more money and time on preventive maintenance) will bring about a strong increase in effectiveness, which—bottom line—creates a higher net productivity. Such considerations can only be made if, in addition to efficiency, particular attention is paid to losses in effectiveness as well.

In all cases, it is necessary to take decisions concerning actions leading to improvement on the basis of facts and figures describing the entire productivity picture. Always remember that world-class manufacturing does not accept any losses at all. That is what management must focus on, and management must have the will to go further than mere window dressing and scratching the surface. Unfortunately, that is often even harder than just opening the wallet. Companies that do take this route discover over and over again: There still lies a nearly unlimited potential for improvement for those who learn to see it and seize it!

Using OEE in the Automotive Industry

Now that you have an overview of some of the concepts of OEE in relation to performance and capacity, let us explore the notion of OEE used as a metric in the automotive industry. In the past,

- There was a limited (if any) assessment of the supplier's capacity.
- Surrogate data were hardly ever used for comparison purposes.
- Capacity verification was conducted too late to make sufficient adjustments.

In other words, the relationship between original equipment manufacturer (OEM) and supplier was based on the concept of "trust me."

In the new era of manufacturing processes, the industry has begun to recognize that early capacity planning is an absolute must for both OEMs and suppliers. Therefore, the trend for the last 1–2 years has been to use the OEE as the "data-driven" methodology for capacity verification by being able to:

- Incorporate capacity analysis as early as possible.
- Incorporate standardized methods using surrogate data and risk sensitivity.
- Use the OEE as part of the Production Part Approval Process (PPAP) verification process.
- Improve the standard of comparison by using the OEE as a metric to measure the effectiveness of the specific OEM capacity plan.

In the new world of automotive manufacturing, the process of verifying capacity using the OEE takes three approaches that begin as early as possible in the sourcing phase, and it gets very specific toward the PPAP phase. The three approaches are

1. *Verify the supplier's demonstrated OEE in the earliest stages, based on surrogate data.* The supplier here has to demonstrate what he or she has done in the past with similar production specifications. It is a way to prove to the customer that the supplier has the capability based on current methods and resources to fulfill the expectations of the requirements. In essence, in this phase, the surrogate OEE should surpass or equal the required OEE set by the customer. (Demonstrated OEE is based on surrogate data during the early planning and actual data from the PPAP phases. On the other hand, required OEE is the minimum level of efficiency required to meet customer demand.)
2. *Verify the supplier's demonstrated OEE at the Run@Rate phase, using actual data from the Run@Rate study.* It is expected that at this stage the Run@Rate OEE should be greater or equal to the required OEE.
3. *Verify supplier's demonstrated OEE (usually) at the end of the PPAP phase.* This is the last opportunity for the supplier to demonstrate what the real capacity is. At this stage, the capacity verification is expected to be greater or equal to the required OEE.

To be sure, OEE is the product of availability (A), performance efficiency (PE), and quality rate (QR). In other words, OEE is indeed a measure of the overall effectiveness of a machine or a process. Another way of looking at this is to think of it as the percentage of time a machine or process is producing good parts during its scheduled operating pattern. The formulas for A, PE, and QR are

$$\text{Availability} = \frac{\text{Operating Time}}{\text{Net Available Time}}$$

where

Operating time = Running time
Net available time = Net operating time

$$\text{Performance Efficiency} = \frac{(\text{Total Parts Run}) \times (\text{Net Ideal Cycle Time})}{\text{Operating Time}}$$

where

(Total parts run) × (net ideal cycle time) = Actual output
Operating time = target output

$$\text{Quality rate} = \frac{(\text{Total Parts Run}) - (\text{Total Defects})}{\text{Total Parts Run}}$$

where

(Total parts run) − (total defects) = Good output
Total parts run = Actual output

Therefore,

$$\text{OEE} = \frac{(\text{Net Ideal Cycle Time}) \times (\text{Total Good Parts})}{\text{Net Available Time}}$$

It is interesting to note that in order to calculate the OEE in a simplified manner, you need only the net ideal cycle time (NICT), good parts produced (GPP), and net available time (NAT). The formula, then, for the OEE is

$$\text{OEE} = \frac{\text{GPP}}{(\text{NAT} \div \text{NICT})}$$

In other words, OEE is the ratio of the GPP (actual) divided by the maximum parts possible (theoretical). Remember that the demonstrated OEE is based on GPP from actual performance data (GPP/Maximum Possible

Parts (MPP)). On the other hand, Required OEE is the customer demand as defined in their requirements of yearly, monthly, weekly, or daily production volumes (Customer demand/Maximum Possible Parts). From this, we can algebraically see that:

$$\text{OEE} = \frac{\text{GPP}}{\text{MPP}} = \frac{(\text{Total Parts Run}) - (\text{Total defects})}{(\text{NAT}) \div (\text{NICT})}$$

Additional helpful information may be the Takt time and the cycle time.

Takt time is the *theoretical* time, and the Tact is the *actual* time. The calculations for both are the same. However, the data calculating the values will be different. [Takt is a word of German origin that means "rhythm," "pace"]

$$\text{Takt time} = \frac{\text{Available Time}}{\text{Customer Requirements}}$$

where
  Available time = run time if everything is perfect
  Customer requirement = customer demand
  Cycle time = total operating time. That is the time it takes a machine to produce a product and includes the load and unloading of the part in the machine

By using the ratio of Cycle time and Takt time, you can figure out how many operators you need:

$$\text{Number of operators needed} = \frac{\text{Total Cycle Time}}{\text{TAKT Time}}$$

Once these numbers have been gathered, then the analysis of measuring the supplier's capacity begins by comparing the required OEE to the OEE that the supplier has demonstrated. This is *not* the traditional approach to evaluating the OEE, so you *must* appreciate the deviation: whereas the traditional approach is more geared to evaluating or comparing the OEE in terms of the 85% world-class benchmark, the new approach is that the OEE of the supplier is used as a benchmark for improvement. That is, whatever the original OEE is, it must improve the demonstrated OEE rather than focusing to reduce the required OEE.

Another way of saying this is: the first priority is to address the demonstrated OEE, and then focus on the required OEE. The first is a calculated number based on existing parameters, and it indicates the potential OEE of a supplier. The second is specific to the customer requirement. That is a drastic change in thinking, however an excellent one. The reason is that now the supplier has plenty of time and opportunity to improve the demonstrated OEE before job 1.

Selecting the Appropriate Process for Demonstrated OEE

The selection of demonstrated OEE is based on choosing the appropriate and applicable process that is aligned with the planned process for the new part. Things to consider:

- Part complexity
- New or old site
- Layout comparison
- Similarity of sizes or cycle times
- Volume considerations
- Operating patterns
- Technology considerations
- Automated versus manual processes

The process selected need not be exactly from a common part or time frame. This is because the intent here is to see if the supplier can actually do what is expected of him. Does he or she have historical data that can demonstrate the capacity and/or performance that is needed? The answer obviously should be a resounding "yes." After all, the minimum information for addressing the OEE at this stage is threefold:

- NICT: This does not include the ramp time.
- NAT: This is for the specific part; it excludes planned downtime; it does not include downtime because of early quit time; and it includes changeover time for shared processes.
- GPP: Includes only the good parts through the process. All of them must actually be observed and verified rather than taken at face value from engineering standards or some other documents.

Demonstrated OEE focuses on trending and variation. That is why the calculation for the capacity uses the average OEE from each process to compare against the required OEE. Also, attention should be given to processes that perform for multiple customers. The allocation percentage obviously will be less than 100% for any process that is shared. The concern here is not that the capacity is fulfilled in total but whether the allocation is enough to support the specific customer.

How to Improve Demonstrated OEE

To improve the demonstrated OEE, the supplier may concentrate his or her efforts on any of the three components of the OEE. For example,

For *availability:*

- Reduce unplanned downtime.
- Identify and eliminate special and common causes.
- Analyze and improve MTBF, MTTR, and other reliability issues.
- Utilize TPM and rapid maintenance teams.
- Prioritize preventive maintenance.

For *efficiency:*

- Improve performance to constraint ideal cycle time.
- Identify the original Takt time and improve the Tact as well as cycle time.
- Assign appropriate resources to identify and close the gap.
- Identify and close the gap of imbalance operations with either appropriate operators or process flows.
- Identify any rogue operations and standardize them.

For *quality:*

- Identify, measure, and improve scrap and rework.
- Identify problem areas and install error or mistake proofing.
- Identify and communicate boundary samples. Ensure that the samples are not victims of degradation because of time or an open environment.
- Allocate appropriate and applicable problem resources and process experts to solve problems that may appear in the development of the OEE. Here, it must be noted that the demonstrated OEE is performed early and, as a result, there is plenty of time to implement OEE enablers

to improve the demonstrated OEE. Furthermore, it is imperative in this stage to realize that each and every action taken in the name of "OEE improvement" must be validated. (Remember: Improving the demonstrated OEE must be the first approach to close any gaps in either capacity or risk.)

Other:

- Understand the value stream so that installation of strategic buffers are placed to prevent shortages.
- Identify the path of communicating metrics and progress for the improvement plans. Remember that the data more likely are surrogate data and as such must be verified for accuracy, and must be taken from a stable process. The initial data are usually based on 25 samples. Also, the initial OEE, if low, may indicate excess available capacity. Therefore, the surrogate selection must be confirmed as appropriate and applicable for the situation. If the initial OEE is over 100%, it may indicate that the metrics are not identified correctly. The OEE should not exceed 100% at any time.
- Prepare simulations with multiple iterations to optimize throughput.
- Apply the knowledge gained from Things Gone Wrong (TGW) and Things Gone Right (TGR) (build on lessons learned).
- Investigate and apply appropriate knowledge from incremental tooling.
- If share equipment is present, investigate possibilities of reducing changeover times and verify that the allocation percentages are possible.
- Review and upgrade PM schedules as needed.
- Prioritize the action items.
- Verify all results, and translate them to OEE in such a way that is understood by everyone concerned.

Reducing Required OEE

Reducing required OEE is the second option of overall improvement. This option is traditionally associated with higher costs. However, if the gaps are identified early on in the process, there is a real opportunity at minimum investment.

One starts the analysis by looking at a single value stream and evaluating each stream for capacity risk. Generally, this analysis is set up on a part-specific basis. In some cases, it is possible to conduct the analysis as a group family of parts or to group similar manufacturing lines. In either case, the required OEE is the minimum level of efficiency that is required to support the demand, based on the defined operating pattern, allocation percentage, and operating parameters. Once all potential losses are identified and evaluated, a process may not be feasible, even if the required OEE is less than 100%. This is because the final determination is when the comparison takes place with the demonstrated OEE. If the demonstrated OEE has never achieved the level of the required OEE, what makes the supplier confident that the required OEE will be achieved? Obviously, at this point, more investigation needs to take place until either the demonstrated OEE is reevaluated or the required OEE is readjusted to the point where both the supplier and the OEM are satisfied. Typical fixes for the required OEE may be in terms of:

- Extra shifts
- Expanded hours
- Overtime
- Machine design
- Reduction of cycle time

Summary

This chapter focused on understanding loss of effectiveness by using the components of OEE. It also addressed the issue of using OEE as a metric for capacity. Chapter 3 will address the traditional method of measuring OEE.

Chapter 3

Measuring OEE: A Traditional Simple Approach

In Chapters 1 and 2, I have discussed some of the basic issues of productivity and effectiveness and the fundamentals of the associated mathematical calculations. This chapter covers the traditional approach of measuring Overall Equipment Effectiveness (OEE).

Measuring OEE is indeed a very important aspect of monitoring how the machines behave and which losses are reducing the effectiveness of which machine. By monitoring and measuring the behavior of machines on a regular basis, problems are identified early, and the effectiveness of the machines is accounted for. Furthermore, by monitoring machines and their behavior, the results of the effort toward improvement are evaluated and adjusted as needed. The evaluation for OEE begins with data collection and proceeds with the evaluation of the data using specific tools and methodologies.

Collecting Data about OEE

In any plant, procedures for manufacturing a product are all based on data that represent the manufacturing process. Because data is the basis of any plan, it should be precise and should measure the intended variables of a process.

Clarify Your Purpose for Collecting Data

Data can be gathered in different forms. First, the purpose of the data must be determined. Only then can a decision be made as to what kind of data would best serve the purpose. There can be many purposes for collecting data in a manufacturing plant. Some common purposes are as follows:

- *Understanding of the actual situation.* Data is collected to check the extent of the dispersion in part sizes coming from a machining process or to examine the percentage of defective parts contained in lots received. When the number of data items increases, they can be arranged statistically for easier understanding. Estimates can then be made concerning the manufacturing process or the machine and the condition of lots received through comparison with specified figures, standard figures, target figures, etc.
- *Analysis.* Analytical data may be used, for example, to examine the relationship between a defect and its cause. Data are collected by examining past results and making new tests. In this case, various statistical methods are used to obtain correct information.
- *Process control.* After investigating product quality, this data can be used to determine whether or not the manufacturing process is normal. Control charts are used in this evaluation, and action is taken on the basis of these data.
- *Regulation.* This data is used, for example, as the basis for raising or lowering the temperature of an electric furnace so that a standardized temperature level may be maintained. Actions are prescribed for each datum, and those measures must be taken accordingly.
- *Acceptance or rejection.* This type of data is used to approve or reject parts and products after inspection. There are two methods: total inspection and sampling. On the basis of the information obtained, a decision on what to do with the parts or products can be made.

It is often hard to obtain data in neat numerical values.

> *Example 1:* Whenever one tries to measure the softness of fabrics, plating luster, or the whiteness of paper, neat numerical figures such as the size or weight of an object are, naturally, impossible to obtain. Suppose you have to determine the softness of the three kinds of fabric. Even if you cannot measure the exact softness, you can arrange the fabrics by degree of softness, and you can thereby obtain excellent data.

Example 2: Similarly, the vibration of a machine or flickering during the projection of a motion picture would be difficult to measure with simple instruments alone, but five persons could test the machine or watch the movie and then report their observations—thus obtaining good data.

As already mentioned, the purpose of collecting data is not to put everything into neat figures but to provide a basis for action. The data can be in any form, but it is generally divided into the following two groups:

1. *Measurement data.* It (which is data that can be classified by size or time) can be categorized to show a number of variables in representing a process or a substance. This data may represent the cycle time of a machine, or it may represent the kilowatts (kW) that a machine consumes.
2. *Countable data.* It can be classified to keep track of two or more categories, such as the "yes" or "no" responses on a questionnaire, acceptable or defective products, or classification of employees as executives, office workers, or factory workers. The criteria used in classifying countable data are often called *attributes*. These data are used frequently by many factory and group leaders with long experience, who can draw appropriate conclusions from them.

Data Collection by Sampling

Data collection and analysis includes several techniques, some of which may be done independently or by the group as a whole:

- Obtaining a sample of the problem's occurrence
- Developing and using tally sheets to collect information
- Displaying and analyzing data in graphs and charts

Data is gathered before working on problems, in order to provide information to ensure that everything is within allowable tolerances and under control. Data can also be analyzed when things go out of control to find out what went wrong. Ideally, 100% of the department's output would be checked for problems. However, this is usually not feasible because of the time and cost that would be involved in such checking. Representative data can be acquired by closely inspecting carefully chosen samples, rather than the entire output. This will allow the use of a smaller number of individual pieces to find the problems or the quality of a lot.

What Is Sampling?

Sampling is making a generalization about a class based on a study of a few of its members.

- Testing a small amount of water from a swimming pool to check chlorinization of the whole pool.
- Polling 5% of TV viewers to see what programs are most frequently watched in the United States.
- Checking 10% of aspirin tablets for quality.

When Is Sampling Necessary?

Sampling is necessary when products or processes are too numerous to test them all. It would be impossible to test every screw that a five-stage screw machine produces. Also, it would be impossible for a pharmaceutical company to test all aspirin tablets. That is because when testing an aspirin, the product is destroyed. It is important to be very careful when sampling because, historically, most faulty generalizations are the result of improper sampling. The science of sampling is very sophisticated. Whole books have been written on the subject. Nevertheless, it is possible to draw some reasonable conclusions after learning some commonsense rules. Once the data is gathered, the most important thing to do first is to determine whether or not the data represent typical conditions. The problem can be stated as follows:

1. Will the data gathered reveal the facts? (This depends on the sampling method.)
2. Are the data collected, analyzed, and compared in such a way as to reveal the facts? (This depends on how data is processed.)

Why Collect Data?

Data is collected in order to identify the eventual "actionable root cause" of the problem through a thorough analysis using specific tools. Looking for root causes, first, you must look at the process to see if some data exist; second, you determine their usability and applicability to your study; and third, you decide what kind of additional data you will need to complete the study. If current data do not exist, then you evaluate your study based on the following three options:

1. *Technical*—What problems with the machinery or materials caused the failure?
2. *Detection*—What detection system failed (or did not exist) that should have caught the problem before it escaped?
3. *Systemic*—What flaws in the manufacturing process and management systems caused the problem and allowed it to continue unaddressed?

Root cause analysis requires a "systems" solution.

It is not enough to blame a problem on human error. You have to discover what, in your system, allowed the error to occur. You must consider all three areas in order to have a thorough root cause analysis.

Basic Tools for Evaluating Data

The following sections describe some of the basic tools used in evaluating data: brainstorming, check sheets, histograms, Pareto diagrams, cause-and-effect diagrams, scatter diagrams, run charts, Is-Is Not analysis, 5-Why Analysis, and process flow diagrams.

Brainstorming to Gather Ideas

Brainstorming is a group creativity technique designed to:

- Identify problems and generate a large number of ideas for the solution to a specific problem.
- Suggest causes of quality problems.
- Suggest ways to improve process.
- Suggest ways of implementing improvement activities.

The method was first popularized in the late 1930s by Alex F. Osborn. His original idea was that groups could double their creative output by using the method of brainstorming. Today, brainstorming is used primarily for three things:

1. To generate ideas by using the thinking capacity of a group of people in a specific area, with common ownership of a specific concern.
2. To encourage creative thinking by applying the principle of synergy—in other words, "the whole is greater than the sum of its parts."
3. To encourage team building, enhance enjoyment of work-related activities, and improve morale.

Procedures for Brainstorming

There are two steps in every brainstorming session. The first one is the *generation* of ideas, and the second is the *evaluation* of ideas. Both steps are independent of one another. The procedure for the first step—generation—is as follows:

1. Each person makes a list of ideas.
2. Sitting in a circle, each person takes a turn reading one idea at a time, starting at the top of their list.
3. As the ideas are read, they should be displayed so that all participants can see them.
4. If an idea has been read by someone else, read the next idea on your list, and so on.
5. The leader then asks each person, in turn, if he or she has any new ideas that they had not thought of before. It is very likely that, on seeing other people's ideas, the group will think of more ideas. This is called *Piggybacking*.
6. The leader continues asking each person, in turn, if there are any more ideas, until the group cannot think of any more ideas.
7. Reflection. Make sure that no personal attacks are vented at the end of the session.

The second step is evaluating the ideas generated by step one. Evaluation of ideas is similar to the generation step in the sense that the same people are involved, but more structure and rules characterize this activity. The focus is on the ideas themselves and not on the person who suggested that idea.

Rules for Brainstorming

The rules are very simple and straightforward. They are

- Do not criticize (by word or gesture) anyone's ideas—attack the idea, *not* the person who suggested it.
- Do not discuss or evaluate any ideas, except possibly for clarification.
- Do not hesitate to suggest an idea because you think it sounds "dumb." Many times, these ideas are the ones that lead to the solutions.
- Limit each team member to only one idea at a time.
- Do not allow negativism.

- Do not allow one or two individuals to dominate the brainstorming session: Everyone must get involved for maximum effectiveness.
- Do not have a gripe session.

Using Check Sheets to Collect Data

There are various ways to gather data. However, the easiest and most prevalent method is the use of *check sheets*. They are so widely used because each sheet is specifically designed for the collection of a specific set of data. Check sheets are simple, systematic ways to collect and organize data. Check sheets come in various forms; each application requires its own check sheet. They are used in many different ways for many different reasons. For example, golfers use them to save time and effort in keeping track of scores.

For industrial applications, two diverse examples of check sheets follow. To find out the number of defects a machine produces each day for one week, the check sheet might look like Table 3.1.

If the intent is to find out the number of defects per shift per day, the same check sheet might look like Table 3.2.

In some cases where the defect location is important, a check sheet may look like a "picture of the product" with the locations of the defect identified.

Designing a Check Sheet

As mentioned earlier, each individual need for data will require a specifically designed data-gathering sheet. Ideas on formatting can be drawn from Tables 3.1 and 3.2, but each check sheet should be different in order to collect the data required. The usual steps in designing a check sheet are

Table 3.1 Check Sheet—Number of Defects by Day

Check sheet—number of defects						
Date:						
Observer:						
	Monday	Tuesday	Wednesday	Thursday	Friday	Total
M/c 1						
M/c 2						
M/c 3						
M/c n						

Table 3.2 Check List—Number of Defects by Shift

Check sheet—number of defects																	
Date:																	
Observer:																	
Day	Monday			Tuesday			Wednesday			Thursday			Friday			Total		
Shift	1	2	3	1	2	3	1	2	3	1	2	3	1	2	3	1	2	3

1. Decide how you want to organize your data. Think of combinations of headings that will give the most useful information. Headings to consider might include
   - Frequency of occurrence
   - Places the problem occurs
   - Shift
   - Defect
   - Department
   - Employee
   - Machine
2. Specify the time period. This may range from a few hours to several months; also, think about factors that may affect the problem, such as
   - Start-up procedures
   - Specific reliability & maintain-ability (R&M) problems
   - Job change
   - Seasonal differences
3. Lay out the check sheet on paper; modify if necessary.

In deciding how to organize the data, do not pick the easiest method of data collection, but the one that is most likely to portray the problem in a realistic fashion from which it can be solved. The check sheet could be identifying a problem with a machine, with an employee or a part, in a specific department, or in many other ways or locations. The time period specified may be partly determined by how you organize the data. By planning the time period properly, you may be able to compare changes in a problem situation by comparing the data from one period to another. This way you could see the problem improving, worsening, or remaining the same.

Check Sheet for Measurable (Continuous) Data

Obtaining data for a process by checking off the size, weight, or diameter will produce a distribution of the measurement. For example, to find out how a machining operation is being performed, the check sheet might look like Table 3.3.

Table 3.3 Check Sheet—Part XYZ Outside Diameter

Check sheet		
Part XYZ outside diameter		
Machine: Number 21		Date: July 10, 2008
Operator: L. Horn		Inspector: Jessica S.
Measurement	*Frequency*	*Total*
1.55–1.64		
1.65–1.74	I	1
1.75–1.84	II	2
1.85–1.94	IIII	4
1.95–2.04	IIIII I	6
2.05–2.14	IIIII IIIII III	13
2.15–2.24	IIIII IIIII IIIII IIIII I	21
2.25–2.34	IIIII IIIII IIIII IIIII IIIII I	26
2.35–2.44	IIIII IIIII IIIII I	16
2.45–2.54	IIIII IIIII I	11
2.55–2.64	IIIII III	8
2.65–2.74	IIII	4
2.75–2.84	III	3
2.85–2.94	III	3
2.95–3.04		
3.05–3.14		
		Total: 120

In designing Table 3.3, it is important to keep in mind what the data will be used for. The sheet was designed to collect the outside diameter dimensions for part # XYZ, which is run on Machine 21 on the shift when Mr. L. Horn (a fictitious name) is the machine operator. This information constitutes the heading and part of the information needed to design the check sheet.

The other information needed to design the check sheet is the details of necessary data. In this case, the dimension was measured in increments of 0.10. Data could have been collected just as easily if the measurement recording increments were designed at 0.020. The measurement numbers would have been one-half as many, and the distribution produced by the data marks would have been higher and narrower.

The shape that the data makes on the completed check sheet represents the distribution of the data. If the check sheet is turned sideways, it can be seen that the data on this check sheet forms a bell-shaped curve (the normal distribution curve). A distribution that resembles a bell, whether it is flat or tall and thin, is called a *normal distribution*. The distribution could also turn out to be shaped differently. For example, the distribution, instead of being single-peaked and smooth, could have two or multiple peaks, or even be shifted in one way or another.

There are many inferences drawn from the shape and the location of the distribution. These inferences indicate what and where specific problems are likely to exist in the process. The location and identification of the problems in turn lead to their solution. A follow-up data collection, after the problems have been identified and corrected, should produce a distribution that resembles a normal bell curve. If the new data shows a distribution that is something other than a normal bell curve, either the right problem was not corrected or another problem (or problems) still exists.

Check Sheet for Attribute Counting

In situations where a product has various attributes or defects, a check sheet can be designed to collect data that will identify which defects are the most prevalent or where they occur on the product. A properly designed check sheet will show the number of defects in a product and how that number relates to the numbers of other defects that are found. A defect location check sheet will show which defects occur where on a given substance, and in what frequency. For example, in a situation where a certain process runs badly and produces various defects that cause the rejection of a large number of the parts, you might want to find the most frequent defect so that the

Table 3.4 Check Sheet for Attributing Counting

Check sheet		
Part number: 123 various defects		
Machine #: B		Date: July 10, 2008
Shift: 1		Inspector: Jamey
Type of defect	*Check*	*Total*
Failure defect	|||||	5
Scratch	||||| ||||	9
Incomplete	||||| ||||| ||	12
Burr	||||	4
Other	||||| ||	7
		Total: 42

problem can be remedied. This requires the design of a check sheet from which the data, when collected, will validate that information. A sample check sheet is shown in Table 3.4.

If you know that there are several types of defects occurring that cause the rejection of a product, and you want to find the location on the product where the rejects occur, you should design a check sheet that will allow you to find that out.

Example

Let us assume you are involved in assembling automobiles. The final inspector consistently finds many surface defects in the various external metal panels on the car. You might design a check sheet to show the metal panels and then mark the defect on the panel.

By finding the defects that occur most frequently and the exact location on the car where they occur, you should be able to trace the problem back to the place in the assembly process where it originates. With the proof provided on the check sheet, the existence of the problem can then be demonstrated, and it can then be rectified.

Using Histograms to Chart Data

After completing the check sheet and seeing the resulting data, the problem can often be identified without further work. At other times, the data must be shown as a graph or chart. There are many ways of charting

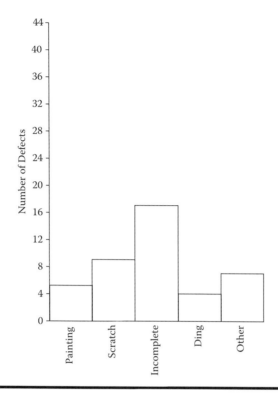

Figure 3.1   A typical histogram.

data; however, here, we will deal with one of the most simple methods—the histogram. A typical histogram is shown in Figure 3.1. Here, we will explain why histograms are useful, how to construct one, and how to interpret one.

Why Histograms Are Useful

First of all, a histogram is nothing more than a presentation of data in a pictorial form. Sometimes, histograms are called *bar charts*. The need for constructing histograms varies, but in an industrial setting, the need is usually to better convey to others the facts that the data represent. When conveying newly found information regarding a production process, it is very important that concise, easily understandable proof be shown to interested individuals. Histograms will readily allow them to see the information that the data contains.

How to Construct a Histogram

The check sheet that counts the amounts of the various defects found on part no. 123 (Table 3.4) can be used as the basis in constructing a histogram.

If you take that check sheet and turn it sideways, you obtain what already resembles a histogram.

When constructing a histogram, it is customary to scale the x-axis (horizontal axis) to indicate the different variables. The variables in this case are Painting Defect, Scratch, Incomplete, Ding (Burr), and Other. For other histograms, the variables might be different defects, different sizes, different colors, different items, etc.

Conversely, the y-axis (vertical axis) represents the amount of each of the variables. The amounts for each variable must be in the same increment in order to be in the same histogram. In other words, one variable cannot be measured in days, when others are measured in minutes. The unit of measure must be the same for all. Only in this way can all the variables be pictorially compared to each other. The data from the check sheet, then, can be represented in a histogram.

The length of line (space) that each variable represents on the horizontal (x) axis must be the same for each variable. The only variation on the chart can be in the vertical (y) axis. It is not important how narrow or how wide the spaces are on the x-axis, as long as they are the same for each variable. It does not matter how large the increments of measure are on the y-axis, as long as they are the same for all variables.

What a Histogram Shows

As stated earlier, a histogram is constructed for the purpose of pictorially representing a set of data. The reason for showing a histogram to someone is to more quickly convey what the data indicates. For that reason, the bars within a histogram are usually darkened (or colored) to show a better contrast on paper.

Using Pareto Diagrams to Chart Data

The Pareto principle states that there are "the important few" that represent the many. This principle applies to many things in life in general as well as in a company setting. Here are some examples:

- Company sales to only 20% of customers account for 80% of all sales.
- In purchasing, a small percentage of the purchase orders accounts for the bulk of the dollars of purchase.
- In inventory control, a small percentage of the catalog items accounts for most of the dollar inventory.

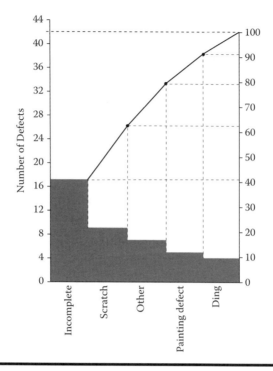

Figure 3.2 A typical Pareto diagram.

- In personnel relations, a small percentage of the employees accounts for most of the absenteeism.
- In machine or equipment, a small percentage of the equipment accounts for most of the delays and defects.

The principle is applicable universally. In fact, it is so common that it is called the 80:20 rule (80% of all problems come from 20% of the sources). Another way of putting this is that the "vital few" are more important than the "trivial many." A typical Pareto diagram is shown in Figure 3.2.

How to Use Pareto Diagrams

A major use of the Pareto principle is in deciding which quality improvement program to undertake. Here, the principle has so wide an application that no practical approach to quality improvement is possible without it. Improvement can be justified only for the few vital projects that have the greatest impact. It is these projects that contain the bulk of the opportunity for improvement of failure rates, quality costs, downtime, process yields, etc.

The "vital few" are identified through a *Pareto analysis*, by the use of *Pareto diagrams*. When applied to solving a problem, the Pareto diagram consists of a listing of contributions to the problem in the order of their importance. The use of Pareto diagrams in the course of problem-solving efforts can be extensive. They can be used to

- Identify the more serious problems.
- Dive deeper into a problem.
- Clarify the problem by representing information pictorially.
- Make management presentations.
- Make before-and-after pictorial comparisons.
- Track a problem and make sure it has been solved.
- Display progress pictorially to other employees.

How to Create a Pareto Diagram

The first step is to organize the data into a table by ranking the variables by their frequency and calculating the percentages of each, relative to all variables. In a manner similar to the construction of the histogram, construct a vertical and a horizontal axis:

- *On the horizontal axis*, include the variables from the tally sheet, and list the variable with the largest number of defects on the left side. Then, in descending order, list the remaining variables.
- *On the vertical scale*, indicate numbers that will represent the frequencies of defects. The bottom number should be zero, and the top of the scale should be the total number of frequencies.

Then, darken (or color) the bars in the graph, as done on the histogram.

However, unlike the histogram, the Pareto diagram will also show the relative percentages of each type of defect. Construct a second vertical axis on the right side of the far-right variable. Calibrate the right-side vertical axis in increments of 10%, starting with 0% at the bottom and ending with 100% at the top.

As you proceed to solve the problems, it might be useful to know the percentage of total identified problems being eliminated. To do that, you have to construct a cumulative line. You merely plot points on the Pareto diagram and draw a line that sums each column with the previous one—starting from the top of the first column.

At this point, your basic Pareto diagram is complete. It shows the relative frequencies of the various types of defects. Using this diagram as the basis, it is clear that to gain the most advantage, the problem to address first is the one closer to the vertical axis (the furthest left characteristic).

Addressing that problem and solving it will reduce the total defect rate by 40.5%. If the objective has been set as reducing the number of defects, the clear course to follow is to solve the problem that creates the most defects first, the second-most defects next, and so forth in that order, until all defect problems have been solved and eliminated.

With the Pareto diagram, you must be careful how you label the *x*-axis. By being very careful with the criteria, you can identify different priorities for prevalent problems. For example, if you begin your analysis with defects in a machine, your objective is to identify the most common defect. On the other hand, if you change the criteria to cost, now you may find that the order of priority is quite different because your objective is different. From this, you can conclude that it is extremely important to be certain of exactly what your objectives are and to ensure that the criteria you set out to investigate matches those objectives precisely.

Using Cause-and-Effect Diagrams to Display Data

When a problem is identified, it is important to collect, organize, and display data to provide pertinent information about the problem. However, in most instances, these steps do not provide the root cause or the solution of the problem. Therefore, a cause-and-effect (CE) tool quite often allows you to come up with a true solution, or solutions, to an identified problem. A typical CE analysis is shown in Figure 3.3.

All problems that are indicated (whether in a work environment or any other) lend themselves to being solved by the use of a CE diagram, or a fishbone diagram, or the Ishikawa diagram. A CE diagram helps you sort out the causes of dispersion or variation and organize all the mutual relationships.

An identified problem is often described in terms of the symptoms that it exhibits. The objective is to treat the cause of the problem, not the symptoms. Otherwise, the problem may reappear through the identification of some other symptom.

Example

You may replace the four tires on your car when it starts to run rough on your way home from work. The next morning, you find that your car will not even start. It is clear that you have been treating a *symptom* of the problem and not *the problem itself.*

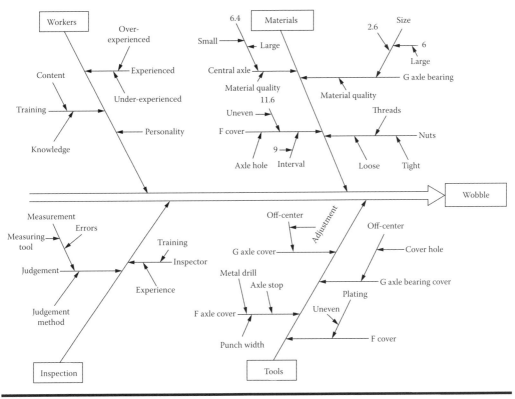

Figure 3.3 A typical cause-and-effect analysis diagram.

Therefore, it is very important that the problem is identified properly. In many cases, this is accomplished by collecting data and grouping it in such a way that a problem becomes evident.

Once a problem has been identified, you need to list causes that have an effect on that problem. At this stage, it is almost always necessary to accomplish that with a group of people. Identifying causes requires a fair amount of knowledge about the problem and the process where it exists. In most cases, the combined knowledge of people who are involved in the process provides enough information to identify all causes and also the solution to the problem. A group meeting produces the best results when conducted in a formal brainstorming format (as described earlier in this chapter).

Types of CE Relationships

Determining the real causes of problems is sometimes simple; more often, though, it is quite difficult. The following information on pinpointing causes of problems provides a good overview of the methods that should be considered. There are several kinds of CE relationships, which fall under two broad categories:

1. *Simple CE relationships.* Here, the causes are obvious because there is generally a one-to-one relationship between cause and effect.

    Example:
    Cause: There is oil on floor.
    Effect: Someone slips and falls.

2. *Complex CE relationships.* Here, the causes are more complicated, and there may be two or more causes at work.

    Example:
    Causes: Untrained operators; flawed material; damaged machine.
    Effect: Defective unit.

    – *Hidden cause.* These causes may be covered by something else, even though they may appear obvious.

    Example:
    Apparent cause: Flu bug.
    Effect: Excessive absenteeism.
    Real cause: Poor supervisor.

    – *Causal chain.* These causes may be affecting or triggering other causes and, therefore, create a chain of causes for the effect. In essence, they become compound causes.

    Example:
    Causes may be multiple:
      – Defective voltage regulator
      – Undependable temperature gauge
      – Too low machine temperature
    Effect: Poor paint job.

Common Mistakes in Analyzing the Cause

Many people tend to greatly oversimplify when they look for causes. There are many mistakes people make in establishing causal relationships; some of the most common mistakes are

- Reasoning that because there is a *correlation* between variables, there must also be *causation*.

    Example: There is a correlation between high training for sales and productivity. However, the cause of high sales may be that the product is very unique and customers want it regardless of who the salesperson is.

- Reasoning that because A happens *before* B, A was the *cause* of B.

    Example: The department hires a new supervisor. The next week production falls off. *Faulty conclusion:* The new supervisor caused the product halt.

People want life very simple: one cause, one effect. But real life is not like that. A cause may have multiple effects just as a symptom may have multiple causes.

- Looking for only one cause.

    Example: An electronics company experiences an increase in defective units. *Faulty conclusion:* The operators need training.

- Mistaking a symptom for a problem, and then treating the symptom.

    Example: A machine shaft rubs on the housing, causing an unbearable squeak. The supervisor greases the shaft and temporarily halts the noise. *Faulty conclusion*: The cause of the problem was the squeak.

Constructing a CE Diagram

If you take the relationship of CE as meaning that one influences the other, you can diagram that relationship as follows:

Cause ⟶ Effect

When applying this relationship to problem solving in your organization, the effect would be called a *problem*, and the problem would normally have several causes affecting it. In a manufacturing environment, virtually all causes fall under six broad categories that influence a problem. These are usually included as the basic branches to the cause side of the diagram, and they are

1. Manpower
2. Machine (equipment)
3. Method

4. Material
5. Measurement
6. Mother nature (environment)

The specific steps for the actual construction are

- *Step 1: Decide on the problem.* This is something you want to understand: It is the effect. It is the problem at hand.
- *Step 2: Write the problem (the effect) on the right side.* Draw a broad arrow from the left side to the right side.
- *Step 3: Write the main factors that may be causing this effect,* directing a branch arrow to the main arrow. It is recommended to group the major possible causal factors of dispersion into such items as the six categories just mentioned. Each individual group will form a branch.
- *Step 4:* Now, onto each of these branch items, *write the detailed factors that may be regarded as the causes.* These will be like twigs. Onto each of these, write in even more detailed factors, making smaller twigs. If you keep the following in mind, you cannot help but find the cause of the problem.
- *Step 5:* Finally, one must check to make certain that all the items that may be causing dispersion are included in the diagram. If they are, and the relationships of causes to effects are properly illustrated, then the diagram is complete.

Types of CE Diagrams

Depending on the problem and the process or context where it originates, the format of the CE diagram may vary. For example, the main categories of causes may not be "Manpower," "Materials," "Method," "Measurement," and so on. The process itself may be sequential, where a series of actions have to follow each other in a certain order. Generally, all CE diagrams fall into three categories:

1. *Dispersion and analysis type:* This diagram was presented earlier in the discussion of how to construct a CE diagram.
2. *Production process classification type:* This is a representation of the sequential steps in a process. This diagram is constructed similarly by starting on the left side with the sequential steps and ending on the right with the problem. An example process might be the baking of

bread. Another example may be sequential stamping, and so on. The construction of the production process classification type diagram is similar to that of the dispersion analysis type in that causal branches are similarly attributed to the steps of the process, and possible causes of the problem are brought out.

3. *Cause enumeration type:* This is simply a listing of causes that influence a problem. This listing should be performed in a group brainstorming environment and should produce a list of causes that will point to the correct solution or solutions to the problem.

In utilizing any of the three types of CE diagrams, you must remember that the solutions to the problem will originate from the listed or diagrammed causes that have been identified. The solution to the problem will only be as good, or correct, as the causes that have been identified.

Using Scatter Diagrams to Show Data

Scatter diagrams depict a pictorial relationship between cause and effect, between a dependent variable and an independent variable (for the "true" numerical relationship one has to do a correlation analysis). A typical *x*-and-*y*-axes scatter diagram is shown in Figure 3.4.

Basically, a scatter diagram is a grouping of plotted points on a two-axis graph. The pattern that the points form indicates if there is any connection between the two variables. The major objective of making a scatter diagram

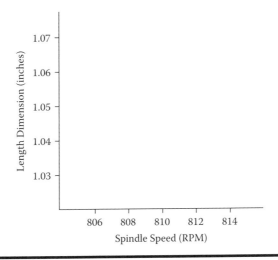

Figure 3.4 A typical *x*-and-*y*- axes scatter diagram.

is to find out if there is a correlation between two variables and, if there is, how much one variable influences the other.

By finding out whether one variable influences another and the extent of that influence, you may prove that a CE relationship exists and take appropriate measures to rectify the causal variable. To construct a scatter diagram, follow these steps:

1. *Select two variables that seem to be related in a CE relationship.*

The length dimension on a part you are machining on the lathe seems to vary as different spindle speeds are used in cutting the part off.
   In this example, the cause would seem to be the spindle speed and the undesirable effect, a varying length dimension.

2. *Construct a vertical axis* that will encompass all the values that you have obtained for the length dimension. Similarly, *construct a horizontal axis* that represents all the measured values for the spindle speeds.
3. *Plot all the measured values for the length dimensions at the different spindle speeds.*

In the sample scatter diagram relating the length of the product to the spindle speed, it can be seen that as the spindle speed increases, the length of the part increases (see Figure 3.5). The scatter diagram shows that a relationship exists between the two variables; that one causes the other to change. Knowing this, you can proceed to control the causal variable.

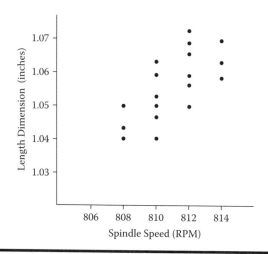

Figure 3.5   Scatter diagram for length dimension and spindle speed.

Reading Scatter Diagrams

In the sample scatter diagram relating the length of the product to the spindle speed, it can be seen that as the spindle speed increases, the length of the part increases (see Figure 3.5). The scatter diagram shows that a relationship exists between the two variables; that one causes the other to change. Knowing this, you can proceed to control the causal variable (spindle speed) and be assured that the part length will also be controlled to some extent in the same way.

Scatter diagrams can be varied. The way the plotted points fall depends on the collected data. The shape and the direction of the shape that they form determine whether the two variables have a relation at all and, if so, what the relation is and how strong it is. The following paragraphs describe the basic, general shapes of scatter diagrams.

Scatter Diagrams Showing Positive Correlation

Figure 3.6 shows a scatter diagram with a positive relationship. This diagram shows a distinct, positive correlation between X and Y:

- If X is increased, Y will also increase.
- If X is controlled, Y will be controlled.

The closeness of the plotted points to an imaginary line drawn through them indicates the degree of correlation that the Y variable has to the X variable. In this case, the value of X will quite accurately determine the value of Y.

Scatter Diagrams Showing Possible Positive Correlation

Figure 3.7 is a scatter diagram showing a possible positive correlation between two variables. The plotted dots are in a more dispersed shape, but they still lie in a general direction from the lower left to the upper right. This

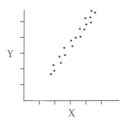

Figure 3.6 Scatter diagram—positive relationship.

Figure 3.7 Scatter diagram showing possible positive correlation between two variables.

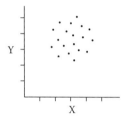

Figure 3.8 Scatter diagram showing no correlation between two variables.

pattern of plotted points indicates that as *X* is changed, *Y* will also change somewhat, usually in the same direction. One cannot, in this case, be sure that changing *X* will equally change *Y*. This point pattern indicates that *X* does have some influence on how *Y* will change, but that other factors are present that will also influence how *Y* changes.

Scatter Diagrams Showing No Correlation

Figure 3.8 is a scatter diagram showing no correlation between two variables. Dots that seem to fall all over the two-axis plot or form a circular pattern indicate that *Y* is not at all influenced by *X*. If *X* increases, the value of *Y* will increase as often as it will decrease. This pattern does indicate that both *X* and *Y* change, but that they change independently of each other. In a case like this, some other variable (cause) must be found that does influence the variable *Y* (effect).

Scatter Diagrams Showing Possible Negative Correlation

Figure 3.9 is a scatter diagram showing a possible negative correlation. The plotted dots again form a dispersed shape that generally lies in the same direction, from the upper left to the lower right. The correlation of *X* and *Y* is similar to that in Figure 3.7. However, in this case, the direction is

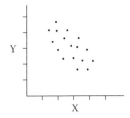

Figure 3.9 Possible negative correlation.

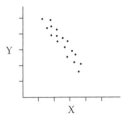

Figure 3.10 Negative correlation.

different: as X increases, Y is likely to decrease. This relationship is referred to as *possible negative correlation*.

Scatter Diagrams Showing Negative Correlation

Figure 3.10 is a scatter diagram showing a negative correlation. The dots form a tight pattern approximating a wide line. The pattern lies in a distinct direction from the upper left to the lower right. There is definite correlation between X and Y: if X increases, Y will decrease. When a tight pattern, as shown, is obtained, you can be assured that there exists a CE relationship between the two variables.

Using Run Charts to Display Data

A run chart is a graph showing incident rates over time (see Figure 3.11). It may show if incidents are increasing or decreasing. For example, a run chart may be initiated to find out what is happening to the parts per million (ppm) over time.

The obvious benefit here is that it is a visual picture showing the team the direction it is headed. Run charts show a series of results over time. You can use them to determine if a pattern exists and if a run is increasing, staying the same, or decreasing.

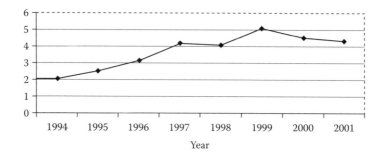

Figure 3.11 A typical run chart.

Table 3.5 A typical Form for Is-Is Not Analysis

Problem	Is	Is Not	Need more information "Parking Lot"
What	Object:		
	Defect:		
Where			
When			
How big			

Using Is-Is Not Analysis to Compare Data

Is-Is Not is a simple but powerful diagnostic tool to help you eliminate incorrect root causes of your problem. A typical form for this is shown in Table 3.5.

You have probably done many informal Is-Is Not analyses in solving simple household problems.

Example

If you were sitting at home reading a book one night (undoubtedly a book on root cause analysis) and your reading light suddenly went out, you might conduct several simple Is-Is Not experiments to determine the root cause. You might start by asking, "Did other lights go out at the same time?" If not, you could say, "No, the problem is not with other lights, it is only with this light." This would save you the trouble of calling the electric company, checking all your circuit breakers, etc.

Table 3.6. A Typical Is-Is Not Matrix

Reading light goes out

Category	Problem description			Potential root cause theories			
	Is	Is Not		External power failure	Burnt-out bulb	Blown circuit breaker	Shorted lamp fixture
What?							
(Part or process)	Reading light	Other lights, appliances		N	Y	N	Y
(Defect or concern)	Goes out	Dim light		N	Y	N	Y
	Will not relight	Flickering light					
Who?							
(Customer/operation)	Reader	Other occupants		N	Y	N	Y
Where?							
(In process)	Light fixture	Light bulb		N	N	N	Y
		Circuit breaker					
		Outside power source					
When?							
(Date or time)	10 min ago	Earlier		Y	Y	Y	Y
	Continuing	Intermittent					
(1st observed)							
How many?	One lamp	More lights and appliances		N	Y	N	Y

You might then try a new light bulb, asking the question: "Is the problem with the light bulb?" If you tried a new bulb and it also did not work, you would say "No, the problem is not with the light bulb." This would save you the expense of throwing out a perfectly good light bulb. You would then go on to conduct further experiments on the wall plug, lamp, etc., eliminating further potential root causes, until you had the real root cause isolated.

Before beginning an Is-Is Not analysis in your work, ask these questions:

- Is the cause of the problem or deviation known?
- Do you need to know the cause of the problem or deviation?
- If the answers are "no" to question 1 and "yes" to question 2, an Is-Is Not analysis may be appropriate.

To facilitate your Is-Is Not analysis, make sure you break down questions into categories. A typical approach to getting the facts and filling the form is to use both the Is and Is Not investigative questions to identify and categorize as much information as you can about your problem. Remember to look for the closest relationship in answering the Is Not questions—category, form, function, etc. The typical approach is

- What is the problem: the object with the defect? (What object could you expect to have the defect but does not?)
- What is the defect? (What defects could you expect to see with this object, but do not?)
- Where on the object do you see the defect? (Where on the object might you see a defect, but do not?)
- Where geographically do you have the problem? (Where geographically could you have this problem, but do not?)
- Where in the process (operation, product life cycle) do you see the problem? (Where in the process (operation, product life cycle) could you see this problem, but do not?)
- When did you see the problem for the first time? (calendar/clock/shift) (When could you have seen this problem for the first time, but did not?)
- When else are you seeing the problem? (When else could you see the problem, but do not?)
- How many instances of this problem do you have: how many objects have this defect? (How many instances of this problem could you have—how many objects could have this defect—but do not?)

- How many defects do you have? (How many defects could you have, but do not?)
- What is the trend of the problem? (What trend could you see with this problem, but do not?)

Note that in the "What" category, you should identify both the object of the problem and the defect. The "Parking Lot" column allows you to note where you need more information.

Determining the root cause using Is-Is Not data can be used to systematically eliminate unlikely causes and to validate root causes. It is important to have the most current and accurate data and information possible. The best candidates for root cause will be theories able to support why both Is and Is Not are true. Concise questioning is an important technique for determining the root cause. Theories can be created by examining the unique characteristics associated with the problem and changes (i.e., modifications, revisions, and adjustments) that have been performed before the problem's occurrence.

The binary question should be responded to with a "Yes" or "No." The question may be used by the Team Leader or Facilitator to confirm that the theory can explain why both the Is and Is Not are true. The question may be stated as follows: Does the theory explain why both the Is and Is Not are true?

- Y = Yes, the theory does explain why both are true.
- N = No, the theory does not explain why both are true.

A matrix is developed with the Is-Is Not vertically listed on the left and the theories across the top to the right (see Figure 3.6 for a sample). One theory should be tested at a time against each Is-Is Not pair. The team should briefly discuss each point so as not to rush through and create inaccuracies. If the team has reached an impasse at one particular point, it should proceed with the other Is-Is Not pairs and return to the point in question.

There can be more than one root cause. Once you have eliminated potential root causes using Is-Is Not, you can take a deep dive into the real root causes. One of the most effective techniques for uncovering the real root cause is the 5-Why Analysis.

Using 5-Why Analysis to Identify the Root Cause of a Problem

One of the most common techniques to "deep dive" into a problem (see Figure 3.12) is 5-Why Analysis.

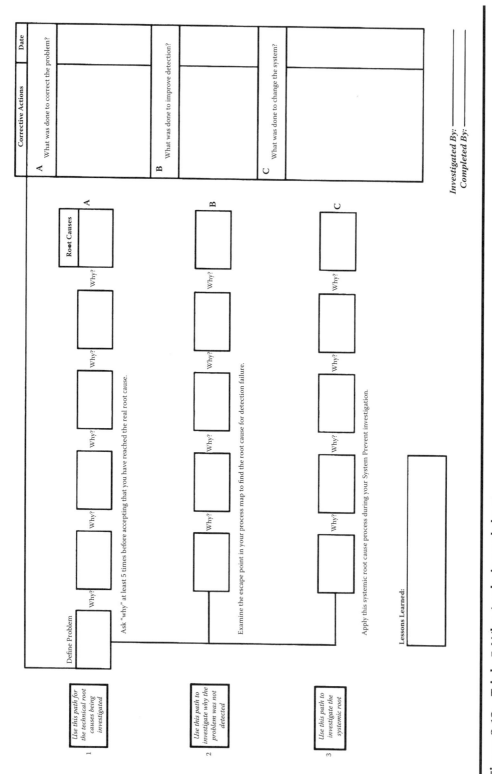

Figure 3.12 Triple 5-Why Analysis worksheet.

Fundamentally, it is an exercise involving a series of questions. In the process of the 5-Why, one keeps asking "Why" questions until the root cause is arrived at. That is a deep dive into one root cause. You know when you have reached that point, when you have "no answer" for the specific "why."

Example

1. Why were you late for work this morning?

    Answer. My car would not start.

2. Why would not your car start?

    Answer: The starter would not turn the engine over.

3. Why would not the starter turn the engine over?

    Answer: The battery was too weak to engage the starter. I had to get a friend to jump start my car.

4. Why was the battery too weak to start the car?

    Answer: The alternator was not producing enough energy to keep the battery charged.

5. Why was the alternator output low?

    Answer: The alternator drive belt was slipping, which my mechanic corrected by tightening the belt.

Repeatedly asking "why" (also known as a 5-Why Analysis) is designed to help you take a deep dive into one potential problem. To do the analysis effectively, you must continue to dive deeply into the same question (and not go off on tangents), and you must ask "why" at least 5 times. It is unlikely you will reach the real root cause in fewer than 5 questions. It is important that your 5-Why Analysis be based on fact (see the preceding example). If you reach a speculative answer, stop and verify the answer before proceeding.

Sometimes your answers to a 5-Why question may give you multiple answers. It is okay to branch off in more than one direction as long as you follow each branch to the root cause. 5-Why Analysis should be used to reach the real root cause of technical, detection, and systemic failures. In other words, you must ask

- Why did the failure occur?
- Why was it not detected before it got to the customer?
- Why were your management systems inadequate to prevent the failure?

Using Process Flow Diagrams to Visualize Processes

This flow diagram is not to be confused with the value stream diagram that was discussed in Chapter 1. This process flow diagram, generally, is a micro diagram of the process (see Figure 1.1). In other words, it is a visual graphic to break down a process into *all* (value and nonvalue) individual steps or activities. It is this characteristic that makes this tool—quite often—the first tool of choice in the "root cause analysis."

Furthermore, it establishes the current state of the machine or the process. Therefore, when you map your process, you should always look for:

- Control points (places in the process where compliance with a standard is monitored)
- Escape points (the earliest point in the process, closest to the root cause, where the problem could have been detected, but was not)
- Potential for human error (where could a human mistake allow a process to falter?)
- Detecting and correcting escape points in your process

In evaluating the process map, try to find ways of eliminating potential errors in your process, by considering prevention systems in the process or at minimum, detection systems. Prevention helps the system avoid problems, whereas detection prevents escape to the customer.

Example

- Does the system automatically shut down when a problem is detected?
- Does your detection system assume your part is good or bad? Detection systems should be set up to reject all parts that cannot prove they are good (rather than accept all parts unless proved bad).
- Have parts and processes been designed to minimize the potential for human error? For example, has the process been designed so that it is only possible to feed the correct end of the part into the machine first?

Finally, the benefit of the process flow diagram (mapping) is that it provides a visual rendering of the process and makes the problem-solving team thoroughly evaluate all the steps needed to complete a process. It acts as the starting point for identifying waste in the process and "weak points" where errors may occur.

Using Kanban to Collect and Analyze Equipment Repair Data

Traditionally, Kanban is used to minimize inventory in a manufacturing process. However, Kanban may also be used to collect and analyze data in repair situations. It is based on the supermarket system. That is, as stock is used, it is replenished as needed. In the manufacturing Kanban system, a machine shop supplies components to final assembly. Assembly is a manual operation with little setup and produces in lot sizes of one, to customer requirements. Machining is more automated and has significant setup costs. Machining produces in batches to amortize the setup and sequence parts to minimize tool changes.

A small quantity of each part is maintained at machining. By observing the quantities, the machinists know what products need to be made. The key points for any Kanban system in reference to machines and equipment are

- Stockpoints
- A withdrawal signal
- Immediate feedback
- Frequent replenishment

These points are very important in planning maintenance as well as reordering spare parts on time appropriately. It is precisely these points that demand accurate data, as well as the next topic of Kaizen, which I have incorporated in this discussion.

Using Kaizen to Improve OEE

Kaizen was created in Japan following World War II. The word *Kaizen* means *continuous improvement*. It comes from the Japanese words *Kai* meaning "school" and *Zen* meaning "wisdom." Kaizen is a system that involves every employee—from upper management to the cleaning crew. Everyone is encouraged to come up with suggestions for small improvements on a regular basis. This is not a once-a-month or once-a-year activity; instead, it is *continuous*.

In Japanese companies, such as Toyota and Canon, a total of 60 to 70 suggestions per employee per year are written down, shared, and implemented. In most cases, these are not ideas for major changes. Kaizen is based on making little changes on a regular basis to always improve productivity, improve safety, improve effectiveness, while reducing waste.

Suggestions are not limited to a specific area such as production or marketing. Kaizen is based on making changes *anywhere* that improvements can be made. In the machine and equipment domain, Western philosophy may be summarized as

> "if it isn't broke, don't fix it."

In contrast, the Kaizen philosophy is to:

> "do it better, make it better, and improve it even if it isn't broken, because if you don't, you can't compete with those who do."

Kaizen in Japan is a system of improvement that includes both home and business life. It even includes social activities. It is a concept that is applied in every aspect of a person's life.

In business, Kaizen encompasses many of the components of Japanese businesses that have been seen as a part of their success. Quality circles, automation, suggestion systems, just-in-time delivery, Kanban, and 5S are all included within the Kaizen system of running a business. Kaizen involves *setting* standards and then continually *improving* those standards. To support the higher standards, Kaizen also involves providing the training, materials, and supervision that are needed for employees to achieve the higher standards and maintain their ability to meet those standards on an ongoing basis.

In the area of machines and equipment, Kaizen plays a major role because it involves the local force for improvement. The local force is

- Operators
- Technicians and engineers
- Maintenance
- Supervisor

The primary responsibility of the local force is to see that the processes and the systems, as given, are run in a stable and predictable condition. That means that they have the responsibility to find, correct, and prevent any special cause from happening.

The secondary responsibility is to give management ideas on how to change processes for the better. Typical things that the local workforce may do to start the improvement process are to ask questions such as:

- What was involved?
- What has changed?
- Is the change good or bad?
- When did it occur?
- Where did it occur?
- Who was affected or involved?
- How big is the change?

On the other hand, it also recognizes the responsibility of management to see that the local workforce has good processes, and good systems that are continually improving forever!

Using Control Charts to Show Variations in Processes

Just as the Kanban and Kaizen generate data, control charts are used for information about the process and the machine. Specifically, control charts are part of the Statistical Process Control (SPC) methodology to identify variation. A typical control chart is shown in Figure 3.13.

In reference to machines, that variation may be due to

- People—abilities, training, motivation.

(*Consider:* The training they have received and motivation they are receiving—Is it adequate?)

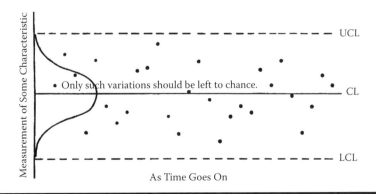

Figure 3.13 A typical control chart.

- Environment—motivation, effectiveness, maintenance.

(*Consider:* Is the physical environment motivating? Are things well laid out for good work? Is the company culture motivating?)

- Materials—practice, usability, variability.

(*Consider:* Properties of particular materials—Are they usable? Are they the right (appropriate and applicable) material for this particular machine to run on? Is the material focusing on reducing variability? Is the cost of the material accounting for the decision or is the cost evaluated based on Life-Cycle Costing?)

- Measurement—accuracy, repeatability, usability.

(*Consider:* Is the measurement system accurate? Is it repeatable, reproducible? Is it easy to use? Have linearity and bias analyses been performed?)

- Machine (itself)—ability, usability.

(*Consider:* Does the machine have the capability? Has it been set up with the operator in mind? Is it maintained well?)

- Methods—clarity, usability, conformance.

(*Consider:* Are the methods applicable, usable, well thought out? Can the people conform to them? Are they actionable?)

Why a control chart? In using control charts, it is important to remember that they will tell *when* to look for problems in the process. They cannot tell *where* to look or *what* the cause of problems is. Updating the control chart with each new point will signal whether the process is operating under control or out of control. If the process is operating under control but exhibiting a trend that will shortly make it out of control, the process should be investigated, and the problem rectified before the process gets out of control.

To find out *where* to look for *what* cause is the most important aspect of statistical process control. Without finding out the problem and rectifying it, the whole effort of developing and maintaining control charts is senseless. To locate the problem causing the process to go out of control, an analysis must be made of the data on the control chart. By interpreting the pattern of points on the chart, one should be able to trace what changes have taken place in the production process.

Control charting has been in use as a tool for monitoring the quality of manufactured products for some time, primarily by quality control and quality assurance employees. Through this application, the format of often-used control charts has been somewhat standardized. Samples of calculation work sheets and control charts for both variable data and attribute data are shown in Figures 3.14 and 3.15.

There are many control charts you can use. Some of the most common are

- *Basic charts for variable data* in which each point represents the most recent data, including X-Bar and R charts, X-Bar and S charts, X-Bar and S-squared charts, Median and Range charts, and individual charts based on X and MR.
- *Basic charts for attribute data,* including P, NP, U, and C charts.
- *Time-weighted charts* in which the points plotted are calculated from both current and historical data, including MA, EWMA, and CuSum charts.

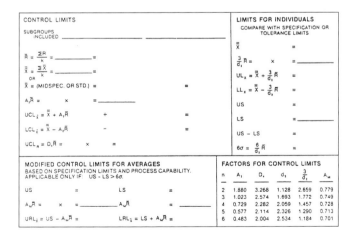

Figure 3.14 Calculation work sheet for variable chart.

Defectives	Defects
Defectives ≅ Count 1 if at least one defect	Defect ≅ Count all the defects
n* = Number in subgroup Statistic = p = fraction (per cent) Defective $$P = \frac{\text{Number of Defectives in Subgroup}}{\text{Number in Subgroup}}$$ $$\bar{p} = \frac{\text{Total Number Defectives}}{\text{Total Number Inspected}}$$ $$CL = \bar{p} \pm \frac{3\sqrt{\bar{p}(1-\bar{p})}}{\sqrt{n}}$$ Note: Use when subgroup size varies – can use when subgroup is constant	n* = Number in subgroup Statistic = u = Number of Defects per Unit $$u = \frac{\text{Total of defects in Subgroup}}{\text{Number In Subgroup}}$$ $$\bar{u} = \frac{\text{Total Number Defects}}{\text{Total Number Inspected}}$$ $$CL = \bar{u} \pm \frac{3\sqrt{\bar{u}}}{\sqrt{n}}$$ Note: Can be used when subgroup varies and Number in subgroup can be greater than one.
n* = Number in subgroup Statistic = pn = Number of Defectives pn = Number of defectives in Subgroup $$\overline{pn} = \frac{\text{Total Number of Defectives}}{\text{Total Number of Subgroups}}$$ $$CL = \overline{pn} \pm 3\sqrt{\overline{pn}(1-\bar{p})} \text{ where}$$ $$\bar{p} = \frac{\text{Total Number Defectives}}{\text{Total Number Inspected}}$$ Note: Use when Subgroup size is constant	n* = Number in subgroup Statistic = c = Number of Defects in subgroup $$\bar{c} = \frac{\text{Total Number of Defects}}{\text{Total Number of Subgroups}}$$ $$CL = \bar{c} \pm 3\sqrt{\bar{c}}$$ Note: 1) Use when subgroup size is constant. 2) Number in subgroup can be greater than one. 3) If number in subgroup is one, the "c" chart and "u" chart are identical.

Figure 3.15 Calculation work sheet for attribute charts.

- *Multivariate charts*, designed for situations where multiple correlated measurements are collected.
- *ARIMA control charts* for autocorrelated data in which the samples collected from one time period to the next are not independent.
- *Toolwear charts* for monitoring data that is expected to follow a trend line and not remain constant at a fixed level.
- *Acceptance control charts* for high C_{pk} processes, where the control limits are placed at a fixed distance from the specification limits rather than the center line of the chart.
- *CuScore charts*, which are designed to detect specific types of patterns when they occur.

However, for machine and equipment, the predominant charts are variable data charts and attribute data charts, which are described in the following sections.

Variable Data

As discussed earlier, variable data can take any value within a given range, and it is restricted to a particular value because of the limitations of the measuring device. When variable data are collected, then you may use:

- *X-Bar chart*: This is a chart of the average of the measurements in a sample. The X chart indicates when a change in central tendency (location) has occurred.
- *Median chart*: This is a chart of the median value of the sample. Its advantage over the X chart is that no computation is required; however, the plotted chart shows greater variability.
- *R chart*: This is a chart of the range of the measurements in a sample. The chart of R values indicates when a significant change in the variability occurs.
- *S chart*: This is a chart of the standard deviations of each sample. This chart is becoming more widely used because computers and calculators with average and standard deviation calculations are available. It is used when high sensitivity of the samples is required.
- *Chart for individuals*: This is a chart where the sample size is limited to one sample because of the cost or homogeneity of the sample. The reading is plotted, and a moving range is used for the MR chart.

Attribute Data

As discussed earlier, this can take only a particular set of values. For example, the number of defectives in a lot must take integer values such as 0, 1, etc. Values other than integers can exist as averages.

- *P chart*: This is a chart of the percent or fraction defective in the sample. The P chart is to the trend of the fraction defective as the X chart is to the trend of the averages. The sample is variable.
- *C chart*: This is a chart of the number of defects per unit where the size is constant. An example is the number of defects per square inch in a protected layer, or the amount of defects per shift on a particular machine. The defects are simply counted and plotted on a chart.

- *NP chart:* This is a chart of the number of defective parts found in the sample. The sample is constant.
- *U chart:* This is a chart of the number of defects per unit, where the sample size varies, that is, is not constant, as in the preceding C chart.

Steps for Control Charting a Production Process

In general terms, the steps are:

- *Step 1: Select the items that should be controlled.* First, decide which problems are to be dealt with and for what purpose. On the basis of this decision, it should be clear what data will be needed.
- *Step 2: Decide which type of control chart to use,* and which chart within that type will be appropriate.
- *Step 3: Make a control chart for process analysis.* Take data for a certain period of time or use data from the past in making the chart. If any points are abnormal, investigate the cause and take action. The cause of the change in quality is studied by rearranging the subgroupings, stratified data, and so on.
- *Step 4: Construct a control chart for process control.* Assume that action has been taken to deal with the cause of the quality change, and the production process is controlled. Now, see if the product satisfies the standards for this state. On the basis of these conclusions, standardize the working methods (or reform them if needed). Extend the control lines of the chart at stable situation, and continue plotting the daily data.
- *Step 5: Control the production process.* If the standardized working methods are being maintained, the control chart should show a controlled state. If an abnormality appears on the chart, investigate the cause immediately, and take appropriate action.
- *Step 6: Recompute the control lines.* If the equipment or the working methods are changed, the control lines must be recomputed. If control over the production process is accomplished smoothly, the quality level of the control chart should keep improving. In this case, make periodic reviews of the control lines. The following rules should be observed in recomputing the control lines:
  - First and foremost, you must know the cause of the change. If you change the control limits without knowing why the change occurred, you will find yourself changing limits all the time. That is not acceptable.

- Second, you should never have specifications and control limits on the same chart. Control limits are calculated based on what the process is doing; therefore, they are process dependent. On the other hand, specifications are what the customer wants and, therefore, are customer dependent.
- When you are recomputing, do not include data on points that indicate an abnormality and for which the cause has been found and corrected.
- Do not include data on abnormal points for which the cause cannot be found or no action can be taken.

Interpreting Control Charts

When you look at the control chart, there are four possible states that the process is in. They are

1. *Ideal state:* The process is "in control," or 100% of the product is conforming.
2. *Threshold state:* The process is "in control," but some nonconforming product is produced.
3. *Brink of chaos:* The process is "out of control," and some nonconforming product is produced.
4. *Chaos:* The process is "out of control," and 100% nonconforming product is produced.

The control charts ideally then help because

- They determine the process limits by detecting nonrandom variations.
- They monitor the process to determine when it is nonrandom—something changing.
- They are used with a small sample instead of 100% testing.

The most common ways to detect this change in variation are to look for:

- Out-of-control points (these are points beyond the upper or lower limit).
- Length of run or trend. How many points are there in a run? For example, if the number of points is 10, 20, 30, or 40, then the times crossing the center should be 2–9, 5–6, 9–12, or 13–28, respectively. If the times crossing the central line are more or less than those given here, then significant changes are taking place in the process.

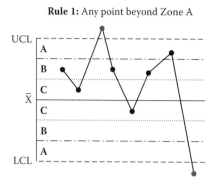
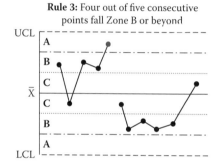
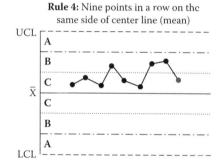

Figure 3.16 Zone control.

- Trends. Watch for a run of seven (7) points up or down. The points must cross the center line. If they do not, then it is called a run.
- Zone control. Points in one of the three zones. Each zone reflects one standard deviation from the center line. See Figure 3.16.

Summary

In this chapter, the measurement of OEE has been discussed, focusing on data collection, basic tools, and introducing Kanban, Kaizen, as well as control charts for purposes of both collecting and evaluating data of processes and machinery.

Chapter 4 continues with measuring OEE; however, more advanced topics such as reliability issues will be covered.

Chapter 4
Measuring OEE: The Reliability Approach

This chapter continues the discussion of measuring Overall Equipment Effectiveness (OEE), focusing on some advanced techniques using reliability.

Reliability is the probability that machinery or equipment can perform continuously, without failure, for a specified interval of time when operating under stated conditions. *Increased reliability* implies less failure of the machinery and, consequently, less downtime and loss of production. In other words, it is a statistical measure of equipment or component performance. It is normally measured as a function of time or number of cycles. For ease of interpretation, and in order to avoid calculating probabilities of occurrence, these can be defined in general terms as follows:

> The standard statistical measure of performance is the amount of time of operation or operating cycles divided by the number of failures or machine events.

Reliability can be measured against a time reference point. This point is identified as the operating time or the mission time for the machine or equipment. The mission time can be defined as the amount of uptime the machinery or equipment is scheduled to operate to meet the production requirements. Mission time is

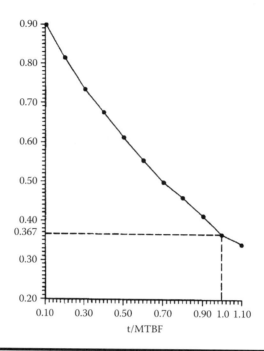

Figure 4.1 Point estimate for reliability.

generally specified by the user and needs to be communicated and clarified by the supplier during the specification development phase of the machinery or equipment. Figure 4.1 shows a point estimate for reliability.

Calculating Reliability Point Measurement

Reliability can be expressed in terms of the following equation:

$$R_{(t)} = e^{(-t/\text{MTBF})}$$

where
$R_{(t)}$ = Reliability point estimate
t = Operating time or mission time of the equipment or machinery
$MTBF$ = Mean time between failures

Note: This calculation can only be performed when the machine has reached the bottom of the bathtub curve.

Example 1

A water pump is scheduled (mission time) to operate for 100 h. The MTBF for this pump is also rated at 100 h, and the mean time to repair (MTTR) is 2 h. The probability that the pump will not fail during the mission time is

$$R_{(t)} = e^{\left(\frac{-t}{MTBF}\right)} = R_{(100)} = e^{\left(\frac{-100}{100}\right)} 0.37 \text{ or } 37\%.$$

This means that the reliability that the water pump will have a 37% chance of not breaking down during the 100 h mission time.

On the other hand, the unreliability of the pump can be calculated as

$$R_{unreliability} = 1 - R$$

$$R = 1 - .37$$

$$R_{unreliability} = 0.63 \text{ or } 63\%$$

This calculation illustrates that the pump has a 63% chance of failing during the 100 h of operation.

Example 2

A water pump is scheduled to operate for 100 h. The MTBF for this pump is rated at 150 h; the M1TR for this pump is 1 h. What is the probability that this water pump will not fail during the mission time? Perform the appropriate calculations as follows. If necessary, refer to the previous example.

$$R_{(t)} = e^{\left(\frac{-t}{MTBF}\right)} = R_{(100)} = e^{\left(\frac{-100}{150}\right)} = 0.51.$$

This pump has a 51% probability of not breaking down during the 100 h mission time. The probability of breakdown is 49%.

Measuring MTBE (Mean Time between Events)

Events are all incidents (regardless of reasons) when equipment is not available to produce parts at specified conditions when scheduled, or is not capable of producing parts or perform scheduled operations at specified conditions. Examples of "events" are

- Machine failure (unscheduled)
- Machine process failure (unscheduled)

- Machine faults (unscheduled)
- Component failures (unscheduled)
- Stops for breaks and lunch (scheduled)
- Quality checks (scheduled or unscheduled)
- Team meetings (scheduled)

Any time that a machine stops during scheduled production, for any reason, it is considered an event. That is measured by the mean time between events (MTBE). Most organizations utilize events to calculate equipment plant floor performance. The event is generated due to the use of electronic monitoring devices on the machine. These data collection sources can only identify whether the equipment is on or off.

MTBF can be expressed in terms of the following equation:

$$MTBF = [\text{Total operating time}/N]$$

where

Total operating time = Total scheduled production time when machinery or equipment is powered and producing parts.

N = Total number of downtime events, scheduled and unscheduled.

Example 1

The total operating time for a machine is 550 h. In addition, the machine experiences two failures, two tool changes, two quality checks, one FTPM meeting, and five lunch breaks. Based on this information, the MTBE is

$$MTBF = [\text{Total operating time}/N] = 550/12 = 45.8 \text{ h}$$

Example 2

What is the MTBF of a machine that has a total operating time of 760 h and experiences two scheduled tool changes, two PM meetings, two quality checks, two lubrication periods, two failed limit switches, one broken belt, one broken damp, three blown fuses, one failed servo board, and two defective cables? Perform the appropriate calculations as follows. If necessary, refer to the previous example.

$$MTBF = [\text{Total operating time}/N] = 760/18 = 42.2 \text{ h}$$

Measuring MTBF (Mean Time between Failures)

A failure is an event when machinery or equipment is not available to produce parts at specified conditions when scheduled, or is not capable of

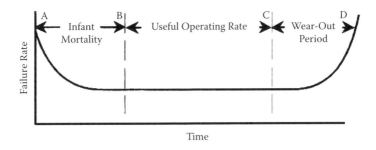

Figure 4.2 Typical bathtub curve.

producing parts or scheduled operations to specifications. For every failure, an action is required.

Failures are a subset of events. These are events that can be attributed to unreliability due to marginal equipment design, installation, or operation beyond useful life of the equipment (e.g., see the bathtub curve in Figure 4.2). The measure of the failure is defined as the MTBF and is a measure of the equipment or component's performance. When MTBF increases, then it can be said that the reliability of the system is increasing.

For some equipment and components, MTBF is not a good indicator because it is a time-based measurement. Therefore, a mean cycle between failures (MCBF) can be used for these items. A definition of MCBF is the average cycles between failure occurrences. Items that can be measured by MCBF may include pumps, cylinders, valves, and others that typically cycle within a certain time period.

Data Collection to Monitor Equipment Performance

Data collection can be described as a process that allows the customer and the equipment supplier to monitor the performance of the equipment. This monitoring is generally implemented during the operation and support phase of the equipment's life cycle, but can be started in the early stages of the design phase. The purpose of data collection is to:

- Look at the existing plant data relating to failures and corrective actions to remove failures.
- Create a feedback system to remove failures in future designs of the equipment.

- Demonstrate achieved levels of machinery and component reliability and maintainability (R&M) performance.
- Identify generic and application-specific problems by comparing achieved levels of R&M to expected levels.

One of the keys to continuous improvement of machinery R&M is a solid data collection process. Specifically, data collection should

- Occur during the design, build/install, and operational phase of the machinery life cycle.
- Involve collecting data used in calculating R&M metrics and solving problems by identifying root causes and taking corrective actions.
- Concentrate on machinery and component levels performance within your organization's manufacturing environment.
- Enable machinery and component improvements to occur.

It is important to note that successful data collection is not the sole driving force behind continuous improvement activities. It merely helps the R&M team to identify which problem-solving tools can be used to correct the root cause of the failure event. Ultimately, continuous improvement activities are successfully achieved when the root cause is identified and eliminated.

The following benefits can be realized when a good data collection process is in place:

- Reduction of design time and research for proven component reliability
- Assurance that proven parts, processes, and designs are specified
- Assistance for quality assurance and purchasing in obtaining acceptable parts and components from vendors
- Assurance that products are capable of meeting specifications
- Reduction of time and expense of testing and design reviews
- Rapid release to production
- Improvement of capability to estimate reliability at the design stage
- Simplified and streamlined decision-making process

The following two basic data collection methods can be utilized within the customer's environment:

1. Manual recording
2. Direct machine monitoring (Electronic)

These are described in more detail in the following sections.

Method #1: Manual Recording of Data

The manual method of recording machinery data is the basis for a solid data collection process. Field personnel typically record machinery R&M data on standardized data forms. These three tools are typically utilized to capture machinery failure data:

- A failure report form
- A failure analysis and corrective action report form
- A universal failure tag

R&M Continuous Improvement Activities

Subsequent to entering data on standardized forms, the field personnel or a designated data entry clerk analyzes the data forms and enters only the significant machinery data into the computer (electronic) database system (a computerized maintenance management system, or CMMS). The data forms are then filed and stored in a central data collection location. Data forms provide critical information to the following entities:

- Plant engineering
- Automation engineering
- Safety
- Tool and die
- Manufacturing engineering
- Union skilled trades
- Maintenance supervision/planners
- Production
- Plant services

Data forms provide the preceding entities with the information that they need to

- Formulate and perform corrective actions.
- Redesign.

- Improve the production process.
- Improve future product design.

Structure of a Manual Recording

A good manual recording system should comprise the following:

- *Master Data List*—This list should provide a summary of the entire manual recording system. Specifically, it should contain machinery and component identification numbers and a brief description of the machinery.
- *Machinery Data File*—This file should contain separate data (information) sheets on each machine. Specifically, it should list all of the components that comprise each machine.
- *Component Data File*—This file should contain separate data (information) sheets on each component of the machinery that is contained in the Machinery Data File.

Failure Report Form

Figure 4.3 shows a typical failure report form that can be used to capture failure data for a failed machine. It is typically used in conjunction with a Failure Analysis and Corrective Action Report Form and a Universal Failure Tag, which collectively completes the manual documentation process of a failure.

Method #2: Direct Machine Monitoring (Electronic)

Direct machine monitoring is the process by which error codes are generated by the failed machine and fed directly to a programmable logic controller (PLC). The data from the PLC is entered into a database. The information in the database is then used to generate reports that can be used to track R&M activities. The following conceptual graphic illustrates the process of electronic machine monitoring:

$$\text{Machine Failure} \rightarrow \text{PLC} \rightarrow \text{Database} \rightarrow \text{Report}$$

Direct Machine Monitoring

Most customers (companies) currently utilize one or both of the following electronic machine monitoring systems at their manufacturing facilities to gather important machinery data:

- PFIS (Plant Floor Information System)
- CMMS (Computerized Maintenance Management System)

When using these different types of electronic monitoring systems, the best data that can be captured is when the machine transitions from the automatic mode to the manual mode. This transition is classified as an event and can be calculated using the MTBE formula.

Project Name (if applicable) --------------   Affected Machine   -----------------

Project ID # (if applicable) --------------   Failure: ---------------------------

FAILURE REPORT FORM

Identification Information

Plant Area:	Department:	Shift:	Date & Time of Failure:
Machine Name:	Machine ID #:	Subsystem Name:	Subsystem ID #:

Failure Review Board (ERB) Members:

Function and Performance Requirements

Recorded by:	Title:		Date:
	Function	*Performance Requirements*	

Failure Data

Recorded by:	Title:		Date:

Description of Failure: Environmental Conditions:

Failure Occurred During:

☐ Inspection ☐ Production ☐ Maintenance ☐ Shipping ☐ Field Use

Symptoms(s):   Observation at time of failure:

☐ Excessive vibration	☐ Excessive noise	☐ Arced	☐ Frayed	☐ Stripped
☐ High fuel consumption	☐ Out of balance	☐ Bent	☐ Loose	☐ Worn
☐ High oil consumption	☐ Overheating	☐ Cracked	☐ Open	☐ Corroded
☐ Inoperative	☐ RPM out of limits	☐ Galled	☐ Plugged	☐ Dented
☐ Interference	☐ Unstable operation	☐ Chafed	☐ Ruptured	☐ Chipped
☐ Leakage	☐ Metal in oil	☐ Peeled	☐ Split	☐ Other ----
☐ Incorrect display	☐ Pressure out of limits	☐ Eroded	☐ Sheared	-------------------
☐ Low performance	☐ Other	☐ Shorted	☐ Burned	-------------------

Effect(s) of failure:

Figure 4.3 A typical failure report.

Maintenance System Data

A good source of R&M data comes from maintenance activities. Maintenance activities are a good source because maintenance personnel are closely involved with machinery PM (preventive maintenance) and respond to machinery failures. The flowchart in Figure 4.4 illustrates how maintenance

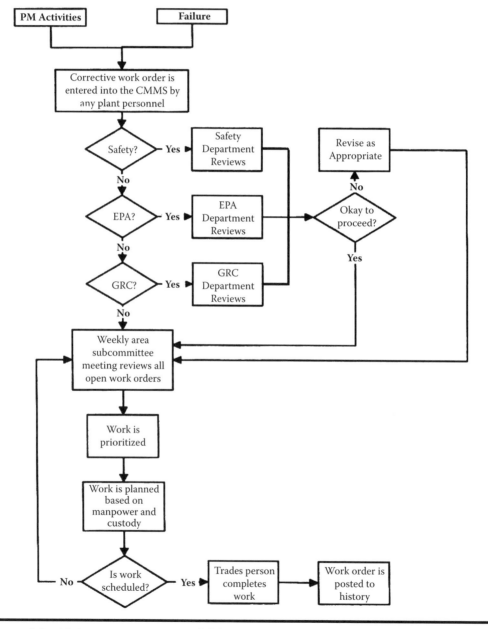

Figure 4.4 Maintenance system activities.

activities (whether related to PM work or an actual failure) are handled by maintenance personnel in a manufacturing environment.

Field History/Service Reports

Field history/service reports are a good source of R&M data that can be used to assist the R&M team with improving R&M parameters. Data from the manufacturer is valuable because it represents the customer's observation of the reliability of the machinery produced. The supplier or manufacturer of the machinery should be able to provide the R&M team with adequate R&M data relating to the machinery they build.

Figure 4.5 illustrates the process by which machinery operation data flows from the field to the analytical tools that are utilized to assess the R&M data.

Component Supplier Failure Data

The R&M team should receive the following data from the machinery and component supplier or manufacturer:

- *Written quality control and reliability program plans,* which should show how the supplier or manufacturer ensures a satisfactory product (conforming to customer's specifications)
- *Quality control records* covering the specific run of the product
- *Inspection reports of the final product,* by lot or batch, showing the lot number or identification, the number inspected, and the number of defects or nonconforming units found (and removed)
- *Reliability test documents* showing the test procedure, environmental conditions, stresses applied, type of test equipment, length of test, number tested, number failed, and causes of failure

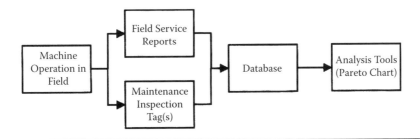

Figure 4.5   Field history report process.

FRACAS: Failure Reporting, Analysis, and Corrective Action System

To reduce the impact of infant mortality and expedite the process of continuous improvement, hardware problems occurring during manufacture and testing at OEM facilities must be tracked by the customer's engineer. The FRACAS approach is an effective closed-loop process that is used to identify failures and their root causes and invoke a corrective action procedure to ensure the failures do not reoccur. A typical FRACAS generic flowchart is shown in Figure 4.6.

The FRACAS system is used for repairable equipment and is used by manufacturing companies, original equipment manufacturers (OEMs), and component suppliers. When components fail and are replaced by like components, the reliability of the equipment remains static. However, when failure causes are identified and actions are taken to remove the cause agents, the equipment's reliability characteristics will improve. The cause agents are typically identified as

- Improperly applied components
- Components overstressed by external stimulation
- Defective manufactured parts
- Insufficient part quality
- Design deficiency

The degree of improvement is a function of an aggressive "Test, Analyze, and Fix" concept employed by the OEM or component supplier. The result of the approach is generally a reliability growth curve. The following sections describe the steps to FRACAS.

Step 1: Problem Investigation Responsibilities of Machinery & Equipment OEMs during Manufacturing and Runoff

- Upon removal of suspected failed component, complete the Failure Tag.
- Notify the customer's Manufacturing Team or R&M Department at the plant.
- Handle the failed component in a manner that will prevent any additional damage.
- Return the component to the approved destination.
- Make sure the OEM notifies the supplier of the returning defective component.
- Track and document the root cause of failure.

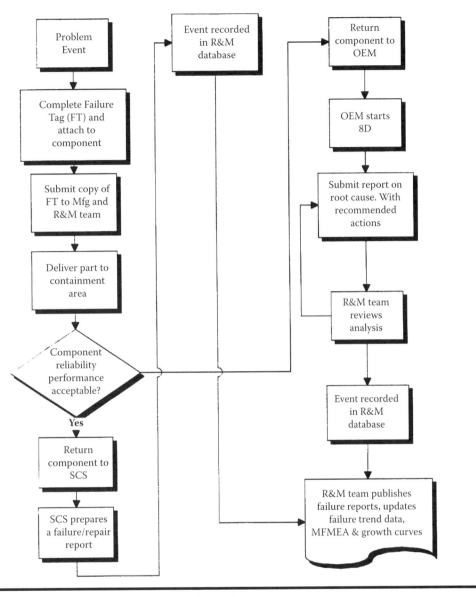

Figure 4.6 A typical FRACAS generic flowchart.

Step 2: Failure Investigation Responsibility of Component Manufacturer or Supplier

- Make arrangements with the local supplier that the component is returned to the supplier.
- Inform the R&M team that the component has been received by the OEM.
- If mode of failure is obvious, then develop a one-page report outlining the failure investigation findings.

- If mode of failure is not obvious, then start the 8D process to determine root cause of the failure.
- If additional information is required, then contact plant for additional failure information.
- Suggest corrective or preventive maintenance activities for future removal of failure.
- Present the failure analysis report outlining the findings of the failure investigation.

Step 3: Problem Responsibility of Customer's Manufacturing R&M Team

- Gather and investigate any problems that have occurred on the equipment.
- Present the failure tag to the customer's R&M team for review.
- Manage the part containment area.
- Return warranty parts to the supplier for replacement.
- Coordinate the problem investigation, and input data into a database.
- Maintain the R&M database.
- Publish periodic reports relating to failures and resolutions.
- Generate reliability growth curves for the various pieces of equipment under investigation.

Failure tagging is the driving force behind a FRACAS. Without tagging, a FRACAS would not be successful. Therefore, it is important to establish an effective tagging process that captures the essential failure data vital to R&M continuous improvement activities.

A Universal Failure Tag is used when a failed part needs to be returned to the supplier or the manufacturer. It is typically completed by a Skilled Trades person (see Figure 4.7).

Subsequently, the failed part is tagged and is placed in a holding area. The tag outlines the essential failure data so that the supplier or the manufacturer has a clear understanding of the failure. A copy of the tag should be sent to the entity (work unit) that experienced the failure, the personnel who are responsible for recording the failure, and the supplier or the manufacturer of the failed part. A copy of the Universal Failure Tag is also sent to the personnel responsible for recording the failure data so that the data from the tag can be manually recorded on a Failure Data Report Form or entered into the CMMS database.

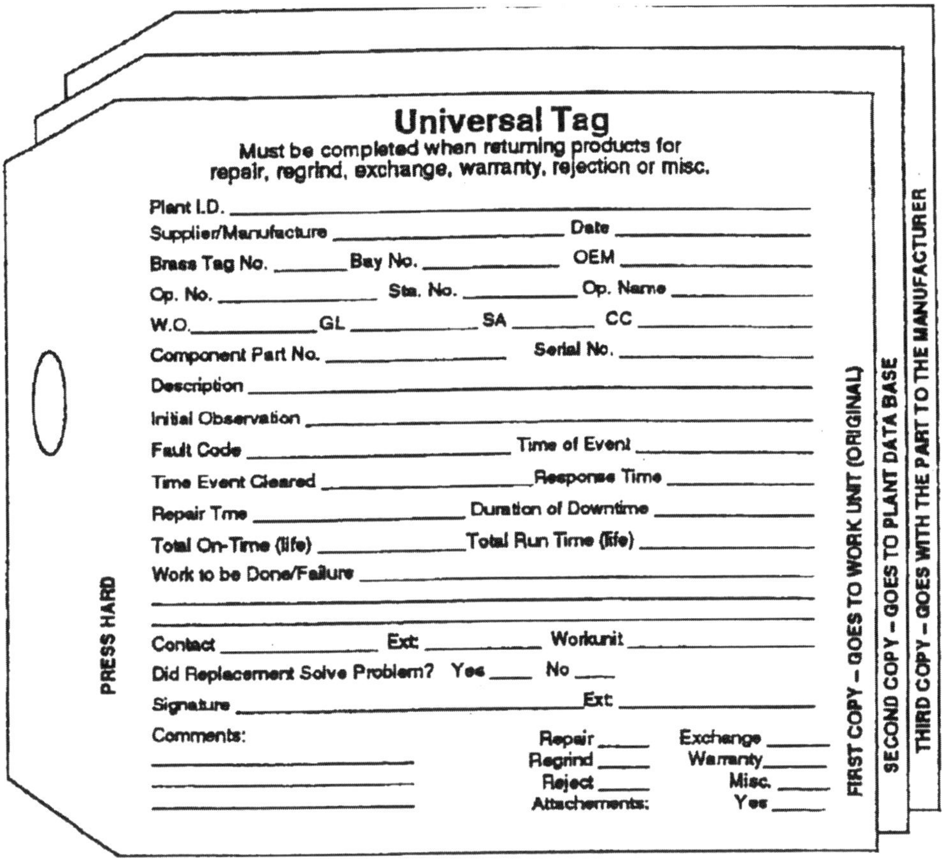

Figure 4.7 A typical universal tag.

Summary

This chapter discussed the process of *measuring* OEE from a reliability perspective. Specifically, it focused on differentiating between events, failures, and concerns about data collection. Chapter 5 discusses how to *improve* the OEE of existing machinery.

Chapter 5

Improving OEE on Existing Machinery

Once you have measured overall equipment efficiency (OEE), the next step is to establish it as the benchmark. At that point, you may be concerned with improving the OEE, if it does not meet the expected requirements.

When machines fail or produce products that do not meet customers' expectations, there are many ways and approaches to fix them. This chapter addresses some of the most common and simple methodologies and some of the traditional tools.

Quite often, the analysis will start with data collection and some usage of the tools discussed in Chapter 3. However, in most cases, two approaches are used right away: the Pareto analysis and the 8-Discipline approach. The chapter begins with a detailed discussion of these two approaches and then follows with a description of other approaches: quick changeover, mistake proofing, P-M analysis, finite element analysis (FEA), failure mode and effect analysis (FMEA), fault analysis, and equipment failure mode and effect analysis (EFMEA).

Using Pareto Analysis to Identify What Needs Improvement Most

Pareto analysis enables the reliability and maintainability (R&M) team to ensure that continuous improvement activities are concentrated where they will do the most good. In other words, it communicates to the R&M team where it needs to focus improvement activities. By way of a definition (for improvement on existing machinery), Pareto analysis is an illustrative and analytical tool that is used to identify, list, and rank the Top 10 list of failures associated with a particular machine.

Often, Pareto analysis is underestimated as an effective tool for facilitating continuous improvement activities. This judgment is usually a result of the perception that Pareto analysis is too simple to critically analyze failure data. Pareto analysis facilitates consensus among all of the entities involved with R&M continuous improvement activities. It tends to correct and clarify any assumptions relating to machinery failure data.

Figure 5.1 is an example of a Pareto analysis chart that identifies and ranks the Top 10 failure list associated with a pump. Notice that there are only eight failures recorded for a Top 10 list. The R&M team should list as many as are available for the equipment under investigation.

Using the 8-Discipline (8D) Process to Resolve Machinery Failures or Problems

The 8D process is another analytical tool that can be utilized to effectively and quickly resolve a failure or problem when its cause is unknown

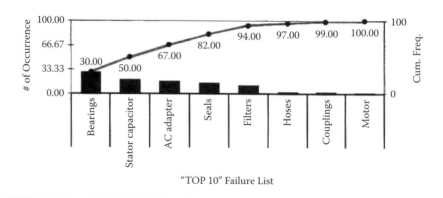

Figure 5.1 Pareto chart—identifying and ranking the Top 10 failures.

Table 5.1 The 8D Generic Report Form

8D Report form				
Equipment name:	Subsystem Name:		Date:	
Failure no.:	Failure title:	Date opened:	Champion:	
(D2) Failure description:		Lines:	Product name:	
		Date(s):	Date(s):	
		(D1) Team members/activity/phone:		
(D3) Containment actions:		% Effective:	Effective date	
			MFG.	Retail
(D4) Root cause(s): Definition:				% Contrib./cause
(D5) Corrective actions:				% Effective
(D6) Implementation of permanent corrective actions:				Effective date
			MFG.	Retail
(D7) Actions to prevent recurrence:				
Reported by:	Last change/ status:		Concurrence:	Date:
(D8) Congratulate your team.				

(see Table 5.1 for the reporting form). It is a reactive eight-step process that should be followed from the moment a failure or problem becomes apparent. It was developed by Ford Motor Company in the early 1980s and continues to be quite successful as a problem-solving tool in many industries throughout the world.* The 8D process simultaneously involves three functions:

- It is a problem-solving tool.
- It is a standard eight-step approach for the organization.
- It is a reporting format.

The eight steps with their deliverables are shown in Table 5.2.

---
* See Ford Motor Company (1987, 2008) and Stamatis (2003) for more detailed information on this process.

Table 5.2 The 8D Process

Step 0: Prepare for the 8D process
■ Has the emergency response action (ERA) been evaluated?
■ If the ERA has been identified and implemented, has it been verified?
■ If the ERA has been identified and implemented, has it been validated?
■ Is there a need for an 8D approach?
Step 1: Establish a team
■ Does the team have the support to accomplish the goal?
■ Does everyone understand the goal? Agree on the focus?
■ Does the leader have the authority to be effective?
■ Are the people most knowledgeable about the problem/product/process on the team?
■ Does the team know to whom they will be accountable?
■ What is the deadline?
■ When do we need to start?
Step 2: Describe the problem
■ Have the problem characteristics been identified?
■ Has the problem been defined in measurable terms? Have trends been evaluated? What is the overall magnitude of the problem?
■ Has the problem been subdivided into its smallest parts?
■ What is the current impact on safety, people, cost productivity, customers, reputation, etc.?
■ What is the urgency and seriousness of the problem?
■ Is there a final, concise problem statement from which to measure results?
Step 3: Interim containment actions (ICA)—problems and symptoms
■ Has containment been documented, implemented, and verified to be effective?
■ Has relevant process data been collected during containment actions?
■ Have customers (internal and external) confirmed effectiveness of containment?
Step 4: Identify all potential root causes
■ Has a thorough list of several or more potential root causes been documented and numbered?
■ Are potential root causes substantiated by data?

Table 5.2 The 8D Process (*Continued*)

■ Has the team identified why the defect escaped and the problem was not detected?
Step 5: Select the real root cause and permanent corrective action (PCA)
■ Has every root cause listed in step 4 been accepted or rejected based on logic?
■ Has the team's reason for accepting and rejecting been documented and numbered?
■ Has the cause been verified? Can this root cause turn on or turn off 100% of the problem?
■ Is there a permanent corrective action identified for every real root cause?
■ Is the "permanent" corrective action really permanent?
■ What will be the future impact of the PCA? What problems could it cause?
■ What resources will we need (people, capital, equipment, time, space, etc.)?
■ Are there any legal, safety, or environmental concerns?
■ When must the PCA be complete?
Step 6: Implement and validate permanent action
■ Have FMEAs, control plans, work instructions, training, etc., been updated?
■ Was the removal of containment actions documented?
■ Has the team determined that the PCAs do not cause negative side effects?
■ Does subsequent data support the conclusion that the problem has 100% disappeared?
■ Have the champion and customer concurred that the cause has been eliminated?
Step 7: System prevent
■ Has the team identified the management systems that failed?
■ Has the team corrected the management systems that allowed the problem to occur and escape?
■ Was the solution applied to similar processes?
■ Does the organization's FMEA process anticipate and prevent similar problems?
■ Have lessons learned been documented and communicated?
Step 8: Celebrate success!
■ Was the team appropriately recognized after management approval of the completed 8D?
■ Was the success communicated throughout the organization?

Although the 8D is a very powerful methodology in its entirety, it will not be effective unless and until you go through step 7 in a very detailed fashion to institutionalize the solutions. The institutionalizing begins with a series of questions to evaluate the solution and also to understand why the failure occurred in the first place. A typical question format may be based on the fundamental question of "Why did our system allow this failure?" An extended format of questioning may be more in line with Table 5.3.

As mentioned in the introduction to this chapter, in addition to these two most popular approaches for improvement (i.e., Pareto analysis and the 8D process), there are other tools and methodologies that may be used. Some of the most common tools and methodologies are described in the rest of this chapter.

Implementing a Quick Changeover to Improve Process Flexibility

Changeover is the amount of time that elapses from when a changeover begins until the first piece is completed. This can be a mold change, die, machine change, tooling, and so on.

When an organization has determined that improvement is a "way of life" in all aspects of the organization, the implication is straightforward. That is, the organization is forced to recognize the importance of flexibility in improving its effectiveness to meet customer demands. Continuous improvement efforts have historically been concentrated on increasing output and reducing unplanned downtime. However, little attention has been paid to R&M, especially to the amount of time necessary to complete changeovers. In today's globalization competition, there is a demand for efficiency as well as a new, higher level of manufacturing flexibility.

To improve process flexibility and reduce inventories, you must focus your attention on the time taken to complete changeover work. If you concentrate on removing Muda (waste), it will permit you to have shorter lead time and give you the time the process is available for new production business.

When an organization takes up the mantle of improvement, the vision for this organization should be for all employees to be committed to work together in quick changeover, to eliminate downtime, improve process flexibility, and eliminate waste (Muda). Changeovers should be quicker and easier so that everyone, not just a few people, will have the ability to perform

Table 5.3   A Typical Extended Questionnaire for Machine Failures

Feasibility
■ Was this potential issue identified during advanced product quality planning (APQP)?
■ Did we have a complete understanding of the customer requirements and functionality of the product?
■ Did we quote appropriately, including all costs of the customer requirements?
■ Did we include a cross-functional team during the quoting process to make sure all areas of the potential launch were covered?
■ Was cost estimated incorrectly? Cost of prevention implementation versus cost of corrections during production (life of program)?
■ Did we have the appropriate (realistic) resources needed to execute a flawless launch? If no, why not?
■ Are we forcing the operators or equipment to perform in an unrealistic cycle time with the equipment or conditions provided?
■ Did we identify previous failures of past or current processes during APQP of this program?
Design
■ Was this potential issue identified during APQP?
■ Did we identify previous failures of past or current processes during APQP of this program?
■ Did we fail to consider potential failures from our suppliers (purchased components)?
■ Did we have a complete understanding of the customer requirements and functionality of the product?
■ Did we fail to achieve customer approval for design changes?
■ Did we fail to track and log design changes? Is this identified on DFMEA?
■ When implementing a change to the tool, did we identify and communicate to all areas of operation and equipment that the change may affect (inspection gages, warehouse, etc.)?
■ Was cost estimated incorrectly? Cost of prevention implementation versus cost of corrections during production (life of program)?
■ Did we fail to consider other risks in all areas of product functionality or process before making the system improvement or design change?

(Continued)

Table 5.3 A Typical Extended Questionnaire for Machine Failures (*Continued*)

■ Have we identified all internal interfaces?
■ Are there any intercompany interfaces?
■ Did we fail to utilize manufacturing (operator, etc.) suggestions or recommendations?
■ Have we performed testing simulations in realistic conditions?
■ What other areas or products could have the same failure but do not?
Timing
■ Was this potential issue identified during APQP?
■ Is our employee changeover rate out of control?
■ Are disciplines in place for the area that caused the delay in the program without a justifiable cause?
■ Is this an input to another process?
■ Are we causing inconsistency in the workflow by frequent management changes?
■ Did we have a complete understanding of the customer requirements and functionality of the product?
■ Did we fail to clearly state our expectations to Manufacturing?
Implementation
■ Was this potential issue identified during APQP?
■ Did we have a complete understanding of the customer requirements and functionality of the product?
■ Is this an input to another process?
■ Have we identified all internal interfaces?
■ Are there any intercompany interfaces?
■ Did we fail to assess the impact of system improvements?
■ Did we fail to utilize Manufacturing's (operator, etc.) suggestions or recommendations?
■ Did we fail to consider other risks in all areas of product functionality or process before making the system improvement or design change?
■ Have Preventive Maintenance disciplines occurred, and if so, did they occur as scheduled?

Table 5.3 A Typical Extended Questionnaire for Machine Failures (*Continued*)

- When implementing a change to the tool, did we identify and communicate to all areas of operation and equipment that the change may affect (inspection gages, warehouse, etc.)?
- What other areas or products could have the same failure but do not?
- Did we fail to verify the operator's complete understanding of process?

Process

- Was this potential issue identified during APQP?
- Is this an input to another process?
- Have we identified all internal interfaces?
- Are there any intercompany interfaces?
- Did we fail to test a representative amount of samples from the lot?
- Have we performed testing simulations in realistic conditions?
- Did we fail to consider potential failures from our suppliers (purchased components)?
- Have preventive maintenance disciplines occurred, and if so, did they occur as scheduled?
- What other areas or products could have the same failure but do not?
- Did we fail to act on out-of-control conditions (key measurables, internal audits)?
- Did we fail to analyze scrap from this process?
- Did we fail to consider other risks in all areas of product functionality or process before making the system improvement or design change?
- Did we have the appropriate (realistic) resources needed to execute a flawless launch? If no, why not?
- Were all areas of the product or process flow identified on the PFMEA? (handling of the product from area to area, lot traceability, etc.)
- Are we forcing the operators or equipment to perform in an unrealistic cycle time with the equipment or conditions provided?
- Is our employee changeover rate out of control?
- Did we determine and analyze product and process failures from sample runs during APQP?
- What other areas or products could have the same failure but do not?

(*Continued*)

Table 5.3 A Typical Extended Questionnaire for Machine Failures (*Continued*)

■ Did we fail to act on out-of-control conditions (key measurables, internal audits)?
■ Did we fail to analyze scrap from this process?
■ What other areas or products could have the same failure but do not?
Logistics
■ Was this potential issue identified during APQP?
■ Were all areas of the product or process flow identified on the PFMEA? (handling of the product from area to area, lot traceability, etc.)
■ Is this an input to another process?
■ Have we identified all internal interfaces?
■ Are there any intercompany interfaces?
■ Are previous failures identified on the PFMEA?
■ Did we quote appropriately, including all costs related to customer requirements?
■ Did we include a cross-functional team during the quoting process to make sure all areas of the potential launch were covered?
■ Did we have a contingency plan in place and did we provide a complete understanding of the contingency plan to all shifts, including IT systems?
■ Did we fail to test a representative amount of samples from the lot?
■ Have we performed testing simulations in realistic conditions?
■ Did we fail to consider potential failures from our suppliers (purchased components)?
■ Did we fail to assess the impact of system improvements?
Off-Standard operations (repair/rework/end of run control/start-up procedure/ engineering change cut-off and start-up, etc.)
■ Was this potential issue identified during APQP?
■ Were all areas of the product or process flow identified on the PFMEA? (handling of the product from area to area, lot traceability, etc.)
■ Is this an input to another process?
■ Have we identified all internal interfaces?
■ Are there any intercompany interfaces?
■ Are previous failures identified on the PFMEA?

Table 5.3 A Typical Extended Questionnaire for Machine Failures (*Continued*)

■ Did we fail to utilize manufacturing (operator, etc.) suggestions or recommendations?
■ Is our employee changeover rate out of control?
■ Did we provide necessary product knowledge to Manufacturing?
■ Did we determine and analyze product and process failures from sample runs during APQP?
■ Was cost estimated incorrectly? Cost of prevention implementation versus cost of corrections during production (life of program)?
■ When implementing a change to the tool, did we identify and communicate to all areas of operation and equipment that the change may affect (inspection gages, warehouse, etc.)?
■ Did we provide complete training or procedures needed?
■ Did we fail to consider other risks in all areas of product functionality or process before making the system improvement or design change?
■ Have preventive maintenance disciplines occurred, and if so did they occur as scheduled?
Lean
■ Was this potential issue identified during APQP?
■ Did we identify previous failures of past or current processes during APQP of this program?
■ Was cost estimated incorrectly? Cost of prevention implementation versus cost of corrections during production (life of program)?
■ Are we causing inconsistency in the workflow by frequent management changes?
■ Are we forcing the operators or equipment to perform in an unrealistic cycle time with the equipment or conditions provided?
■ Did we have a contingency plan in place and did we provide a complete understanding of the contingency plan to all shifts, including IT systems?
■ Have preventive maintenance disciplines occurred, and if so, did they occur as scheduled?
■ What other areas or products could have the same failure but do not?
■ Did we fail to act on out-of-control conditions (key measurables, internal audits)?
■ Did we fail to analyze scrap from this process?

(*Continued*)

Table 5.3 A Typical Extended Questionnaire for Machine Failures (*Continued*)

■ Did we fail to consider other risks in all areas of product functionality or process before making the system improvement or design change?
Packaging and shipping verification
■ Was this potential issue identified during APQP?
■ Did we identify previous failures of past or current processes during APQP of this program?
■ Were all areas of the product or process flow identified on the PFMEA? (handling of the product from area to area, lot traceability, etc.)
■ Is this an input to another process?
■ Have we identified all internal interfaces?
■ Are there any intercompany interfaces?
■ Are previous failures identified on the PFMEA?
■ Did we have a complete understanding of the customer requirements and functionality of the product?
■ Have we performed testing simulations in realistic conditions (shake test, test run, etc.)?
■ Did we fail to consider potential failures from our suppliers (purchased components)?
■ Did we quote appropriately, including all costs of the customer requirements?
Plant tryout
■ Was this potential issue identified during APQP?
■ Did we identify previous failures of past or current processes during APQP of this program?
■ Were all areas of the product or process flow identified on the PFMEA? (handling of the product from area to area, lot traceability, etc.)
■ Have we identified all internal interfaces?
■ Are there any intercompany interfaces?
■ Are previous failures identified on the PFMEA?
■ Did we have a contingency plan in place and did we provide a complete understanding of the contingency plan to all shifts, including IT systems?
■ Did we fail to utilize manufacturing (operator, etc.) suggestions or recommendations?

Table 5.3 A Typical Extended Questionnaire for Machine Failures (*Continued*)

- Did we fail to test an appropriate amount of samples as representative of the lot?
- Have we performed testing simulations in realistic conditions?
- Did we fail to assess the impact of system improvements?
- Have we made any system or design changes?
- Did we fail to consider other risks in all areas of product functionality or process before making the system improvement or design change?
- Did we provide necessary product knowledge to Manufacturing?
- Did we determine and analyze product and process failures from sample runs during APQP?
- Did we have a complete understanding of the customer requirements and functionality of the product?
- Did we fail to clearly state our expectations to Manufacturing?
- Are disciplines in place for the area that caused the delay in the program without a justifiable cause?
- Did we have the appropriate (realistic) resources needed to execute a flawless launch? If no, why not?
- Did we provide complete training or procedures needed?
- Are we forcing the operators or equipment to perform in an unrealistic cycle time with the equipment or conditions provided?
- Did we fail to act on out-of-control conditions (key measurables, internal audits)?
- Did we fail to analyze scrap from this process?

Customer return system

- Was this potential issue identified during APQP?
- Did we identify previous failures of past or current processes during APQP of this program?
- Were all areas of the product or process flow identified on the PFMEA? (handling of the product from area to area, lot traceability, etc.)
- Is this an input to another process?
- Have we identified all internal interfaces?
- Are there any intercompany interfaces?
- Are previous failures identified on the PFMEA?

(*Continued*)

Table 5.3 A Typical Extended Questionnaire for Machine Failures (*Continued*)

■ Did we determine and analyze product and process failures from sample runs during APQP?
■ Did we fail to clearly state our expectations to Manufacturing?
■ Are disciplines in place for the area that caused the delay in the program without a justifiable cause?
■ Did we provide necessary product knowledge to Manufacturing?
■ Did we fail to test a representative amount of samples from the lot?
■ Did we remove containment practices prior to verifying the corrective action?
■ Did we fail to isolate the suspect product?
■ Did we have the appropriate labeling system in place to identify all phases of the containment?
Communications/contacts
■ Was this potential issue identified during APQP?
■ Did we identify previous failures of past or current processes during APQP of this program?
■ Have we identified all internal interfaces?
■ Are there any intercompany interfaces?
■ Did we fail to utilize manufacturing (operator, etc.) suggestions or recommendations?
■ Is our employee changeover rate out of control?
■ What kind of attitude are we representing to Manufacturing?
■ Did we provide necessary product knowledge to Manufacturing?
■ Are disciplines in place for the area that caused the delay in the program without a justifiable cause?
■ Are we causing inconsistency in the workflow by frequent management changes?
■ Did we provide complete training or procedures needed?
■ Did we have a complete understanding of the customer requirements and functionality of the product?
■ Did we fail to consider other risks in all areas of product functionality or process before making the system improvement or design change?

changeovers. If you accomplish these goals, you can look to the future to adapt, to meet the changing needs of your customers, and to make a difference in the global market.

Benefits of Quick Changeover

The specific benefits for a fast changeover are that by reducing your downtime in changeovers, the penalty of lost production is less. You have the ability to make only daily requirements of each part. You are now making small lots, instead of a couple of days' worth. You have reduced your inventory and saved money—and by saving money, you save jobs. In addition, by making smaller lot sizes, you improve overall quality because you have better control of your process, including the machine or equipment. This is accomplished because the smaller lot sizes reveal quality issues faster. With determining the quality of parts in a large batch is more difficult. You also have more parts to scrap if something is wrong with the parts. Quality is Job 1, and you want to be the lowest-cost producer. If you accomplish this, you have secured your future.

The rules to quick changeover are that there are no rules. The only exception to this is safety. Safety will not be compromised for any reason. There is no changeover anywhere that will compromise the safety of any employee. In addition to safety, there are two essential parts to setup:

1. *Internal setup:* These are elements of the changeover that can be done while the machine is idle.
2. *External setup:* These are elements that can take place while the machine is operating.

Shifting internal elements to external elements is the most basic concept for quick changeover.

To begin the process, you must:

1. Select a changeover process that needs improvement.
2. Select your team. This should include Manufacturing Tech, Coordinator, Advisor, Engineers (it is important to have their buy-in), the entire production team, and members from skilled trades.
3. Choose a changeover that hampers productivity, and select a part with the largest inventory.
4. Videotape the current changeover procedures to isolate each element of the changeover process, so the team can review and replay the study of the operation.

5. Make a list of the current changeover procedures.
6. Make a detailed analysis of the changeover from the preparation of the changeover to the restart of production.
7. Do a Kaizen time-study analysis. Have the team set a target for a reasonable changeover goal. Once the goal is accomplished, raise the standard to a new level.
8. Record the process. Have the team display a visual changeover chart. The team should update the chart daily to notice improvements and focus on changeover times that need action.

Roles and Responsibilities (R&R)

As in all methodologies, in the changeover process, there are requirements for the appropriate roles and responsibilities (R&R). Key R&R are described in the following subsections.

Manufacturing Tech

Get all materials for changeover ready before you empty the machine from the previous run, and make sure tools are returned to the cabinet. Make sure stock and tags are marked correctly to prevent future quality issues.
Reaction Plan:

- Inform Advisor if no material handling is available.
- Make corrections on improperly marked tags.

Coordinator

Inform Manufacturing Tech of changeover in advance of changeover that will take place that day. Open locker for tools needed for changeover. Inform Manufacturing Tech of how many pieces to run before next changeover. Provide stock for Manufacturing Tech for next changeover. Make sure tools are back in cabinet after changeover is complete.
Reaction Plan:

- Revert back to cookie cutter schedule unless emergency dictates an unscheduled changeover.
- Inform Advisor if there are problems opening the locker; if pieces are not known, revert to cookie cutter schedule.

- Inform Advisor if stock is not available.
- Inform Advisor if any missing tools cannot be located.

Advisor

Inform Coordinator at the start of the shift of changeovers that will occur that day. Communicate with the Maintenance Supervisor, MPS (Manufacturing Planning Specialist), Production Superintendent, Maintenance Superintendent, and Business Unit Manager of any unscheduled changeovers.

Reaction Plan:

- Revert back to cookie cutter schedule unless emergency dictates an unscheduled changeover.
- If unable to communicate verbally, page appropriate person.

Maintenance Supervisor

Inform advisor if skilled trades are not available to assist in changeover. Alert maintenance clerk to schedule additional manpower if needed.

Reaction Plan:

- Alert maintenance clerk to schedule additional manpower if needed.
- Inform Advisor if skilled trades are not available to assist in changeover.

Manufacturing Planning Specialist (MPS)

Obtain daily production schedule and inform Advisor how many different batches to run, and make sure they have materials required to complete the changeover.

Reaction Plan:

- Contact Material Control for any changes.
- Revert back to cookie cutter schedule.
- Contact Material Control and inform Advisor and Coordinator of material shortages.

Production Superintendent

Make sure that all needed parties have as much information as possible about the schedules and the stock that needs to be run during the shift. Support the quick changeover processes.
   Reaction Plan:

- Contact appropriate resource person for any needed information.

Maintenance Superintendent

Communicate with the Maintenance Supervisor if there is a problem that would stop the changeover process, and let the appropriate personnel know in a timely manner. Support the quick changeover process.
   Reaction Plan:

- Contact appropriate personnel in a timely manner.

Business Unit Manager

Provide jobs and materials for the department so the quick changeover process can exist and grow. Support the quick changeover process.
   Reaction Plan:

- Contact appropriate vendor for needed materials.

Plant Manager

Provide work for the entire plant and show that upper management supports the quick changeover process that will secure the future.
   Reaction Plan:

- Reviews and approves the changeover process.

Examples of Changeover

Working with different die sizes. If the die footprint is different from job to job, everything from the T-nuts and studs to the finger clamps needs to be changed. One way to address this problem is to make all the dies with the same footprint. This can be quite costly if you have a large number of dies. Another way is a die-clamping system that works with a variety of die sizes. Some stampers locate dies with T-slots and simple spacers or locator pins on magnetic clamps to reduce the time it takes to locate a die.

Changing clamps. To reduce changeover times, look for clamps that can accommodate all die sizes run on a press. For example, hydraulic clamps that fit in existing T-slots can be slid in and out easily. Magnetic clamps that cover the entire ram and bolster can eliminate all tools while accommodating all die sizes without standardization.

Locating needed tools. A die-clamp standard is helpful because it can save time by eliminating decisions such as which stud is needed, which finger clamp will work, and where to place clamps. Shops that alter studs and clamps to make them similar save the most time. Many stampers create a die-change toolkit that contains everything needed, so no time is wasted looking for the right wrench, studs, and T-nuts. Automated clamping mechanisms that require very little adjustment with as few tools as possible also will reduce setup times.

Loading large dies. Line-up blocks mounted to the press bolster or a red stripe painted on the center of the die and center of the press can help an operator line up a die more easily. Die lifters installed into existing T-slots can reduce the effort needed to get dies lined up and do not use up precious die space. Die lifters can be used in conjunction with hydraulic or manual die loading and hydraulic or magnetic clamping.

Recognizing component wear and tear. T-nuts, studs, nuts and bolts, finger clamps, and hydraulic clamps wear out over time and become unusable at inopportune moments. T-nuts can strip, finger clamps will bend with overtightening and normal use, and hydraulic clamps can wear and develop leaks. Even the most organized plants can suddenly find themselves in a pinch. Stampers that have a planned maintenance program or a clamping system with no wearable parts fare the best in reducing this element of changeover time.

Understanding die shoe thickness. If a shop uses a hydraulic mechanical system to clamp dies, the shoe usually has to be the same thickness from die to die for the clamping system to work. Sometimes, the cost of altering the shoe thickness is far greater than the initial clamp investment. This is not an issue for systems that can accommodate any die shoe thickness.

Making ram adjustments. Making ram adjustments is crucial and time consuming. Altering die heights to a small window reduces the time needed to get the next die running, because some rams can adjust at less than 1 in. per 5 min. For example, running two dies back to back with similar shut heights can reduce ram adjustment times.

Moving dies in and out. Reducing a die's total travel distance is a quick way to shave time off changeovers. For example, one stamper uses a die storage unit with multiple-height die racks. It has enough room to put all the dies on the same level as the bolster. The stamper then installed a simple roller system that extended all the way to the press, which eliminated a lift truck or die cart in the die-changing process. This was a vast improvement because the stamper had 26 dies on four levels, 12 ft above floor level, assigned to a single press. As a result, this stamper saved enough time to eliminate an entire production shift on two presses.

Locating the next die. This is another time-consuming process if you do not have a system to track where a specific die should be. This seems simple, but many hours are wasted trying to locate dies. Have a place for each die, and label it. A die-labeling system tracks where a die is at all times. For example, a labeling system can be as simple as this for Die #1XB17:

- 1 is the part number.
- X is the material.
- B is a revision number.
- 17 is the exact storage space the die should be returned to.

This is an inexpensive way to reduce changeover downtime.

Keeping a clean pressroom. The cleaner a pressroom is, the more efficiently it operates. Seeing hydraulic oil on the floor is becoming rarer, but sometimes unusual items end up around presses, such as buckets of oil, clutch parts, broken wrenches, worn-out feeders, and pieces and parts of entirely unidentifiable objects. Catalog noninventory items at least once a year. If you find something that does not belong, get rid of it.

Implementing Mistake Proofing to Prevent or Mitigate Errors

The concept of mistake proofing involves controls or features in the product or process to prevent or mitigate the occurrence of errors. Mistake proofing also requires simple, inexpensive inspection (error detection) at the end of each successive operation to discover and correct defects at the source.

Six Mistake-Proofing Principles

There are six mistake-proofing principles or methods. These are listed in order of preference or precedence in fundamentally addressing mistakes:

1. *Elimination* seeks to eliminate the possibility of error by redesigning the product or process so that the task or part is no longer necessary. For example:
   - Product simplification that avoids a part defect or an assembly error in the first place.
   - Part consolidation that avoids a part defect or an assembly error in the first place.
2. *Replacement* substitutes a more reliable process to improve consistency. For example:
   - Use of robotics or automation that prevents a manual assembly error.
   - Automatic dispensers or applicators to ensure the correct amount of a material, such as an adhesive, is applied.
3. *Prevention* engineers the product or process so that it is impossible to make a mistake at all. For example:
   - Limit switches to ensure a part is correctly placed or fixtured before process is performed.
   - Part features that only allow assembly the correct way.
   - Unique connectors to avoid misconnecting wire harnesses or cables.
   - Part symmetry that avoids incorrect insertion.
4. *Facilitation* employs techniques and combining steps to make work easier to perform. For example:
   - Visual controls including color coding, marking, or labeling parts to facilitate correct assembly.
   - Exaggerated asymmetry to facilitate correct orientation of parts.
   - A staging tray that provides a visual control that all parts were assembled, locating features on parts.
5. *Detection* involves identifying an error before further processing occurs so that the user can quickly correct the problem. For example:
   - Sensors in the production process to identify when parts are incorrectly assembled.
   - Built-in self-test (BIST) capabilities in products.
6. *Mitigation* seeks to minimize the effects of errors. For example:
   - Fuses to prevent overloading circuits resulting from shorts.

- Products designed with low-cost, simple rework procedures when an error is discovered.
- Extra design margin or redundancy in products to compensate for the effects of errors.

When to Conduct Mistake Proofing

Ideally, mistake proofing should be considered during the development of a new product, to maximize opportunities to mistake-proof through design of the product and the process (elimination, replacement, prevention, and facilitation). Once the product is designed and the process is selected, mistake-proofing opportunities are more limited (prevention, facilitation, detection, and mitigation). However, one can still use mistake proofing in an ongoing operating machine.

How to Minimize Human Errors

In dealing with machines or equipment, you must be very careful not to blame the operator for the failures. The reason for not blaming the operator is that failures occur due to systems failure that the operator quite often has nothing to do with. Some ways to minimize the chance for human error are described in the following list.

Blind testing. When hospitals test blood, they take three vials, tag them, and send them to three different places to be tested. None of the testers know who the other testers are. Therefore, the chance of all three testers making the same mistake on a patient's blood sample is virtually zero.

Different criteria. Checking accuracy in more than one way significantly reduces human error. For example, to verify that a box contains the right number of parts, you could have someone count the parts. To ensure accuracy, you could also weigh the box before shipment because the weight will not match the standard if the box contains the wrong number of parts.

Mechanical screens. These include cameras, computers, and other measuring devices that screen parts to ensure compliance to standards. Mechanical screens will catch errors that get by humans, but they are not root cause solutions, because the problem that allows the bad parts to be made still exists.

Error proofing (poka-yoke). Error proofing means redesigning a process so that a particular mistake cannot be made. For example, if an operator is installing a part upside down, you can redesign the part so that it can only be installed right side up. If two parts look so similar that they get mixed up in shipment, you can physically separate them, so that they can not come in contact with each other.

How to Prioritize Mistake-Proofing Opportunities

Mistake-proofing opportunities can be prioritized by performing design and process FMEAs. They can also be developed for every process step in a manufacturing or service process.

Special note: Whereas the generic Japanese term is *poka-yoke*, in the United States we use three terms to say the same thing: *mistake, error, or poka-yoke*. However, quite often in some industries, we make a distinction between *error* and *mistake proofing*. *Error* is reserved for design situations, and *mistake* is reserved for manufacturing or assembly. In Japan, there is no distinction.

Using P-M Analysis to Analyze and Eliminate Chronic Machinery Problems

P-M Analysis is designed to help your total preventive maintenance teams analyze and eliminate chronic problems that have been neglected or unresolved in the past. Through P-M Analysis, teams really get in touch with their equipment. Its unique skill-building process improves technological know-how while delivering solutions to persistent problems. The first four steps of the rigorous eight-step process help teams isolate and understand the root (actionable) causes of defects and failures within main equipment mechanisms and peripheral systems. The final four steps provide a systematic approach for effectively controlling those causes.

A critical concept of P-M Analysis is that thinking about how defects and failures are generated should force us to look at the physical principles involved and to quantify the changes in the relationship between the equipment mechanisms and product parts involved. When a proper physical analysis is carried out, teams are far less likely to overlook important factors or to waste time pursuing unrelated ones. Although not a cure-all,

P-M analysis is indeed a very effective way to reduce chronic losses to zero and raise technological experience and competence in many manufacturing environments.

Chronic quality defects and other chronic losses are hard to eradicate because they typically have multiple, interrelated causes that vary with every occurrence. Common improvement strategies, such as cause-and-effect analysis, are usually ineffective in dealing with such complex problems. P-M Analysis was specially developed to overcome the weakness of traditional methods. It offers a very rigorous eight-step method for ensuring that all possible factors are identified and investigated. Steps for implementing P-M analysis:

1. Clarify the phenomenon.
2. Conduct a physical analysis.
3. Identify constituent conditions.
4. Study the 4Ms and 1E for causal factors.
5. Establish optimal conditions (standard values).
6. Survey causal factors for abnormalities.
7. Determine abnormalities to be addressed.
8. Propose and make improvements.

Using Finite Element Analysis (FEA)* to Improve Product Refinement

FEA was first developed in 1943 by R. Courant, who utilized the Ritz method of numerical analysis and minimization of variational calculus to obtain approximate solutions to vibration systems. Shortly thereafter, a paper published in 1956 by M. J. Turner, R. W. Clough, H. C. Martin, and L. J. Topp established a broader definition of numerical analysis. The paper centered on the "stiffness and deflection of complex structures." By the early 1970s, FEA was limited to expensive mainframe computers generally owned by the aeronautics, automotive, defense, and nuclear industries. Since the rapid decline in the cost of computers and the phenomenal increase in computing power, FEA has been developed to an incredible precision. Present-day supercomputers are now able to produce accurate results for all kinds of parameters.

---

* Some of the material on FEA in this section was adopted from: www.sv.vt.edu/classes/MSE2094_NoteBook/97ClassProj/num/widas/history.html.

FEA consists of a computer model of a material or design that is stressed and analyzed for specific results. It is used in new product design and in existing product refinement. A company is able to verify whether a proposed design will be able to perform to the client's specifications prior to manufacturing or construction. Modifying an existing product or structure will qualify the product or structure for a new service condition. In case of structural failure, FEA may be used to help determine the design modifications to meet the new condition.

There are generally two types of analysis that are used in industry: 2D modeling and 3D modeling. Whereas 2D modeling conserves simplicity and allows the analysis to be run on a relatively normal computer, it tends to yield less accurate results. In contrast, 3D modeling produces more accurate results while sacrificing the ability to run on all but the fastest computers effectively.

With each of these modeling schemes, the programmer can insert numerous algorithms (functions), which may make the system behave linearly or nonlinearly. Linear systems are far less complex and generally do not take into account plastic deformation. Nonlinear systems do account for plastic deformation, and many also are capable of testing a material all the way to fracture.

FEA uses a complex system of points called *nodes* (points at which different elements are jointed together; nodes are the locations where values of unknowns [usually displacements] are to be approximated) that make a grid called a *mesh*. This mesh is programmed to contain the material and structural properties *which* define how the structure will react to certain loading conditions. Nodes are assigned at a certain density throughout the material, depending on the anticipated stress levels of a particular area. Regions that will receive large amounts of stress usually have a higher node density than those that experience little or no stress. Points of interest may consist of the fracture point of previously tested material, fillets, corners, complex detail, and high-stress areas. The mesh acts like a spider web in that from each node, there extends a mesh element to each of the adjacent nodes. This web of vectors is what carries the material properties to the object, creating many elements. A wide range of objective functions (variables within the system) are available for minimization or maximization:

- Mass, volume, temperature
- Strain energy, stress strain
- Force, displacement, velocity, acceleration
- Synthetic (user defined)

There are multiple loading conditions that may be applied to a system. For example:

- Point, pressure, thermal, gravity, and centrifugal static loads
- Thermal loads from solution of heat transfer analysis
- Enforced displacements
- Heat flux and convection
- Point, pressure, and gravity dynamic loads

Each FEA program may come with an element library, or one is constructed over time. Some sample elements are

- Rod elements
- Beam elements
- Plate/shell/composite elements
- Shear panel
- Solid elements
- Spring elements
- Mass elements
- Rigid elements
- Viscous damping elements

Many FEA programs also are equipped with the capability to use multiple materials within the structure such as:

- Isotropic, identical throughout
- Orthotropic, identical at 90°
- General anisotropic, different throughout

Types of Engineering Analysis Using FEA

Structural analysis. Structural analysis consists of linear and nonlinear models. Linear models use simple parameters and assume that the material is not plastically deformed. Nonlinear models consist of stressing the material past its elastic capabilities. The stresses in the material then vary with the amount of deformation.

Vibrational analysis. Vibrational analysis is used to test a material against random vibrations, shock, and impact. Each of these incidences may act on the natural vibrational frequency of the material, which in turn may cause resonance and subsequent failure.

Fatigue analysis. Fatigue analysis helps designers to predict the life of a material or structure by showing the effects of cyclic loading on the specimen. Such analysis can show the areas where crack propagation is most likely to occur. Failure due to fatigue may also show the damage tolerance of the material.

Heat transfer analysis. Heat transfer analysis models the conductivity or thermal fluid dynamics of the material or structure. This may consist of a steady-state or transient transfer. Steady-state transfer refers to constant thermoproperties in the material that yield linear heat diffusion.

Results of finite element analysis. FEA has become a solution to the task of predicting failure due to unknown stresses by showing problem areas in a material and allowing designers to see all of the theoretical stresses within. This method of product design and testing is far superior to the manufacturing costs that would accrue if each sample was actually built and tested.

Using Failure Mode and Effect Analysis (FMEA) to Predict and Prevent Machine Failures

Failure mode and effect analysis (FMEA) is a methodology to predict failures at the design level and figure out an action plan to avoid these failures either in design or process.* EFMEA (described later in this chapter) is an extension of the traditional FMEA and, of course, it is focused exclusively on the specific equipment.

One must not confuse the FMEA with the RCM2: RCM2 = reliability-centered maintenance × remote condition monitoring = cost-effective life cycle asset management strategy. Both methodologies are related; however, the focus of each is different. The FMEA's focus is *failure*; the RCM2's is *cost*. One may use FMEA in figuring out the RCM2, but not the other way around.

Using Fault Tree Analysis to Show the Causes of Machine Failures

Fault tree analysis (FTA) is a graphical evaluation of failures and their causes, as shown in Figure 5.2. It is a top-down method in which each event that may cause the failure is evaluated. The FTA starts with the top-level undesired event and then proceeds through all the known causes that could lead

---
* Stamatis 2003.

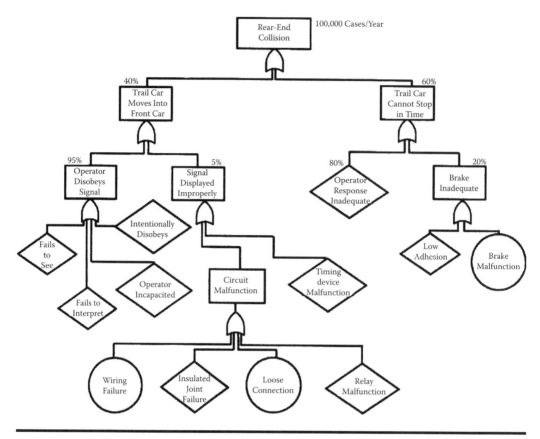

Figure 5.2 An FTA example.

to the top-level event. The top-level undesired event develops a branching process that forms a tree of events.

The purpose of an FTA is to structure a root cause analysis of known failure modes that are not yet fully understood. If it is used in conjunction with both FMEAs and MFMEAs (machinery failure mode and effect analysis), these failure modes are identified in the "Failure Mode" column. The steps that can be implemented to develop an FTA are shown in Table 5.4.

Five basic symbols are used in developing the FTA. These symbols are described and identified in Tables 5.5 and 5.6, respectively:

Using a simple example to demonstrate the concept, consider one from a rear-end collision (see Figure 5.2). Once you become familiar with the symbols, then it is easy to accomplish root-cause analysis based on the flow of the FTA. A typical approach is shown in Figure 5.3.

Table 5.4 FTA Steps

Step	Action	Description
1.	Define the system.	Define the system and any assumptions to be used in the analysis. Define the term of failure. Look to the hierarchical model for details.
2.	Simplify the scope.	Simplify the scope of the analysis by developing simple block diagrams of the machine, inputs and outputs, and interfacing.
3.	Identify the unknown causes of failure.	Identify the top list of failures for the machine. Look at the "Top 10" failure listing. Those failures that are not understood or safety related should be evaluated.
4.	Identify the contributing causes.	Using a series of symbols, arrange the causes of the top-level event. List all the contributing factors of failure. Try to complete the analysis to include both items with lowest replacement units (LRUs), as well as environmental or plant conditions.
5.	Develop the FTA to the lowest event.	Work toward identification of failures down to the lowest level.
6.	Analyze the FTA.	Review the various paths of the failures and which has the highest probability of causing the failure. Look for interrelations between causes of failure.
7.	Determine the corrective actions for lowest events.	Determine which cause relationships need to have design changes for elimination of failure path. Corrective action plans need to be developed for the lowest level of causes.
8.	Document the process.	Document the process to ensure that all corrective actions have been addressed to eliminate the failure path on the FTA.

Table 5.5 Key FTA Symbol Description

FTA symbol description	Definition
Top-level failure/causal event	The failure/causal event symbol (depicted by a rectangle) denotes a top-level event or a subsequent-level event below an AND or OR gate requiring further definition.
OR gate	The OR gate is depicted by a circular polygon with a concave base. When used, it signifies that if any of the subsequent events below it occur, then the next-highest event above it also occurs.
AND gate	The AND gate is similar to the OR gate, consisting of a circular polygon with a straight horizontal base. When used, it signifies that all of the subsequent events below it must occur before the next-highest event above it can occur.
Undeveloped event	The Diamond denotes an undeveloped event. This usually results from a root cause identification that is not within the scope of the I-TA team to address.
Basic event	The Circle denotes an identified root cause that is within the scope of the FI'A team to address. It can also represent basic events that are independent of each other.

Using Equipment Failure Mode and Effect Analysis (EFMEA) to Identify Potential Machine Failures and Causes

EFMEA is the same as machinery failure mode and effect analysis (MFMEA). In either case, whichever is used, it is a systematic approach that applies a tabular method to aid the thought process used by simultaneous engineering teams to identify the machine's potential failure modes, potential effects, and potential causes of the potential failure mode. It is also used to develop corrective action plans that will remove or reduce the impact of the potential failure mode. It may be used in the design phase or as a troubleshooting methodology of existing failures.

In the last several years, most of the OEMs require their equipment suppliers to perform an FMEA. There are two types of FMEAs that will need to be delivered by the supplier:

1. *Machinery:* In which the original equipment manufacturer (OEM) (customer) shall generate this for full and assembly-level systems.

Table 5.6 FTA Symbols

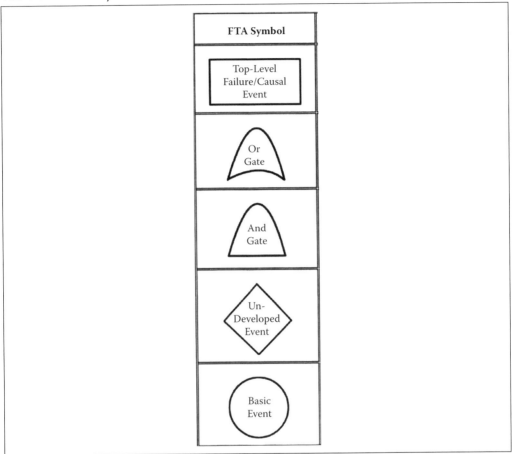

2. *Design:* In which the OEM and supplier shall generate this for transfer mechanisms, spindles, switches, and cylinders exclusive of assembly-level equipment.

For the appropriate FMEA, the supplier must understand the requirements of the customer, without any misinterpretation. To help in this endeavor, the supplier may use a hierarchical model. The basic purpose of the hierarchical model is to divide the machine into subsystems, assemblies, and lowest replaceable units (LRUs). This process assists the design team to review the individual failures, the MTBF, and the reliability levels of the systems. It also assists in the development of a maintenance strategy.

The by-product of this analysis is also very important in the sense that when the design process begins, it is important that the designers attack the

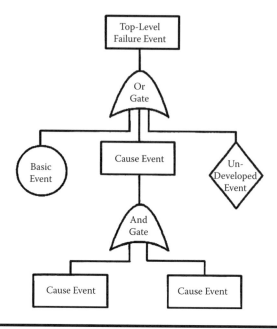

Figure 5.3 A typical approach of root cause analysis using FTA.

weak links in the equipment first. The objective is to remove the failures from these designs, allowing for a reduction in overall system failures.

Equipment availability >95%; MTBF > 100 h; MTTR < 1 h

Subsystem 1: MTBF = 250 h; R = 90%

Subsystem 2: MTBF = 60 h; R = 80%

Subsystem 3: MTBF = 125; R = 99%

Steps to determine the weak link:

- Draw the system and subsystem in a hierarchical model.
- Review past documentation relating to failure modes, warranty reports, or service reports.
- Assign MTBF or failure rate values and reliability values to the system and each subsystem.

Attack the system with low MTBF and reliability values. Hint: Look for new technology and interfacing of electrical/mechanical parts.

In this example, the design team may want to start the EFMEA on subsystem 2 for the improvement process because it has the lowest MBTF and lowest R in the system.

Table 5.7   A Comparison of MFMEA/FTA

Characteristics	FTA	MFMEA
Multiple failure description.	X	
Analysis of noncritical failures.	X	
Identify higher-level events that are caused by lower-level events.	X	
Broader scope in identification of failure mode.		X
Fewer restrictions and easier to follow.	X	
Identification of external influences.	X	
Identification of critical characteristics of failure.		X
Validation format.		X
Quantitative analysis.	X	
Each component characteristic is needed to evaluate failure and effects.		X
Information is limited to basic functions.	X	
Design information is available in detail drawings.		X
Evaluation of alternate design approaches.	X	
Evaluation of failure detection and failure safe.		X
Top-down approach to failure and cause relations.	X	
Bottom-up approach to failure and cause relations.		X

Components of the Form

MFMEA Header Information

1. Enter the MFMEA number in order to track the MFMEA and the future revisions of the document.
2. Identify the proper name of the machine.
3. Enter the equipment supplier's name. The supplier should be responsible for implementing the MFMEA.
4. Enter name, telephone number, company, and engineer responsible for implementing the MFMEA.
5. Enter the model of the equipment.
6. Enter the initial MFMEA date. This MFMEA should be conducted during the Concept and Design phases of the equipment.

7. Enter the original date that the MFMEA was completed.
8. List the names of the MFMEA core team.
9. List all team members' phone numbers (if applicable).
10. List all team members' addresses (if applicable).

System/Subsystem Name Function

Enter the description of the subsystem name being analyzed. Base the analysis from the hierarchical block diagram:

- Describe the design intent of the system or subsystem under investigation.
- Function is identified in terms of what can be measured. If there are multiple functions, list them separately.
- What is the system or subsystem supposed to do?
- Describe the function in verb–noun statements.
- List environmental conditions.
- List all the R&M parameters.
- List all the machine's performance conditions.
- List all other measurable engineering attributes.

Failure Mode

A failure is an event when the equipment or machinery is not capable of producing parts at specific conditions when scheduled, or is not capable of producing parts or performing scheduled operations to specifications. Machinery failure modes can occur in three ways:

1. Component defect (hard failure).
2. Failure observation (potential).
3. Abnormality of performance is equivalent to equipment failure.

Potential Effects

The consequence of a failure mode on the subsystem is described in terms of safety and the BIG 7 losses (listed in Table 5.8).

- Describe the potential effects in terms of downtime, scrap, and safety issues.
- If a functional approach is used, then list the causes first before developing the effects listing. Typical severity descriptions are shown in Table 5.8.

Table 5.8 Typical Severity Descriptions

Effects of Failure	Types of Loss
Downtime	**Breakdowns**—Losses that are a result of a functional loss or function reduction on a piece of machinery requiring maintenance intervention.
	Setup and adjustment—Losses that are a result of set procedures. Adjustments include the amount of time production is stopped to adjust a process or a machine to avoid defect and yield losses, requiring operator or job setter intervention.
	Start-up losses—Losses that occur during the early stages of production after extended shutdowns (weekends, holidays, or between shifts), resulting in decreased yield or increased scrap and defects.
	Idling and minor stoppage—Losses that are a result of minor interruptions in the process flow, such as a process part jammed in a chute or a limit switch sticking, etc., requiring only operator or job setter intervention. Idling is a result of process flow blockage (downstream of the focus operation) or starvation (upstream of the focus operation). Idling can only be resolved by looking at the entire line or system.
	Reduced cycle—Losses that are a result of differences between the ideal cycle time of a piece of machinery and its actual cycle time.
Scrap	**Defective parts/Scrap**—Losses that are a result of process part quality defects resulting in rework, repair, and/or scrap.
	Tooling—Losses that are a result of tooling failures/breakage or deterioration/wear (e.g., cutting tools, fixtures, welding tips, punches, etc.).
Safety	**Safety considerations**—For the MFMEA, this consideration is the priority. It means an immediate life- or limb-threatening hazard or minor hazard to the operator.

Severity Ratings

A severity rating is a rating corresponding to the seriousness of the effects of a potential machinery failure mode. Always remember that a reduction in severity rating index can be affected only through a design change.

- Severity comprises three components:
  1. Safety of the machinery operator (primary concern)
  2. Product scrap
  3. Machinery downtime
- Rating should be established for each effect listed.
- Rate the most serious effect.
- Begin the analysis with the function of the subsystem that will affect safety, government regulations, and downtime of the equipment.

Classification

The classification column is not typically used in the MFMEA process, but it should be addressed if it is related to safety or noncompliance with government regulations.

- Address the failure modes with a severity rating of 9 or 10.
- Failure modes that affect worker safety will require a design change.
- Enter "OS" in the class column. OS (Operator Safety) means that this potential effect of failure is critical and needs to be addressed by the equipment supplier. The difference from the traditional criticality is that this OS is critical to the Operator, not the product from a customer's perspective. The traditional criticality refers to the product characteristic as viewed by the customer. Other notations can be used but should be approved by the equipment user. Typical other designations are high-impact or high-priority items, and significant characteristics.

Potential Causes

The potential causes should be identified as design deficiencies. These could translate as

- Design variations, design margins, environmental or defective components
- Variation during the build or install phases of the equipment that can be corrected or controlled

Identify first-level causes that will cause the failure mode. Data for the development of the potential causes of failure can be obtained from

- Failure logs
- Surrogate MFMEA
- Warranty data
- Concern reports
- Test reports
- Field service reports

Occurrence Ratings

Occurrence is the rating corresponding to the likelihood of the failure mode occurring within a certain period of time. The following should be considered when developing the occurrence ratings:

- Each cause listed requires an occurrence rating.
- Controls can be used that will prevent or minimize the likelihood that the failure cause will occur, but should not be used to estimate the occurrence rating.

Data to establish the occurrence ratings should be obtained from

- Service data
- MTBF data
- Failure logs
- Maintenance records
- Surrogate MFMEAs

Current Controls

Current controls are described as being those items that will be able to detect the failure mode or the causes of failure. Controls can be either design controls or process controls.

A design control is based on tests or other mechanisms used during the design stage to detect failures. Process controls are those used to alert the plant personnel that a failure has occurred.

Current controls are generally described as devices to

- Prevent the cause or mechanism failure mode from occurring.
- Reduce the rate of occurrence of the failure mode.
- Detect the failure mode.
- Detect the failure mode and implement corrective design action.

Detection Rating

Detection rating is the method used to rate the effectiveness of the control in detecting the potential failure mode or cause. The scale for ranking these methods is based on 1–10.

Risk Priority Number (RPN)

The RPN number is a method used by the MFMEA team to rank the various failure modes of the equipment. This ranking allows the team to attack the highest probability of failure and remove it before the equipment leaves the supplier floor.

Special reminder: Ratings and RPNs in themselves have no value or meaning. They should be used only to prioritize the machine potential design weaknesses failure modes for consideration of possible design actions to eliminate the failures or make them maintainable.

The RPN value typically:

- Has no value or meaning
- Is used to prioritize potential design weaknesses (root causes) for consideration of possible design actions
- Is a product of severity, occurrence, and detection
- Formula: RPN = (S) × (O) × (D)

Recommended Actions

Each RPN value should have a recommended action listed.

- Designed to reduce severity, occurrence, and detection ratings.
- Actions should address in order the following concerns:
  - Failure modes with a severity of 9 or 10.
  - Failure mode or cause that has a high severity occurrence rating.
  - Failure mode or cause or design control that has a high RPN rating.
- When a failure mode or cause has a severity rating of 9 or 10, the design action must be considered before the engineering release to eliminate safety concerns.

There is nothing wrong with saying that no action is being taken at this time, if it is appropriate, applicable, and documented.

Date, Responsible Party

Without a specific day and without a responsible person to carry out the recommended actions, you will never have a complete and thorough FMEA. The reason? Without these essential items, there will be no accountability and no ownership. Therefore:

- Document the person, department, and date for completion of the recommended action.
- Always place the responsible party's name in this area.

Actions Taken/Revised RPN

The action taken must be selected from the pool of the recommended actions. Make sure it is one action at a time. If more actions are taken at the same time, you will not know what action caused what change. The evaluation for the revision should be based on the *severity, criticality*, and *detection:*

- After each action has been taken, document the actions.
- Results of an effective MFMEA will reduce or eliminate equipment downtime.
- The supplier is responsible for updating the MFMEA. The MFMEA is a living document. It should reflect the latest design level and latest design actions. Any equipment design changes need to be communicated to the MFMEA team.

Revised RPN

Recalculate S, O, and D after a specific action taken has been completed. A reminder here is that at this stage the MFMEA team needs to review the new RPN and determine if additional design actions are necessary.

Summary

This chapter described how to improve OEE on *existing* machinery, using some traditional tools for improvement and some specific methodologies such as quick changeover, mistake proofing, P-M analysis, FEA, and FTA. Chapter 6 continues the discussion of OEE improvement, but the focus is on *new* machinery.

Chapter 6

Improving OEE on New Machinery: An Overview of Mechanical Reliability

Chapter 5 discussed improving the overall equipment effectiveness (OEE) on *existing* machinery. This chapter continues the discussion, however with a focus on mechanical reliability for *new* machinery.

The overall goal of the reliability and maintainability (R&M) program is focused on three main areas:

1. Increase the machine's uptime
2. Reduce the repair time when a failure occurs
3. Minimize the overall life-cycle costs of the equipment

To fulfill these three basic requirements, this chapter addresses the basic needs of the acquisition and OEM design engineers to review the fundamental tools necessary to reduce failures, reduce maintenance actions, and minimize the overall life-cycle cost of the machine.

One of the primary outcomes of the goals just mentioned is the prioritization of problems. It is very important for the team (including operations, engineering, purchasing, quality, reliability, manufacturing, etc.) to recognize the temptation to address *all* problems just because they have been identified. However, if the team takes that action, it will diminish the effectiveness of the total R&M.

Instead, it should concentrate on the *most important* problems, based on performance, cost, quality, or any characteristic identified on an a priori basis through a variety of tools and methodologies.

There are many tools and methodologies that the team may use. Some are very simple, and some are very advanced. Some may be the same as the ones discussed in Chapter 3. In any case, this chapter selectively discusses the frequent methodologies that are not too simple or too complicated. To give you an idea of the variety of tools and methodologies you may use, here's a short list:

- Brainstorming
- Affinity diagrams
- Force-field analysis
- Quality Function Deployment
- Kano model
- Benchmarking
- Machine failure mode effects analysis (MFMEA)
- Pugh concept
- Axiomatic designs
- Geometric Dimensioning and Tolerance
- Design of experiments (DOE)
- TRIZ
- There are many more.

As mentioned, this chapter discusses *some* of these tools and methodologies; it also introduces such topics as mechanical components, design margins for electrical and mechanical components, reliability, MFMEA, steps for successful application engineering, and what is needed to address the testing of machinery. Let us begin with a general discussion of design.

Understanding Equipment Design Variables

The design variables are very important for the equipment designer to understand. The equipment supplier should complete an equipment analysis to ensure that all variables to the design have been identified. Typical issues of this analysis are

- **Piece-to-piece variation:** This includes the variations that are developed from one component to another as placed on the equipment. For example, a manufacturer produces a cylinder, and it has a mean time between failures (MTBF) of 100 h; the same cylinder may be manufactured later and have an MTBF of 80 h.
- **Dimensional strength degradation:** This relates to the degradation of the component, subsystem, or assembly over time. The duty cycle related to the operation of the equipment must be identified.
- **Customer usage:** This relates to how the customer will use the equipment. Items that should be reviewed are jobs per hour (JPH), maintenance activities, and the skill of the workforce.
- **External environment:** This relates to the environment in which the equipment will be located in the manufacturing operation. Temperature, humidity, ultraviolet lighting, and other environmental parameters should be listed.
- **Internal environment:** This relates to the internal operation of the equipment.
- **Internal temperatures:** This and the location of other components, power regulation, and other internal environmental issues should be identified.

Developing Design Input Requirements

The design input requirements are developed by the customer and are translated into specifications for the equipment. These specifications may include

- Jobs per hour (JPH)
- Mean time between failures (MTBF)
- Mean time to repair (MTTR)
- Availability
- Maintainability issues

and other reliability parameters. Typical concerns here are described in the following paragraphs.

Equipment under Design

The equipment profile is a description of the environmental and operating conditions in which the equipment must operate. This profile is important to the equipment design because it will outline certain design characteristics that may develop into failure modes once in operation. The equipment profile should identify

- Factory requirements such as JPH, duty cycles, maintenance characteristics, product being produced, yearly operating hours, etc.
- Environmental analysis
- Customer's needs, translated into customer specifications
- "Top 10" failure analysis
- Equipment performance profile statement

Ideal Function of Equipment

The ideal function of the equipment under design relates to translating the customer's specifications into the actual equipment on the customer's floor. The equipment at this point should meet all design input requirements listed by the customer.

Control Factors

Design control factors are used to offset the design variables. Design control factors may be strength of materials, cylinder sizing, safety margins, and types of oils for lubrication or other controls that can be used to minimize the impact of the design variables. It must be remembered that these design controls must be analyzed from the overall cost standpoint of the equipment. To control the MTBF of a cylinder, each cylinder should really be tested to MTBF values. This might prove to be an effective design control; however, the cost of implementing these tests would outweigh the benefits.

Failure States of Equipment

The designer should evaluate the various sources of failures that may be established with the equipment under design. The starting point is

- "Top 10" failure analysis
- MFMEA
- Past designs

- Warranty reports
- Service reports
- Field reports
- Fault tree analysis

Implementing Factory Requirements

Factory requirements include the overall view of the manufacturing conditions of the equipment's location for the equipment design team. These requirements should include data from the following sources.

Jobs per Hour (JPH)

The machine cycle time or jobs per hour is a basic parameter for the design of the equipment, but it also drives secondary issues such as coolant, air, electrical usage, waste generated, and machine speed relative to other machines on the line. Fast machines can be scheduled for maintenance online, whereas bottleneck machines must have high reliability built in to ensure low failure rates.

Required Quality Levels

Knowledge of the process capabilities of the machine will identify the amount of scrap that will be produced by it. This level of performance is generally copied from other successful machine designs instead of fixing the quality problems on the floor.

Duty Cycle Operating Patterns

The percentage of time the equipment operates is identified as the duty cycle. The patterns of operation vary from plant to plant and from plant area to plant area. The design team should evaluate these patterns and include them as part of the design.

Operator Attention

How will the machine be staffed? The design team should look at loading of raw materials and minor maintenance activities.

Maintenance Required

The amount of preventive maintenance (PM) is a function of the design, time, and number of people available to perform maintenance activities.

Management

The design team should evaluate what machines will be required to be operational in order to meet production requirements. The PM activities and crisis maintenance will be focused on bottleneck operations. For example, a cylinder block line showed that three machines were critical and required an availability of 80%. If these three machines did not maintain 80% availability, the line could not achieve the desired production rates. Keeping these three machines at an 80% availability rate was more cost effective than spending hundreds of thousands of dollars on noncritical areas.

Conducting an Environmental Analysis before Installing New Equipment

It is important to develop a set of environmental parameters for the equipment. This includes listing the various conditions the equipment will be subjected to while in the plant. These factors could be a source of failure, and could extend the repair time needed for equipment on the plant floor. The environmental factors could include

- Temperature
- Mechanical shock
- Immersion or splash
- Electrical noise
- Electromagnetic fields
- Ultraviolet radiation
- Humidity
- Corrosive materials
- Pressure or vacuum
- Contamination and its sources
- Vibration
- Utility services

Table 6.1   A Typical Environmental Data Sheet

Environmental Factors	Range	Rate of Change
Relative humidity	5%–95% R.H.	+ 10% R.H.
Electrical power	110–120 VAC	+ 10 VAC
Pneumatic pressure	45–75 psi	>5 psi
Hydraulic pressure	75 psi	+ 3 psi

Table 6.1 shows how a typical environmental data sheet could look.

Understanding Possible Environmental Failures to Prevent Them from Occurring

Environmental conditions and the consequent stress applied can cause components and equipment to fail. Typical effects of the environment and the causes of the failures are listed in Table 6.2.

The "Top 10 List" of Equipment Failures and Causes

While still in the concept phase of the equipment, it is generally a good practice to come up with a "Top 10" failure list, which can provide valuable information on the design of the new equipment. (Of course, this list will be based either on historical or surrogate data.) The failure list allows the designer to see the major sources of downtime associated with the current equipment.

Once the "Top 10" list is compiled, a root cause analysis should be conducted on each of the failure modes. When the root cause of failure is known, the designer can then modify the design of the new equipment to eliminate the major source of downtime associated with the existing equipment. A typical "Top 10" failure list for a pump may resemble the one in Table 6.3.

Developing an Equipment Performance Profile

When beginning the equipment design process, a clear understanding of the equipment's intent should be developed. This statement is commonly referred to as an *equipment performance profile statement*. A solid performance profile statement should address

Table 6.2 The Environment and the Cause of the Failure

Environment	Principle Effects	Typical Induced Failures
High temperature	• Thermal aging • Oxidation • Structural change • Chemical reaction • Softening, melting, or sublimation • Viscosity reduction or evaporation • Physical expansion	• Insulation failures • Structural failures • Loss of lubrication properties • Increased mechanical stress • Increased wear on moving parts
Low temperature	• Increased viscosity and solidification • Ice formation • Embrittlement • Physical contraction	• Loss of lubrication properties • Alteration of electrical properties • Loss of mechanical strength, cracking, or fracture • Structural failures • Increased wear on moving parts
Thermal shock	• Mechanical stress	• Structural collapse • Weakening of electrical components • Wear • Reduced function

High relative humidity	• Moisture absorption
	• Chemical reaction, i.e.,
	  – Corrosion
	  – Electrolysis
		• Swelling, rupture of container, physical breakdown; loss of electrical strength
		• Loss of mechanical strength, interference with function, loss of electrical properties, increased conductivity of insulators

Low relative humidity	Desiccation • Embrittlement • Granulation	• Loss of mechanical strength, structural collapse, alteration of electrical properties, "dusting"

Table 6.2 The Environment and the Cause of the Failure (*Continued*)

Environment	Principle Effects	Typical Induced Failures
High pressure	• Compression	• Structural collapse • Interference with function • Penetration of sealing
Sand and dirt	• Abrasion • Clogging	• Increased wear • Interference with function • Alterations of electrical properties
Ozone	• Chemical reaction • Cracking • Embrittlement • Granulation • Reduced dielectric strength of air	• Rapid oxidation of electrical properties • Loss of mechanical strength • Reduced function • Insulation breakdown resulting in arcing

Table 6.3 An Example of a "Top 10" Failure List for a Pump

Top 10 Failure List		
Failure Mode	Downtime (in hours)	Root Cause
Does not pump water	6	Broken impeller, blown fuse
Reduced pressure	4	Leaky seal
Overpressure	3	Dirty filter, wrong filter
High contamination	2	Dirty filter, wrong filter
Overheating	1	Dirty filters, dirty intake line
Leaky outlet	1	Defective seal
Excessive noise	1	Defective bearing

- Environment
- Standards
- Safety
- Quality
- Jobs per hour
- Duty cycle

- Hours of operation
- Cycle time

Example

The following is an example of an equipment performance profile statement:

The proposed machine design will perform the operation described in Drawing XW73-67A30 Operation 10 to an aluminum blank, with cycle time of 34.6 s. The machine will produce no more than 1% scrap (assuming that the raw materials conform to specifications) and will exhibit a C_{pk} of 1.6.

The machine will be located at the Southgate Stamping Plant, which has no air conditioning or tempering, and is located 4200 ft above sea level. The normal temperature range for the plant is between 50°F and 120°F, which can change by 30°F in an 8 h period. The humidity range for the plant is from 20% to 95% noncondensing. The machine can operate in a condensing environment without damage to the equipment or affecting the safety of personnel.

Electrical power to the machine will comply with IEC346 for stability, variability, and noise. Input voltage will be 480 VAC ± 10%. The machine will require 15 cfm of compressed air at a minimum of 88 psi. A total of 200 gal per minute of coolant at 80°F maximum will be required.

The maximum noise level from the machine will not exceed an average of 82 dbA for a period of 2 min. Peak noise levels will not exceed 100 dbA.

The machine will require no operator attention between startup and stop, nor will any planned or preventive maintenance actions be required during the automatic operation of the machine. All planned or preventive maintenance will be performed during machine shutdown and shall not exceed a 4 h period of time to complete the action.

Design Concept

Various tools can be used by the designer to develop a reliable and maintainable piece of equipment. These tools should be used in the early design stage of the equipment in order to take advantage of three factors:

1. Reusability: parts and/or equipment that have been used in the past. For example, a part was used in model year (MY) 2002. In MYs 2003–2007, that part and/or equipment was discontinued. However, in MY 2008, it was used as an active part of a new subassembly.
2. Carryover: parts and/or equipment that have been used in the preceding model year. For example, a part was used in 2007 and in 2008.
3. Reduction in complexity (*simplicity*): a part and/or equipment that is very complicated and has many redundant systems unnecessarily. It is a very

complicated design with low reliability. The idea here is to redesign the same part or equipment with better reliability and less complicated design.

Design Simplicity

One of the most fundamental characteristics of improved reliability and maintainability is to reduce the complexity of the equipment. Remember "KISS": "Keep it simple, Stanley." Components that can be removed in the design process will ensure that part count in the equipment is reduced, thus generating a reduction in potential failures of the equipment.

The design simplification also relates to maintenance action for the equipment. Reducing the number of components will reduce maintenance times, thus resulting in an increase in overall equipment availability.

A classic example of design simplification can be found with Landis Grinding Machines. In the early 1990s, design simplification was applied to reduce the complexity and parts count of the grinder. The new design employed a direct-drive motor to turn the cam. This reduced the need for two pulleys, a drive belt, and the associated supporting mounting structures. This action improved the overall reliability of the grinder, reduced maintenance actions, and eliminated limited-life items (i.e., the drive belts).

Design simplification can also be associated with reduction in the overall number of machines that will be used in a process. One automobile company reported an engine program at a Spain facility that was able to review past failure data from other engine plants. As a result, the program was able to reduce the number of machines from 176 to 129, a 27% reduction. An increase in reliability and reduced maintenance actions were seen through this program. Another benefit of the program was reduction of overall program costs.

As an additional benefit to design simplification, *standardization* is quite often an outcome. Standardization is a concept in reliability and maintainability which, when followed, reduces the number of different components used on a specific machine within a particular plant. Standardization can affect the overall operation of the plant in three ways:

- Reduces maintenance time due to skilled resources' familiarity with component and removal/replacement methods
- Increases the probability of having the correct spare parts in the plant
- Reduces plant's inventory of spare parts

An example of this action was employed at an engine plant of a major U.S. automotive company. Allen–Bradley was the supplier of all stainless steel proximity switches throughout all of the machining departments. This ensured that there were only two different sensor configurations used in the entire installed base of approximately 18,000 sensors. Other standardized AC and servo motors are being used to reduce maintenance time and inventory and promote other cost-saving activities.

Commonality of Design

Commonality of design refers to the concepts of using "like designs" of proven reliability in the design and development of new equipment. The main function of commonality is a reduction in design time and costs.

Lambda is a supplier of DC power supplies, and uses the same packaging techniques and modularized design approach to create a family of products. Power supplies that demand larger output voltage will have the same engineering and design techniques that applied to lower-voltage ones. The same thermal analysis, stress analysis, and other tools are applied to all power supplies.

An extension of this commonality principle, we find *modular design* to be also a factor in new machinery. This is a concept that applies the grouping of physical and functionally distinct units to facilitate removal and replacement. Modular design concepts can be applied to electrical black boxes, printed circuit boards, and other quick attach/detach electrical components.

An example of electrical modular design is the removal and replacement of servo drive boards in the control cabinet. The board plugs into a back plane, and with the removal of two fasteners, the board unplugs. All cable connections to the board are quick disconnects, resulting in reduced downtime of the servo system.

Mechanical components can also be designed with modular design in mind. A gear box can be dislodged by removing five fasteners. The new gear box can be replaced, and the equipment can return to normal operation with a short maintenance downtime. The results of modular design are

- Less downtime of equipment
- Less training of personnel
- Reduced maintenance actions

Improving Mechanical Reliability by Understanding Mechanical Failures

Mechanical reliability is defined as engineering in the design intent of a machine or component function rather than engineering out failures.

The aim of mechanical reliability focuses on the understanding of failure mechanisms at the part and material level. The broad range of mechanical failures are described in this section.

Designing Equipment to Prevent and Maintain against Fatigue or Fracture

Fatigue is destruction that is caused by repeated mechanical stress applied to the component. This is identified as the critical stress level of the component. The damage that is created by these critical stress levels deforms the internal structure of the metal; therefore, the metal cannot return to its original prestressed condition when the stresses are removed.

Critical stresses are cumulative; repeated or cyclical stresses above the critical stress level will eventually result in failure of the component. Fatigue is very important when designing equipment that will be subjected to

- Repeated load application
- Wave loading
- Vibration

Temperature and corrosion are the main factors that can accelerate the effects of fatigue. Higher temperatures accelerate crack growth by maintaining critical energy levels. Corrosion can also be a factor in increased fatigue failures.

Designing against Fatigue

When designing against fatigue, the following information may be useful to the design team:

- Develop a complete understanding of the critical stress levels of the components being used in the equipment.
- Control stress distribution by paying special attention to holes, corners, and fillets. These locations may cause vibration, and additional simulations—such as finite element analysis—need to be completed by the designer.
- Design for "failsafe mode" into the mechanical design so that other components will be able to take the load when a fatigue failure is observed. This "failsafe mode" should be a long enough period of time so that equipment can remain running while the repair is being completed.

- Design for ease of inspection and maintenance in order to observe the fatigue.
- Ensure that all areas that could develop into cracks have proper fillets applied, surface treatment (to relieve surface stresses) done, and/or increased surface toughness.
- Establish manufacturing inspections that will detect these areas of fatigue to ensure that they are not delivered to the customer.

Maintenance Action against Fatigue

Several maintenance actions can be developed to detect components that have a high degree of fatigue. These actions are as follows:

- Visual inspection
- Nondestructive testing such as x-ray, dye penetration, or acoustic emission testing
- Monitoring vibration spectrum
- Scheduled replacement before the end of the fatigue life

Preventing and Minimizing Equipment Wear

Wear can be defined as the removal of materials from the surfaces of components as a result of these surfaces coming into contact with other components or materials. Wear can occur in a variety of mechanisms, and more than one mechanism can result in the wear of a component. Table 6.4 shows several types of wear.

Methods to Reduce Wear

The main methods that can be used to reduce wear are

- Minimize the contact of vibrating surfaces.
- Select materials and surface treatments that are wear resistant or self-lubricating.
- Design efficient lubrication systems and ease of access for lubrication when necessary.

If a plain bearing shows signs of adhesive wear at one end, oil film thickness and likely shaft deflection or misalignment should be checked. If the failure is identified as abrasive wear, then the lubrications and surfaces should be checked for contamination or wear debris. In serious wear

Table 6.4 Types of Wear Found in Equipment

Wear Type	Description
Adhesive	Found when smooth surfaces rub against each other. Generally caused by high points on the surface and relative motion, causing heating and dragging on the surfaces. Results in particles being scraped and broken, generating wear debris
Fretting	Similar to adhesive wear but occurs between surfaces subject to small oscillatory movements. The movements do not allow wear debris from escaping, thus creating oxidation. Repeated movement over the same area will generate surface fatigue and corrosion, leading to failure
Abrasive	Generated when a soft surface comes in contact with a hard surface. The wear action is often a cutting action, with the displacement of the soft materials to the sides of the groove on the soft materials
Fluid erosion	Caused by fluids coming into contact with surfaces with sufficient energy. If the fluids contain solid materials, then the wear is accelerated
Corrosive	Materials are removed from the surface by electrolytic action. Corrosive materials may wear away the protective coating, thus resulting in accelerated wear actions

problems, a design change or operational limitations might be needed to eliminate the wear problem. Other changes, such as materials, surface treatment, or lubrication may be sufficient. It is also important to have proper lubrication filtration to reduce contamination.

Maintenance Actions against Wear

Maintenance actions should be established for high-wear items in the equipment. These are

- When wear is observed, classify the wear item into one of the described categories and identify the root cause of the wear.
- Look for atmospheric conditions that could result in wear.
- Look for contamination in lubricants, filtration devices, and other areas that may increase wear failure rates.
- Schedule monitoring of lubricants for contamination.

Preventing and Maintaining against Equipment Corrosion

Strength degradation can be caused by corrosion. Corrosion occurs when ferrous metal is placed in damp locations. It can be accelerated by chemical change contamination, such as salt in the atmosphere, resulting in corrosion of certain metals. The primary cause of corrosion is oxidation. Stress corrosion is caused by a combination of tensile strength and corrosion damage. Corrosion creates surface weakness, leading to crack formation. Further corrosion attracts the weakened cracked tips where the metal is in a chemically active state, and where high temperatures generate accelerated chemical actions. This combination of effects causes increased corrosion of the materials, resulting in failures. The design methods that can be identified to alert the team to corrosion are

- Select materials that are appropriate for the environment and the application.
- Protect the surface of the materials.
- Use protective environmental devices such as dryers or desiccators.
- Promote awareness of actions that will generate stress corrosion.

Developing Equipment Safety Margins

Safety margins for mechanical components are generally defined as the amount of strength of a mechanical component relating to the applied stresses (see Figure 6.1).

Generally, a rule-of-thumb in mechanical design with a normally distributed stress-load relationship is that the safety margin should always be greater or equal to three. This will provide for highly reliable mechanical designs. Mechanical reliability can be described as follows:

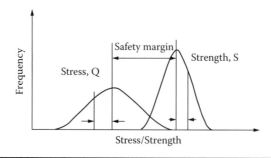

Figure 6.1 Safety margin of stress strength.

- Mechanical Application Static—tension, shear, or torsion applied mainly to structures, housing, and components
- Mechanical Application Dynamic—reversing or irregular stresses mainly applied to components and simple structures; variable stress levels, especially with irregular time events, are most often associated with components of basic materials.

In addition, mechanical reliability analysis can take many forms:

- Stress-load analysis
- Modeling
- Probabilistic estimates
- Coefficients of variability
- Cyclical loads
- Reversing loads
- Material limits
- Simulation
- Finite element analysis

The safety margin can be defined by the ratio of the critical design strength parameters (tensile, yield, etc.) to the anticipated operating stress under normal load conditions. The safety margin equation is

$$SM = \frac{U_{Strength} - U_{Load}}{\sqrt{SV^2 + LV^2}}$$

where
  SM = safety margin
  $U_{Strength}$ = mean strength
  U_{Load} = mean load
  L_V^2 = load variance
  S_V^2 = strength variance

Example

A robot's arm (see Figure 6.2) has a mean strength of 80 kg. The maximum allowable stress applied by end-of-arm tooling is 50 kg. The strength variance is 8 kg, and the stress variance is 7 kg. What is the safety margin for this combination?

$$SM = \frac{U_{Strength} - U_{Load}}{\sqrt{SV^2 + LV^2}} = \frac{(80-50)}{\sqrt{8^2 + 7^2}} = \frac{30}{10.63} = 2.822$$

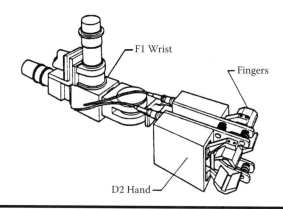

Figure 6.2 A robot's arm.

Example

A robot has a maximum lifting strength capacity of 15 kg. The part to be transferred exhibits a load of 3.5 kg, and the tooling weight load is estimated to be 5.5 kg. The variance for the strength is 1.5 kg, and the variance for the load is 2.0 kg. Calculate the safety margin for this unit. What is the interference for the strength and load? What are your recommendations for this design?

$$SM = \frac{U_{Strength} - U_{Load}}{\sqrt{SV^2 + LV^2}} = \frac{(15-9)}{\sqrt{1.5^2 + 2^2}} = \frac{6}{2.5} = 2.4$$

[A low safety margin may indicate assigning another size robot or redesigning the tooling material.]

Stress Analysis

Stress analysis is an evaluation of the stress levels when a component is operating under actual operating and environmental conditions. The resulting stress analysis defines the maximum anticipated operating stress. This point is identified as the reliability boundary (Rb).

Rb is where the component may operate with a certain failure rate or reliability. For mechanical components, it may be interpreted as a point where operating stress just exceeds the minimum material strength.

Stresses and the Bathtub Curve

Due to high stress areas with components, different shapes of the bathtub curve may be generated (see Figure 6.3).

Stress–Strength Relationship

Equipment can fail in a number of ways. For example, excessive vibration can cause nuts and bolts to become loose and fall off; human error can cause misadjustment of valves. Wear items such as belts or seals can deteriorate, causing mechanical failures of equipment functions. Figure 6.4 shows a normal distribution for stress and strength.

Normal Stress–Strength Relationship

In Figure 6.4, the flatter distribution on the left characterizes *stress*, while the steeper curve on the right characterizes *strength*. Notice that there is no overlapping of the two curves, greatly reducing the probability of failures.

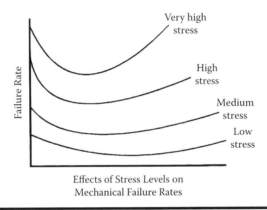

Figure 6.3 Effects of stress levels on mechanical failure rates.

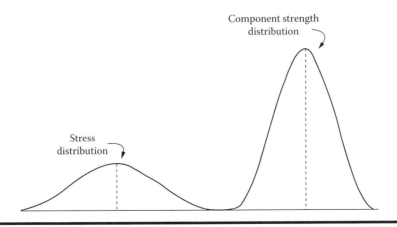

Figure 6.4 Normal stress–strength relationship.

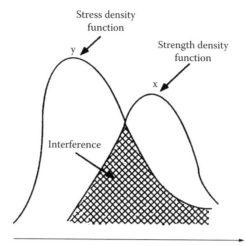

Figure 6.5 Interference location for failure.

When stress exceeds strength, a potential failure mechanism develops. This is shown graphically in Figure 6.5. The shaded overlapping of the stress–strength curve shows the highest probability of failure.

The mechanical strength–load interference is defined as the probability of the strength distribution exceeding a random load distribution observation. The probability of this interference can be observed from the safety margin calculation; and, from a normal distribution Z table (Table 6.5; see also Appendix A), the probability of interference can be obtained. The reliability of the strength–stress relationship can be estimated from the safety margin calculation.

Example of Interference Calculation

The safety margin calculation can now be used to determine the interference and reliability of the components under investigation. The terms of interference and reliability are defined as follows:

Interference: the probability that a random observation from the load distribution exceeds a random observation from the strength distribution.

(The values used in this example are arbitrary. They are used to explain the transformation of the Z-value to Reliability).

$$\text{Reliability: } R = 1 - \text{Interference}$$

$$Z = SM = \frac{U_{Strength} - U_{Stress}}{\sqrt{SV^2 + LV^2}} = \frac{(8000 - 5000)}{\sqrt{(800)^2 + (700)^2}} = \frac{3000}{1063} = 2.822$$

Table 6.5 Z Table—Normal Distribution (Area Beyond Z)

Z	.00	.01	.02	.03	.04	.05	.06	.07	.08	.09
0.0	.5000	.4960	.4920	.4880	.4840	.4801	.4761	.4721	.4681	.4641
.1	.4602	.4562	.4522	.4483	.4443	.4404	.4364	.4325	.4286	.4247
.2	.4207	.4168	.4129	.4090	.4052	.4013	.3974	.3936	.3897	.3859
.3	.3821	.3783	.3745	.3707	.3669	.3632	.3594	.3557	.3520	.3483
.4	.3446	.3481	.3372	.3336	.3300	.3264	.3228	.3192	.3156	.3121
.5	.3085	.3050	.3015	.2981	.2946	.2912	.2877	.2843	.2810	.2776
.6	.2743	.2709	.2676	.2643	.2611	.2578	.2546	.2514	.2483	.2451
.7	.2420	.2389	.2358	.2327	.2296	.2266	.2236	.2206	.2177	.2148
.8	.2119	.2090	.2061	.2033	.2005	.1977	.1949	.1922	.1894	.1867
.9	.1841	.1814	.1788	.1762	.1736	.1711	.1685	.1660	.1635	.1611
1.0	.1587	.1562	.1539	.1515	.1492	.1469	.1446	.1423	.1401	.1379
1.1	.1357	.1335	.1314	.1292	.1271	.1251	.1230	.1210	.1190	.1170
1.2	.1151	.1131	.1112	.1093	.1075	.1056	.1038	.1020	.1003	.0985
1.3	.0968	.0951	.0934	.0918	.0901	.0885	.0869	.0853	.0838	.0823
1.4	.0808	.0793	.0778	.0764	.0749	.0735	.0721	.0708	.0694	.0681
1.5	.0668	.0655	.0643	.0630	.0618	.0606	.0594	.0582	.0571	.0559
1.6	.0548	.0537	.0526	.0516	.0505	.0495	.0485	.0475	.0465	.0455
1.7	.0446	.0436	.0427	.0418	.0409	.0401	.0392	.0384	.0375	.0367
1.8	.0359	.0351	.0344	.0336	.0329	.0322	.0314	.0307	.0301	.0294
1.9	.0287	.0281	.0274	.0268	.0262	.0256	.0250	.0244	.0239	.0233
2.0	.0228	.0222	.0217	.0212	.0207	.0202	.0197	.0192	.0188	.0183
2.1	.0179	.0174	.0170	.0166	.0162	.0158	.0154	.0150	.0146	.0143
2.2	.0139	.0136	.0132	.0129	.0125	.0122	.0119	.0116	.0113	.0110

(Continued)

Table 6.5 Z Table—Normal Distribution (Area Beyond Z) (Continued)

Z	.00	.01	.02	.03	.04	.05	.06	.07	.08	.09
2.3	.0107	.0104	.0102	.0099	.0096	.0094	.0091	.0089	.0087	.0084
2.4	.0082	.0080	.0078	.0075	.0073	.0071	.0069	.0068	.0066	.0064
2.5	.0062	.0060	.0059	.0057	.0055	.0054	.0052	.0051	.0049	.0048
2.6	.0047	.0045	.0044	.0043	.0041	.0040	.0039	.0038	.0037	.0036
2.7	.0035	.0034	.0033	.0032	.0031	.0030	.0029	.0028	.0027	.0026
2.8	.0026	.0025	**.0024**	.0023	.0023	.0022	.0021	.0021	.0020	.0019
2.9	.0019	.0018	.0018	.0017	.0016	.0016	.0015	.0015	.0014	.0014
3.0	.0013	.0013	.0013	.0012	.0012	.0011	.0011	.0011	.0010	.0010

By looking up the value of 2.82 on the Z table (shown in Table 6.5), the interference or alpha area can be obtained. The reliability of the system can then be estimated from the equation

$$R = 1 - \alpha$$

From the preceding example, interference and reliability can be calculated. From the Z table (see Table 6.5), the area above the Z-value is identified as α. The α value indicates that there exists a 0.0024 or 0.24% probability of failure. Reliability can be calculated as

$$R = 1 - \alpha$$
$$R = 1 - 0.0024$$
$$R = 0.9976 \text{ or } 99.76\%$$

From this data, you can see that the strength and the load have a 0.24% probability of failure, which is obviously very low, and the reliability is 99.76%.

In any stress–strength interference (SSI) analysis, stress and strength are considered in the general sense and are not limited to mechanical stress and strength of the material. An important characteristic of SSI analysis is that these causes are random variables. The area of concern between the strength and the stress curves is the location where failures occur, and this location is called *interference*.

One of the key benefits of understanding the SSI relationship is that the designer can visualize that stress and strength are not constant but are subject to variability (see Figure 6.6). By understanding and modeling this

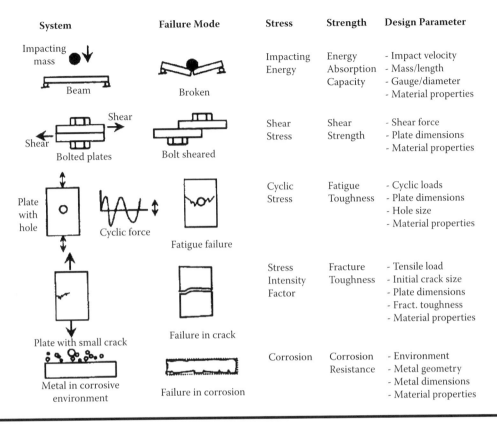

Figure 6.6 Stress–strength interference (SSI) chart.

variability, the design process can be improved. The SSI relationship can be started in the early stages of the design when nominal values of the design parameters, such as dimensions and gauges, are available.

Stress–Strength Process

The process that is used for the implementation of this application is as follows:

Step 1—Identify the key failure modes of the system.
Step 2—Identify the stress and strength characteristics.
Step 3—Identify key parameters associated with stress and strength.
Step 4—Gather key data.
Step 5—Evaluate the data.
Step 6—Make design changes.

Figure 6.6 takes a look at the stress and strength process relating to the SSI design and the typical failure modes associated with it. This example uses

Table 6.6 Typical Mechanical Failure Modes

Failure Modes	Description
Tensile yield strength	Occurs under pure tension. The strength of the material is under the rated stress of the material selected
Ultimate tensile strength	Occurs when the applied stresses exceed the ultimate tensile strength of the material and cause total failure of the structure at cross-sectional points. This is a catastrophic failure
Compressive	Occurs when the compression load is greater than the compression strength of the material. This failure will result in deformation of the materials
Shear loading	Occurs when the applied shear stress is greater than the strength of the material. Generally, this failure occurs on a 45° axis with respect to the principal axis
Creep and stress rupture	Long-term loads generally measured in years cause elastic materials to stretch, even though they are below the normal yield strength of the material. If the load remains, the material stretches (creep) and will normally terminate with a rupture. High temperatures will cause this failure to rupture in a shorter period of time
Fatigue failures	Generally occurs when the material is subjected to repeated loading and unloading over a period of time

a beam connecting an end-of-arm tool to the machine. This beam is connected with fasteners at the machine and at the tool.

During the design process, at least the following potential failure modes (as listed in Table 6.6) should be addressed to ensure there is sufficient material strength and a large-enough safety margin.

Mechanical systems and components obviously can fail for a variety of reasons. To reduce the potential of mechanical failures, the following items should be addressed in design:

- Are high-wear items on the preventive maintenance list—such as backlash in control, linkage inspection and gear fatigue due to wear, excessive tolerance, or incorrect assembly or maintenance?
- Do valves, metering devices, etc., have a visual factory marking to ensure proper setting?

- Do moving parts that seize (such as bearings or slides) have preventive maintenance scheduled to clean or remove contamination, corrosion, or surface damage?
- Are seals that are due to wear or damage on a preventive maintenance list?
- Are loose fasteners due to incorrect tightening, wear, or incorrect torque specifications listed in maintenance manuals or marked with match marks for proper location?
- Is excessive vibration in rotational components due to wear, out-of-balance rotating components, or is resonance measured to reduce potential failure modes?

Selecting the Proper Materials and Components

Selection of proper materials is important when designing a reliable piece of equipment. The equipment designer should take into account

- Relevant properties of the materials
- Application environment of the materials

Table 6.7 shows some examples of what a designer should be aware when considering different material.

Equipment that is design specified with high reliability will contain a vast selection of components ranging from springs to pumps, power transmissions to motors, and other mechanical devices. However, with the advent of new components and technology, little standardization is being designed into the equipment. This will result in weaker reliability of the equipment and additional unscheduled downtime events.

The designer can take several actions to ensure higher reliability:

1. Review all past performance of component data to locate sources of failures in past designs.
2. Make sure all personnel on the design team are aware of the application of equipment in the customer's location. Make sure component suppliers are aware of the application of their components in the equipment.
3. Use bookshelf designs, which are preferred; these designs are tested and have proven their reliability with past performance. This is also an issue of "reusability."

Table 6.7 Typical Material/Component Selection

Type	What to Look for in Selection
Metal alloys	• Fatigue resistant • Corrosion environment compatibility • Surface protection methods • Electrochemical when two dissimilar metals come into contact
Plastics, rubber	• Resistant to chemical attacks • Temperature stability • Ultraviolet radiation • Moisture absorption
Ceramics	• Brittleness • Fracture toughness
Composites, adhesives	• Impact strength • Erosion • Directional strength

4. Place all new technologies and components on a critical list so that the designer can track their performance.
5. Minimize the number of components and component types: this will increase the overall reliability of the equipment.
6. Pay attention to the small things in the equipment design, which will result in fewer design failures in the field.

Designing for Equipment Maintainability

Maintainability in design must be considered to minimize both corrective downtime and lengthen the period between required preventive maintenance actions. Design techniques such as accessibility, modularity, standardization, ability to be repaired in the shortest period of time, and interchangeability should be considered during the design by the supplier's designer. Maintainability should also take into account the end users and their historical preventive plant maintenance activities.

There are several techniques available to improve the overall maintainability of the equipment (see Table 6.8). This improvement takes place during the design stage where maintainability and maintenance

Table 6.8 Benefits of Maintainability in Design

Maintainability Techniques	Benefits
Visual Factory	Reduces MTTR
	Use of Poka-Yoke concepts to reduce mistakes
Quick disconnects of electrical devices	Improves MTTR
Captive hardware	Eliminates lost hardware
Machine diagnostic software	Improves troubleshooting time
	Reduces incorrect repairs
Designing with internal lighting	Improves MTTR
	Improves maintenance conditions
Self-aligning mounting brackets	Improves MTTR
	Eliminates misalignment
Modularity in Design	Reduces spare part inventory
	Reduces training requirements for maintenance personnel
Eye bolts designed into large/heavy subassemblies for lifting	Improves MTTR
	Ease of maintenance
Standardize components	Reduces space part inventory
	Reduces training requirements for maintenance personnel
	Use of "Best-in-Class" components
Design for Accessibility	Reduces removal and replacement time

strategies are developed. There are four rules to follow when designing for maintainability:

1. Include diagnostics that provide quick location of the cause of failure.
2. Allow repairs to be made as quickly as possible.
3. Know how to repair the failure online or offline.
4. Include the ability for the failure to be repaired ontime or offtime.

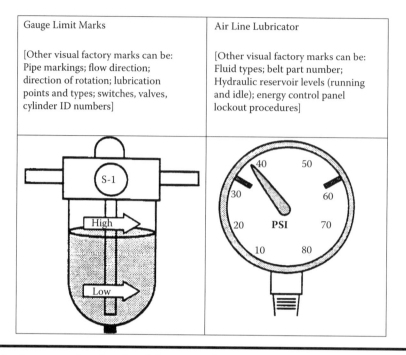

Figure 6.7   An example of a visual factory—for gauges.

Visual Factory

In today's world, most customers of machinery and equipment require some type of a visual factory program to be in place. Visual factory is a method that clearly marks gauges for high and low settings, as well as pipes for fluid content and other such strategies to develop quick and easy maintainability of the equipment. Figure 6.7 shows a simple example of such gauges.

Designing Equipment Components to Ensure Accessibility for Maintenance

During the design phase, the design team should review the location of the various components and the failure rates associated with those components. High-failure-rated components should be located in a position where they are easy to repair. A trade-off must be developed between the design's functionality and the location of the part. Always try to reduce the amount of time for repair. For example, maintenance believes that guarding is increasing the time for maintenance actions; however, no one would think of removing the guarding to make a location more accessible.

Improving OEE on New Machinery: An Overview of Mechanical Reliability ■ 173

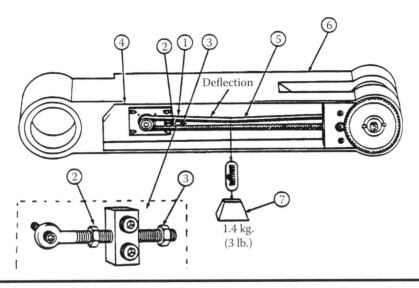

Figure 6.8 A typical design for "belt" accessibility.

Reduce the amount of components that need to be removed when replacing defective components. Do not make the maintenance personnel remove a motor to replace a sensor. Make components with known wear-out items so that they can be isolated and replaced while the machine is running. A hydraulic filter can be isolated and replaced. This can be accomplished by using a bypass or a redundant filter system. The following strategies may be initiated for accessibility in design (see also Figure 6.8):

1. Items that are frequently replaced should be easily accessible without the use of ladders, steps, or other aids. Parts should not be out of the reach of maintenance personnel.
2. All lockout procedures should be followed.
3. Items that are adjusted frequently should have high accessibility.
4. Items that wear out should be situated in an accessible location.
5. Equipment should be laid out with adequate space for maintenance activities. An area large enough to swing a wrench or enough to bring in hoists, lifts, or other items for speedy maintenance activities is necessary.
6. Group high-failure-rate components together to reduce motion time in maintenance activities.
7. Make repairs offline. If the component needs repair, replace it with a known good one and repair the failed component offline. The cost of the spare must be considered when developing this type of accessibility.

Design with Maintenance Requirements in Mind

The ultimate goal is to design out the need for maintenance. For example, if the machine uses bearings that require lubrication every 500 operating hours, using a sealed, prelubricating bearing providing 10,000 operating hours results in significantly improved maintainability and the equipment's overall availability. The maintenance design strategy should include matching the component's failure rate to the preventive maintenance schedules. If the preventive maintenance strategy is one per quarter, then the equipment should operate for one quarter without maintenance.

Many components are overspecified in order to increase reliability and reduce maintainability of the equipment. For example, if a filter is required to be changed every 3 months, then why select a filter that will last for 2 years? A design margin is good, but overspecification may be wasteful and costly for the equipment. If maintenance actions are required, then ensure that failures are easily diagnosed and located, and ensure that spare parts are available for a quick return to service of the machine. Delicate adjustments and calibrations should be removed to reduce the overall maintenance requirements. Maintenance personnel should have the proper training, tools, and materials available for maintenance actions.

Minimize Maintenance Handling

During a machine's design stage, many aspects of its configuration are determined. Therefore, review is important and necessary to ensure that maintenance-handling requirements are reduced to a minimum. For example, a machine may encounter a sensor failure, requiring 15 min to replace the sensor. However, if this sensor is located behind a drive assembly that takes 3 h to remove and replace, an additional 3 h of maintenance is realized. In addition, frequent removal of subassemblies to perform maintenance increases the likelihood of additional equipment failure and subcomponent damage.

The designer should be aware that many failures result from repair activity, such as

1. Parts broken during maintenance actions
2. Parts lost during disassembly and reassembly
3. Foreign objects left in mechanisms after maintenance actions
4. Wrong component removal during maintenance

Design with Maintenance Tools and Equipment in Mind

Quite often when machines are designed, special tools may be required to perform routine maintenance. This is usually the result of the configuration established during design. Therefore, if your machine requires a special tool, be sure to obtain it when the machine is delivered, not when it fails and one must then be obtained from the supplier. *The machinery should be designed to be maintained with the fewest number of tools.* The tool kit used to repair and maintain the equipment should be standard throughout the plant. Equipment should also be designed to ensure that the same tools are used during the repair or maintenance cycle.

Design for Removal and Replacement of Components Requiring Repair

As equipment R&M deployment actively influences machine builders to modify machines complementing today's factories, greater emphasis is placed on defining how repairs can most efficiently be performed. One technique is to define in advance the level of repair maintenance people should perform. Then review the repairs associated with various potential failures to determine the most efficient solution. Often, this is further defined by identifying a machine's "lowest replaceable unit" (LRU).

For example, if the machine has a programmable logic controller (PLC), one possible failure may be a blown output module. Once diagnosed, the output module can be removed and a new one installed. (The failed module can be returned to the OEM for repair or can be repaired in-house if the technical capabilities are present. Removal and replacement provides for quick and easy maintenance. This also means that a supply of modules must be kept in stock to ensure that MTTR values are kept to the lowest levels. Increasing MTTR will reduce availability and cause reduced production.)

Design for Interchangeability and Standardization of Parts

Because of the large number of machines purchased in any organization, a tremendous potential savings is associated with standardizing components when possible and striving to obtain interchangeable parts. For example, if all electrical motors are provided by one manufacturer, the spare parts inventory is reduced because multiple same-size motors are not necessary. In addition, if one machine is out of service, its electrical motors can be reused on other production equipment. Therefore, every effort should be made to consider standardization opportunities.

Review the Working Environment

When machines are installed in various locations by the customer, the plant's environmental aspects must be reviewed to determine their effects on reliability. For example, if the machine is installed in a glass plant, its reliability must be evaluated when considering high temperatures, the presence of sand, and round-the-clock operation. If these conditions are ignored, there is a much greater probability of frequent machine failure.

Use R&M Validation/Verification Techniques

The purpose of R&M validation and verification methods is to adequately demonstrate the performance of the equipment before shipment to the plant. These tests should consider the following information relating to the R&M process:

- Reliability and maintainability
- Vibration analysis or machine condition signature analysis
- Electrical panel thermal characterization
- Fault diagnosis
- Software reliability
- Part-quality requirements

All unscheduled equipment stoppages will be documented, and a root cause analysis will be conducted. Major design defects must be corrected before the test can continue.

Conduct a Component Application Review

The component application review area is designed to ensure that the components used in the design of the equipment are properly applied. This section of the R&M activity should be implemented by the equipment supplier and the standardized component application (SCS) supplier. The following questions should be answered by the design team to ensure proper application of the components in the equipment design:

- Have all functional requirements of the equipment been considered?
- Have all environmental issues relating to the component's application been considered?
- Have all derating criteria for the component been considered?
- Are specific application guidelines for the component identified by the SCS?

The answers to these questions should result in a Component Application Agreement (CAA) letter outlining that the components have been properly applied in the design. The CAA letter should clearly demonstrate a mutually respective and productive partnership between the supplier and the SCS, as well as a thorough engineering analysis of the application of the equipment in the design.

All component application deficiencies should be documented by the SCS and communicated to the customer's acquisition engineer. The root cause of the deficiency should be documented and reported back to the customer within one week with the corrective actions.

Use Control Point Analysis to Measure Failed Component Reliability

A controlled point analysis is the simplest method for measuring failed component reliability. This method involves having all failed components pass through a point before being repaired. These repair points could be hydraulic repair shop, electrical repair shop, or other repair points within the plant (see Figure 6.9).

Failure information is easily placed into a database as the part moves into the repair shop. Reports generated from the various repair locations can then be translated into a plantwide failure report to identify common failures throughout the plant.

Use the "Bucket of Parts" System to Compile Failure Reports

Another method used to collect data relating to the operation of equipment is identified as the "Bucket of Parts" concept (see Figure 6.10). This data collection system allows maintenance personnel to place failed components into selected bins for that component. At regularly scheduled intervals, the component supplier comes to the plant and collects all of the failed components. The supplier also creates a failure data report for the plant.

Figure 6.9 Control point analysis.

Figure 6.10 Bucket of parts.

Collect Failure Data by Tracking Spare Part Utilization

The spare part utilization method is the simplest method to collect failure data relating to failed components being requisitioned from stores. Assuming that all spare parts are being controlled from one central location, this provides for a simple trending of the components usage over time.

Summary

This chapter covered improving OEE for new machinery; specifically, it gave an overview of mechanical reliability, emphasizing some of the design variables and concepts that are needed in improving new machinery, followed by a discussion of topics that relate to mechanical reliability such as fatigue, wear, safety margin, stress–strength relationships, and maintenance considerations.

The next chapter continues the discussion on improving the OEE, focusing specifically on electrical reliability.

Chapter 7

Improving OEE on New Machinery: An Overview of Electrical Reliability

Chapter 6 began the discussion of improving overall equipment effectiveness (OEE) for new machinery, emphasizing *mechanical* reliability. This chapter continues the discussion, focusing on *electrical* reliability.

The reliability of electrical equipment employed in a manufacturing process is a function of three basic areas. These are as follows, and are covered in detail in this chapter:

1. Thermal considerations
2. Electrical device derating
3. Electrical power quality

Consider the Thermal Properties of Electrical Equipment

Heat and cold are powerful agents of chemical and physical deterioration for two very simple reasons:

- The physical properties of almost all known materials are modified greatly by a change in temperature.
- The rate of almost all chemical reactions is influenced markedly by the temperature of the reactants. The familiar rule of thumb for chemical

reactions is that the rate of many reactions doubles for every rise in temperature of 10°; this is equivalent to an activation of energy of about 0.6 eV.

Basically, heat is transferred in three methods away from electrical components: via *radiation, conduction,* and *convection.* One of these three methods (or a combination) can be used to protect electrical components from degradation. High-temperature degradation can be minimized by passive or active heat-transferring techniques:

- Passive techniques use the natural heat sinks to remove the heat.
- Active cooling methods use devices such as heat pumps or refrigeration units to create heat sinks.

When designing to remove the heat from the electrical components in a manufacturing environment, it may be necessary to look at other design alternatives. The designer should look at

- Can the components be compartmentalized to prevent heat transfer?
- Can the walls of the compartment be insulated?
- Can interdepartmental walls and intra walls of the enclosure have air flow applied independently?
- Is a substitute component available that will generate less heat?
- Can the position of the component reduce the amount of heat radiated to other components in the enclosure?

For steady temperature within an enclosure, the amount of heat generated must be equal to the amount of heat removed. Thermal systems (such as conduction cooling, forced convection, blowers, direct or indirect liquid cooling, direct evaporation or evaporation cooling, and radiation cooling) must be capable of handling natural and indirect heat-generating sources.

Passive heat sinks require some means of progressive heat transfer from intermediate sinks to ultimate sinks until the desired heat extraction has been achieved. Thus, when the heat sources are identified, and the heat removal elements selected, they must integrate into an overall heat removal system so that heat is not redistributed within the system. Efficient heat removal systems can greatly improve the overall reliability of the system.

A reduction in the operating temperature of components is a primary method for improving reliability. This is possible by providing a thermal design that reduces heat input to minimally achievable levels and provides low thermal-resistance paths from heat-producing elements to an ultimate heat sink of reasonably low temperature. The thermal design is often as important as the device design in obtaining the necessary performance and reliability characteristics of electrical equipment. Adequate thermal design maintains equipment and parts within their permissible operating temperature limits under operating conditions. Thermal design is an engineering discipline in itself.

Understanding the Benefits of Thermal Analysis

Electrical and electronic components dissipate heat through the conversion of electrical energy into thermal energy (i.e., heat). In order for equipment to operate failure free, it must dissipate its thermal energy. Excessive temperature will result in failure modes, and the causes of those failure modes are the inability of the equipment to reject the temperature increases to the surrounding environment. To minimize the overall effects of temperature, a thermal analysis can be conducted with the equipment. This analysis will assist the designer in developing an adequate heat transfer method.

In most applications, the equipment and its electrical/electronic components are housed in a NEMA 12 electrical enclosure. The National Electrical Manufacturers Association (NEMA) 12 standard will protect the equipment's electrical/electronic components from harsh environmental conditions. The NEMA 12 enclosures are intended to keep the components free from dirt, dust, and dripping of noncorrosive liquids. Because these enclosures are sealed, they must be able to keep the equipment's electrical/electronic components operating at the correct temperatures.

The reliability and maintainability (R&M) design team should request the supplier to provide a thermal analysis to calculate the steady-state internal enclosure temperature based on uniform heat conduction through the panel walls or doors and based on uniform heat transfer from the panel's external surfaces. The following information should be reviewed to determine whether to conduct a thermal analysis:

- Enclosure dimensions (height, width, and depth)
- Whether an external enclosure, such as side, bottom, top, rear, front, etc., is not available for heat transfer
- Maximum external ambient temperature

- Desired internal temperature
- Total amount of heat generated from internal component racks, panels, etc.

The maximum external temperatures are defined as the yearly maximum temperatures within the manufacturing facility. For example, in the case of an engine plant that has tempered air of 90°F (32.3°C), the desired limit of internal temperature for panel operation has been established at 104°F (40°C). This is based on a safety margin that has been established by the Personal Computing (PC) industry. The design team should consult the manufacturer's electrical thermal ratings. Typically, the company expects that, under full-load conditions, the worst-case combination of in-panel ambient heat, plus heat rise, will not exceed 44°C.

If the calculated thermal temperatures exceed the limits of desired performance, then the design team should consider the following option list for thermal design actions:

1. Increase the electrical panel size.
2. Isolate the highest heat-generating components in another enclosure.
3. Increase the overall electrical derating of the components.
4. Design in other passive and active cooling systems.
5. Increase the overall available surface area.

In most cases, on the manufacturing floor, options 2, 4, and 5 are acceptable for the reduction of temperatures. In manufacturing facilities that do not have tempered air, a side mounting of active cooling systems may be evaluated as a design alternative. To develop highly reliable designs, controls for hydraulics and pneumatics should not be mounted in the enclosure. Transformers and other high-heat-generating items should be mounted at least 6 in. away from the panel's surface.

Thermal design actions for the removal of excess heat from a panel are as follows:

- National Electrical Codes and the customer's (if any) electrical/electronic requirements must be strictly followed.
- Electrical/electronic component manufacturer's applications guidelines must be strictly adhered to as well (e.g., spacing between rack-mounted components).
- Air inlets/outlets must not be blocked.
- Preventive maintenance schedules must be followed for all filters.

- No components should be placed directly under heat-rejection fans.
- High-heat areas on panels with small internally mounted muffin fans are designed in the system to generate air circulation. Excessive temperature is a primary cause of both performance and reliability degradation in equipment. Each design must thoroughly evaluate and establish thermal characteristics so that they are consistent with the equipment's reliability requirements.
- Large heat buildups within a panel may require fans or air conditioning to move air within the panel.
- Thermal design should consider *conduction, convection*, and *radiation*.

Conducting a Thermal Analysis

A design engineer should conduct a thermal analysis on a control cabinet to ensure that the thermal characteristics of the panel do not cause premature failures of the electrical system. The thermal analysis is completed as follows:

- Step 1: Develop a list all of the electrical components in the enclosure.
- Step 2: Identify the wattage rating for each component located in the enclosure.
- Step 3: Sum the total wattage for the enclosure.
- Step 4: Add in any external heat-generating sources.
- Step 5: Calculate the surface area of the enclosure that will be available for cooling.
- Step 6: Calculate the thermal rise above the ambient temperature.

Example of a Thermal Analysis: Steps 1, 2, 3, and 4

Thermal Calculation Values			
Internal Component Name	Qty	Individual Wattage Maximum	Total Wattage
Relay	4	2.5	10.0
A18 Contactor	1	1.7	1.7
A25 Contactor	2	2	4.0
PS27 Power supply	1	71	71.0
Monochrome monitor	1	85	85.0
Subtotal wattage			*171.7*

External Component Name	Qty	Individual Wattage Maximum	Total Wattage
Servo transformer	1	450	63.0*
Subtotal wattage			63.0*
Total enclosure wattage			**234.7**

Note: Asterisk (*) indicates that the servo transformer is mounted externally and next to the enclosure. Therefore, only 14% of the total wattage is estimated to radiate into the enclosure.

Step 5: Calculate the enclosure surface area. An electrical enclosure is 5 ft tall, 4 ft wide, and 2 ft deep. The surface area of this enclosure is calculated as follows:

Front and back = 5 ft × 4 ft × 2 = 40 ft²

Sides = 2 ft × 5 ft × 2 = 20 ft²

Enclosure top = 2 ft × 4 ft = 8 ft²

Total Surface Area = **68 ft²**

Step 6: Calculate the thermal rise (ΔT).

Thermal rise ((ΔT) = Thermal resistance (θ_{CA}) cabinet to ambient × power (W).

$$\theta_{CA} = \frac{1}{(\text{Thermal conductivity} \times \text{Cooling area})}$$

$$\theta_{CA} = \frac{1}{(.25 W/\text{degree} F \times \text{Square Footage})}$$

$$\theta_{CA} = \frac{1}{(.25 W/\text{degree} F \times 68\,\text{ft}^2)} = 0.0588$$

The thermal conductivity value for the NEMA 12 enclosure is 0.25 W/degree. If the equipment inside the enclosure generates 234.7 W, then the thermal rise is

$$\Delta T = \theta_{CA} \times \text{Wattage} = 0.0588 \times 234.7 = 13.80°F$$

If the plant's ambient air is 100°F, then the enclosure temperature will see 113.8°F. If the enclosure temperature is specified as 104°F, then the design exceeds the specification by approximately 9.8°F. The enclosure must be increased in size, the load must be reduced, or active cooling techniques need to be applied.

This method is not valid for enclosures that have other means of heat dissipation such as fans, heavier metal, or if the materials were changed. This calculation method is used assuming that the heat is being radiated through convection to the outside air.

Electrical Design/Safety Margins

A design margin in simple terms is the margin that exists between the rated and applied stresses of a component. All components (whether electrical or mechanical) are designed or specified to withstand certain forces (e.g., loads, shears, torque, voltage, current, etc.).

To reduce the probability of part failure, the stress on the part can be reduced or the strength of the part can be increased. If the stress on the part cannot be reduced, then the part's strength can be increased by designing a larger or stronger part into the equipment.

Design margins applied in electrical engineering of the equipment is referred to as *derating*. Mechanical design margins are referred to as *safety margins*.

Derating to Improve Reliability and Availability of Electrical Components

Derating for electrical components essentially focuses on four different stresses:

1. Voltage
2. Current
3. Power
4. Thermal

The applications of derating concepts are generally established by internal design guidelines of the component supplier. For example, Allen–Bradley components being applied to a new piece of equipment will already have a derating factor for the active and passive components.

Another level of derating can be achieved when a power supply is being considered and requires 400 W under full load for circuit operation. A 500 W power supply can be used, thus giving a 25% derating number under full-load operation.

A servo drive requires 50 A for motor control under full load. A 20% design margin applied would suggest that a 60 A drive would be required to provide the necessary margin. A 20% derating factor is a good rule of thumb for electrical components.

These examples increase the overall reliability of electrical components in the equipment under design, resulting in fewer failures and increased availability of the equipment.

An Example of Electrical Stress

During a design review, the question was raised of whether the 24 V power supply for the pallet conveyor is adequately derated. The power supply takes 480 VAC three phase, with a 2 A circuit breaker, and it has a rated output of 10 A. A quick examination of the power supply shows that 24 V power is delivered to the load through three circuit breakers CB2, CB3, and CB4, as shown in Figure 7.1.

Note here that, when these circuit breakers are combined, 11 A of current flow to the load. This situation will probably not happen, but it requires additional investigation. The parts list on Table 7.1 illustrates the components that use 24 V, and the schematic diagram shows the output to the load.

Only one of the two solenoids will be on at any one time (total of two). The electrical derating for this example can be calculated as follows:

$$\% \text{ Derating} = 1 - \frac{I_T}{I_S} = 1 - \frac{6.737}{10.0} = 0.3263 = 32.63\%$$

where
I_T = Total circuit current draw
I_S = Total supply current

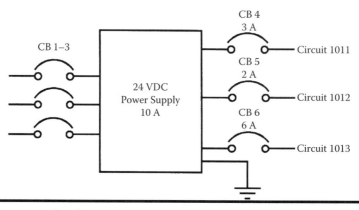

Figure 7.1 An example of a power supply with three circuit breakers.

Table 7.1 Parts List of Components (for Figure 7.1)

Description	Quantity	Current Draw	Total Draw
1011 Circuit			
Safety relay	1	0.090	0.090
Master relay	1	0.120	0.120
Power ON light	1	0.100	0.100
Emergency stop light	1	0.100	0.100
Main air solenoid	1	0.067	0.067
Subtotal			0.477 A
1012 Circuit			
Input card	3	0.015	0.450
Pressure switch	1	0.060	0.060
Prox switch	11	0.010	0.110
24 VDC to switch/relay Contacts	15	0.007	0.110
Subtotal			0.73 A
1013 Circuit			
Relay card	2	0.150	0.300
Motor starter	2	0.050	0.100
Escapement solenoids	4	0.067	0.130
External connection	10	0.500	5.000
Subtotal			5.530 A
Total circuit			**6.737 A**

From this electrical derating analysis, it can be pointed out that the power supply will not be overloaded and that the circuit breakers are generously oversized. The circuit breakers should not be tripped due to false triggers.

Some power supplies require a minimum load to be applied in order to develop proper power source regulation. The designer should check the data sheet for the power supply and evaluate if the load meets these requirements. Generally, the load required is about 20%, and we see that the master relay, power-on lights, and I/O cards come close to this figure and represent a base load that will not change. The designer should conclude that there will be no problems associated with this power supply on this equipment.

Preventing Problems with Electrical Power Quality

Table 7.2 illustrates the potential electrical failure mode and its potential cause and effects.

Preventing Electrical Failures

Electrical stress can be reduced, thus increasing the overall reliability of the component. Two methods to reduce the stresses are *electrical derating* and *thermal analysis*.

The most effective and economical method of protecting electrical components from failure is to eliminate the voltage spikes by clipping them off with surge suppressors. Transient surge suppressors do not suppress surges but rather provide a low impedance path to ground. Often, they do not provide a fast-enough response time, and the resulting voltage transients are present, leading to failed electrical components. In this case, metal-oxide varistors (MOV) should be used. MOVs simply cut off the voltage spikes.

Isolation transformers can be used to reduce electrical harmonics. Dips, sags, or surges can be corrected using voltage regulation devices.

Most motor failures are caused by environmental contamination, which includes dust, coolant, metal shavings, and other forms of grit that abrade winding insulation and attack motor shaft bearings. Filters installed on motor air intake passages must be inspected, cleaned, or replaced regularly. High ambient operating temperatures or motor overload is a destroyer of motor windings. Experience has shown that, for every 10°C increase in winding design operating temperature, winding life is reduced by half.

Motors must be of proper size for the application. An overloaded motor will run hot and fail prematurely. Motor overload protection devices must never be adjusted to solve overloading trip problems. The cause of overloading must be analyzed and corrected. Overvoltage or undervoltage will shorten winding life.

Motor overheating can also be caused by unbalanced three-phase supply voltages, resulting in a significant increase in current and accompanying overheating. Motor insulation should be tested regularly in areas where

Table 7.2 Typical Electrical Failure Modes

Power Problems	Potential Cause	Potential Effects
High voltage, spikes, and surges	• Lightning • Utility grid switching • Heavy industrial equipment	• Equipment failures • System lockups • Data loss
Low voltage, electrical noise	• Arc welders • Electronic equipment switching devices • Motorized equipment • Improper grounding • Fault clearing devices (contactors and relays) • Photocopiers	• Data corruption • Erroneous command functions • Timing signal variations • Changes in processing states • Drive state and buffer changes • Loss of synchronous states • Servomechanism control instability • In-process information loss • Protective circuit activation
Harmonics	• Switch-mode power supplier • Nonlinear loads	• High neutral current • Overheated neutral • Conductors • Overheated distribution and power • Transformers
Voltage fluctuations	• Overburdened distribution networks • Heavy equipment startup • Power line faults • Planned and unplanned brownouts • Unstable operators	• System lockup • Motor overheating • System shutdown • Lamp burnout • Power supply damage • Data corruption and loss • Reduced performance

(Continued)

Table 7.2 Typical Electrical Failure Modes (*Continued*)

Power Problems	Potential Cause	Potential Effects
Power outage and interruptions	• Blackouts • Faulted or overloaded power lines • Backup generator startup • Downed and broken power lines	• System crash or lockup • Loss of data • Loss of production • Control loss • Power supply damage • Loss of communications • Complete shutdown

switching operations create surges. Vibration caused by the motor itself or the driven load will result in both bearing and winding failure. Causes of vibration include misalignment of motor and load, bent shaft, worn bearings, and eccentricity of motor or load-rotating elements. Vibration cause should be determined, and the root causes should be eliminated.

Selecting the Appropriate Electrical Components for the Environment

Electrical component selection is based on the ability of the component to operate under the manufacturing and environmental conditions as stated in the equipment profile. The component supplier must provide evidence of reliable performance of their devices under anticipated environmental conditions, such as

- Heat
- Humidity
- Shock
- Vibration
- Dust
- Contaminants
- Electrical power changes

Off-the-shelf electrical device suppliers should be asked to provide the results of product design tests and/or production acceptance tests as a tool for verification of reliability capabilities. Warranty data information related to electrical devices should be provided as evidence illustrating the ability of that component to operate in a manufacturing environment. If the electrical device supplier has derated the component, then this information should be reviewed by the application engineer to ensure that the electrical device is properly applied. If derating is not provided, then the device should be derated when applied to the machine.

Summary

This chapter addressed improving OEE on new machinery, focusing on electrical reliability. Specifically, the topics of thermal considerations and electrical/safety margins were discussed. The next chapter continues describing how to improve OEE for new machinery, but the focus is on selected methodologies that contribute to design improvement. (For more information on mechanical and electrical reliability, the reader is encouraged to see Stamatis (2003, 1998, 1997); Kececioglu (1991).)

Chapter 8

Improving OEE on New Machinery: Selected Methodologies

Whereas the last chapter described how to improve OEE on new machinery and focused on electrical reliability, this chapter introduces new and old methodologies for innovation techniques that focus on *design improvements*, such as hazard analysis, rate of change of failure (ROCOF) analysis, and others.

Innovation relies on carefully planned measures that encompass people and processes. In fact, innovation is not just about *product invention*; it is about *reinventing business processes* and *building entirely new markets* that meet untapped customer needs (Rogers 2003). Because all innovations are not successful in the marketplace, a business needs many to ensure that a sufficient number will be commercially successful and lead to profitability during the growth and maturity phases of the product life cycle.

In the area of machinery and equipment, this axiom also holds true. It means that organizations should do everything within their power to improve *existing* machinery and/or equipment or to *redesign* that machinery with the most efficient designs. It is certainly much easier to talk about improvements and redesigns than to actually deliver a "better" machine or equipment.

There are many ways to optimize the process of improvement. However, this chapter will describe some of the most-often discussed and easy methodologies for machinery improvements. These include

- TRIZ: the theory of inventive problem solving
- Pugh diagrams
- Geometric dimensioning and tolerancing
- Short-run statistical process control (SPC)
- Measurement system analysis (MSA)
- Capability analysis
- Design of experiments (DOE)
- Hazard analysis and critical control points (HACCP)
- ROCOF (rate of change of failure) analysis
- Calculating reliability growth.

Detailed discussions of each of these methodologies follow, beginning with TRIZ; calculating reliability growth is discussed in the next chapter.

Using the TRIZ Method—The Theory of Inventive Problem Solving to Improve Machine Design

It has been said that TRIZ is one of the components of customer-driven robust innovation. (The other two are QFD and Taguchi.) TRIZ* is a methodology that was developed in Russia in 1926 by Genrich Altshuller, and has been growing all over the world since then.

The foundation of the theory is the realization that contradictions can be methodically resolved through the application of innovative solutions. Terninko, Zusman, and Zlotin (1996) have identified three premises that support the theory. They are

1. The ideal design is a goal.
2. Contradictions help solve problems.
3. The innovative process can be structured systematically.

---

* Because TRIZ is the pronunciation in Russian, many names have been given to this methodology. Some of the most common are TIPS (Theory of Inventive Problem Solving), TSIP (Theory of the Solution of Inventive Problems), and SI (Systematic Innovation).

Striving for a Higher Level of Innovation

TRIZ certainly focuses on innovation, but what is innovation? According to the founder (Altshuller), there are five levels of innovation. They are

- *Level 1: Refers to a simple improvement of a technical system.* It presupposes some knowledge about the system. It is really not an innovation because it does not solve the technical problem.
- *Level 2: An invention that includes the resolution of a technical contradiction.* It presupposes knowledge from different areas within the relevancy of the system at hand. By definition, it is innovative because it solves contradictions.
- *Level 3: An invention containing the resolution of a physical contradiction.* It presupposes knowledge from other industries. By definition, it is innovative because it solves contradictions.
- *Level 4: A new technology containing a breakthrough solution that requires knowledge from different fields of science.* It is somewhat an innovation because it improves a technical system, but it does not solve the technical problem.
- *Level 5: Discovery of new phenomena.* The discovery pushes the existing technology to a higher level.

For Altshuller (1997, pp. 18–19), the benefit of using TRIZ is to help inventors elevate their innovative solutions to Levels 3 and 4.

A Step-by-Step Guide to Using the Core Tools of the TRIZ Methodology

To optimize the levels just described, Altshuller suggests the following tools:

- Segmentation—finding a way to separate one element into smaller elements
- Periodic action—replacing a continuous system with a periodic system
- Standards—structured rules for the synthesis and reconstruction of technical systems
- ARIZ—an algorithm to solve an inventive problem

ARIZ is the core tool of the TRIZ methodology. It consists of nine steps.

Step 1: Analyze the problem. Identify the problem in concise, clear, and simple language. Do *not* use any jargon.

Step 2: Analyze the problem's model. Identify the conflict in relation to the overall problem. A boundary diagram may facilitate this. The idea of this step is to *focus on the conflict.*

Step 3: Formulate the ideal final result (IFR). Here, you identify the physical contradiction. The process is to identify the vague problem and transform it into a *specific* physical problem. [Another clue for the $Y = f(x_1, x_2,..., x_n)$.]*

Step 4: Use outside sources and field resources. If the problem remains, imaginatively interject outside influences to understand the problem better.

Step 5: Use an informational data bank. The use of standards and databases with appropriate information is recommended here to solve the problem.

Step 6: Change or reformulate the problem. If at this stage the problem has not been solved, it is recommended that you go back to the starting point and reformulate the problem with respect to the super system.

Step 7: Analyze the method that removed the physical contradiction. Check whether or not the quality of the solution provides satisfaction. A key question here is: Has the physical contradiction been removed most ideally?

Step 8: Use a found solution. Here, the focus is on interfacing an analysis of adjacent systems. It is also a source for identifying other technical problems.

Step 9: Analyze steps that lead to the solution. This is the ultimate scorecard. This is where the former process is compared to the current one. The analysis has to do with the new gap. Obviously, you should record any deviations for future use.

Using TRIZ in a Design Situation

To actually use TRIZ in a design situation, the reader must be aware not only of the nine steps just mentioned but also the 40 principles that are associated with the methodology, which are included in Appendix J. In this chapter, we list only the 39 engineering principles without any further discussion (see Table 8.1).†

---

* The characterization model that identifies the dependent variable with the independent relationships.
† See Altshuller (1997), Terninko et al. (1996), and other sources in the bibliography for more details.

Table 8.1 The 39 Engineering Parameters

Weight of moving object	Energy spent by nonmoving object
Weight of nonmoving object	Power
Length of moving object	Waste of energy
Length of nonmoving object	Waste of substance
Area of moving object	Loss of information
Area of nonmoving object	Waste of time
Volume of moving object	Amount of substance
Volume of nonmoving object	Reliability
Speed	Accuracy of measurement
Force	Accuracy of manufacturing
Tension, pressure	Harmful factors acting on object
Shape	Harmful side effects
Stability of object	Manufacturability
Strength	Convenience of use
Durability of moving object	Repairability
Durability of nonmoving object	Adaptability
Temperature	Complexity of device
Brightness	Complexity of control
Energy spent by moving object	Level of automation
	Productivity

It is imperative to recognize that using innovative approaches to solve problems especially in R&M (as shown in Figure 8.1), you must use several methods and must recognize that there is a trade-off between effort and difficulty, as shown in Figure 8.2. One of the better tools to determine the appropriate priority is the Pugh diagram (which is discussed in the next section of this chapter).

The key to the TRIZ methodology is to resolve contradictions with the Contradiction Matrix. To map the problem to the principle, Altshuller devised a 39 × 39 Engineering Parameter Matrix (see Figure 8.3 for an example); based on the outcome results (parameter improved versus undesired result), the more probable solution is selected. The numbers inside the boxes are the principles that apply to that contradiction.

198 ■ *The OEE Primer*

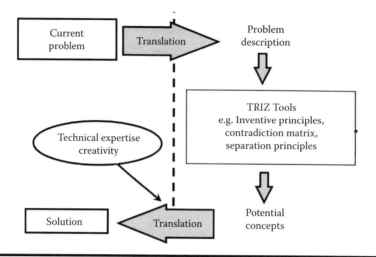

Figure 8.1 Innovation and concept selection.

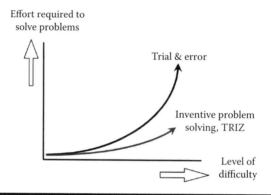

Figure 8.2 Effort required versus level of difficulty.

	Speed	Adaptability	Complexity
Speed		15, 10 26	10, 28 4, 34
Adaptability	35, 10 14		15, 29 37, 28
Complexity	34, 10 28	29, 15 28, 37	

Parameter Improved

Figure 8.3 Contradiction matrix.

Table 8.2 Summary of the 10 Key Steps in Using TRIZ to Design Innovation

1. Identify the current problem.
2. Describe the inventive issue.
3. Define how the system (problem) can be improved.
4. Choose the parameters that you want to improve.
5. Describe the way of improving the chosen parameter.
6. Determine, as specifically as possible, which parameter gets worse (there may be multiple ones).
7. Investigate the inventive principles—even further.
8. Analyze the applications and draw analogies (use analogic thinking).
9. Document all conceptual solutions.
10. Evaluate, synthesize, and come up with the best solution.

Even though the process sounds very complicated, in reality, the TRIZ approach to innovation may be summarized in 10 steps, listed in Table 8.2.

Using Pugh Diagrams to Improve Machine Design

Having generated several concepts, we need to now select the best concept. This step is very important, and cannot be trivialized or brushed off. Therefore, a structured decision-making process is needed. The decision must be based on consensus by a team that is cross functional and willing to have an open mind for the possibility of new or even improved ideas even at this late hour. The final solution should be able to be analyzed and evaluated not only for the criteria set but also against other concepts.

Many tools exist for this evaluation; however, the Pugh diagram is simple and has been accepted by most to be the tool of choice. A typical Pugh diagram is shown in Table 8.3.

Here's a guide to Table 8.3:

- The criteria are the issues that you are interested in optimizing (e.g., speed of response, accuracy, or anything else).
- The datum is the evaluation of the current process for those specific criteria.
- Concepts B, C, and n are the alternatives presented to the current process.

Table 8.3 Pugh Selection Matrix

Criteria	Concept A (Current Process)	Concept B	Concept C	Concept n
Speed of response	Datum	+	S	+
Accuracy		S	S	+
---		−	−	+

- Concept S indicates that the criteria is about *the same* as that of the current process.
- The + sign indicates that the criteria is clearly *better* than the current process.
- The − sign indicates that the criteria is clearly *worse* than the current process.

The actual evaluation is based on the most "+" signs that any concept will have and the elimination of all the worst "−" signs. As part of your evaluation, it is not unusual to combine several best concepts by incorporating the superior characteristics of other concepts. At that point, you select the best concept for pilot.

Special note: For this incorporation to be effective, instead of using "+, S, −" we recommend using "+++, S, ---" to increase the discrimination of the scoring.

You may also use a weighting system to reflect the importance of the criteria. However, if you do that, be careful with the use of numbers. They can give a false sense of precision and often lead to much debate within the team.

Using Geometric Dimensioning and Tolerancing (GD&T) to Improve Machine Design

GD&T is a technical database through which product design and production personnel can communicate via drawings to provide a uniform interpretation of the requirements for making a product. It replaces confusing and inconsistent notes and datum lines with *standard symbols* that refer to a universal code: the American Society of Mechanical Engineers (ASME) Y14.5M-1994 and Y14.5-2009 Dimensioning and Tolerancing Standard (ASME,

1994, 2009), an international symbolic engineering language. This standard improves crossfunctional communication and helps avoid costly changes to drawings after initial release.

GD&T allows design engineering to communicate the requirements of a part more clearly, and should eliminate the need for implied and/or ambiguous part definition. For a very cursory summary of the new revised Y14.5-2009 see appendix M.

An Overview of the Key Concepts of GD&T

It is beyond the scope of this book to delineate all the characteristics of GD&T. However, because of the important role it plays in translating customer requirements through drawings in the tooling and equipment industry, the following short list of characteristics or concepts is presented.*

Bonus Tolerance: When the maximum material condition (MMC) or the least material condition (LMC) modifiers are used with a geometric specification as a produced part's size departs from its stated size limit, an additional amount of geometric tolerance is permitted.

Composite Position: Using this design tool, it is possible to allow greater tolerance on a pattern of holes to the boundary of a part while maintaining tighter control of the spacing within the pattern of holes.

Datum Reference Frame: The datum reference frame defines the location and/or orientation of a part feature in relation to other specific part features. It is made up of three mutually perpendicular planes that correspond to the X, Y, and Z axes. The concept replaces the use of implied data, which often leads to erroneous interpretations. Data may then be used by production, but it is intended to identify correct inspection setup procedures.

Diametral Tolerance Zone: This is a concept that creates a cylindrical tolerance zone for the axis of round holes, pins, studs, and shafts. Although there are several applications for this type of tolerance, it is expressly intended to allow additional tolerance for round mating features.

Functional Specification/Gauging: If the maximum material condition (MMC) modifier is used, we are able to make a functional design

---

* For more information on GD&T, see ASME Y14.5M-1994 (ASME, 1994) and ASME Y14.5M-2009 (ASME, 2009).

specification. That is, bonus tolerance is permissible as long as the part will perform its intended function. We are also able to verify part conformance by a simple go/no-go gauging process when feasible.

GD&T Controls: There are 13 control symbols that are used to define various characteristics and replace the typical title block or note specifications found in drawings. All 13 of these are clearly defined in the standard, thereby eliminating any ambiguity.

Material Conditions: There are drawing symbols that are used to identify three material conditions:

1. LMC: least material condition
2. MMC: maximum material condition
3. RFS: regardless of feature size

When appropriately applied, these material conditions more clearly define the function of a part and can allow additional tolerance or datum shift when permissible in the design.

Projected Tolerance Zone: This is a concept that is generally applied to tapped holes and dowel pinholes to ensure a noninterference fit during assembly. Without this design consideration, it is difficult to obtain tight-fitting or controlled assembly conditions due to orientation limitations.

Rules: There are three rules; two deal with the application of the material condition modifiers. Rule 1 is by far the most significant, and delineates the concept of form, also being controlled by size limits.

Virtual Condition: This is a term used to describe the worst-case accumulative effect of size and any applicable geometric tolerances. This condition is used in mating part design to ensure part fit, to do stack-up analysis, and also to design functional gauging and fixturing.

Zero Tolerance: Using zero tolerance with the MMC modifier in a geometric specification assures you that any part you buy will function properly.

Using Short-Run Statistical Process Control (SPC) to Improve Machine Design

In the 1980s, SPC was reintroduced in the manufacturing industry with both enthusiasm and very good results in process improvement. However, when SPC was tried in the tooling and equipment industry, it met with difficulty and provided some questionable results. These results were experienced for two reasons:

1. Because SPC was misunderstood
2. Because in tooling and equipment organizations there is an issue of the "short run."

Let us look at these in a bit more detail.

Clarifying Misunderstandings of What SPC Is

First, let us examine the issue of misunderstanding. When someone hears the term "SPC," *charts and statistics* are automatically assumed and envisioned. That may or may not be correct. SPC is *a bundle of tools* to be used in identifying, defining, correcting, evaluating, and monitoring a process for some kind of improvement. The improvement is organization dependent and may be defined differently for each organization as well as each process. Typical improvement expectations are reduced waste, improved efficiency, improved cycle time, and so on.

To measure the improvement in an organization, one may use qualitative and/or quantitative tools (both are part of SPC):

- Some of the *qualitative* tools are
  - Brainstorming
  - Affinity charts
  - Process flow charts
  - Cause-and-effect diagrams
- Some of the *quantitative* tools are
  - Check sheets
  - Control charts
  - Scatter plots
  - General statistics
- Advanced statistics

What is important about both categories of tools is the fact that they can be used in any organization, given appropriate and applicable data. Therefore, in the tooling industry, there are situations for which SPC is appropriate and quite useful. By the same token, however, *there are some limitations*, which bring us to the second point—limited production.

Understanding the Problems of Using SPC for Short-Run Production Cycles

A short run is an environment in which there are a large number of jobs per operator in a production cycle—typically a week or month—with each job involving different products. In essence, a short run involves the production of very few products of the same type, termed a *limited production run*. An extreme case of a short production run is the one-of-a-kind product.

When a short run is used, traditional SPC is not applicable. Therefore, the analysis should focus on the *process* itself as opposed to the *product*. When that happens, modifications are in order. For example, because short runs involve less than the recommended number of pieces for capability, the acceptability criteria are usually modified. It is beyond the scope of this book to identify and discuss in detail all of the specific tools used in short-run SPC*; however, some of the short-run SPC common tools are

- Traditional X-bar and R charts
- Traditional individual (X) chart
- Traditional individual (X) chart and moving range
- Traditional attribute charts (p, c, np, and u)
- Code value charts (evaluating the deviations of the target rather than the actual value)
- Stabilized control charts (multiple characteristic charts with independent units)
- The exact method as defined and developed by Hillier (1969) standardized control charts
- Demerit control charts

It is important that these tools be applied when the appropriate data have been defined and collected.

Using Measurement System Analysis (MSA) to Improve Machine Design

Measurement plays a significant role in helping a facility accomplish its mission. Therefore, the *quality* of the measurement systems that produce those measurements is important. For example, too much variation in a

---
* See Stamatis (2003), (2003a,b) Griffith (1996), Pyzdek (1992), Hillier (1969), and others.

measurement system being used for SPC may mask important variations in the production process. A typical measuring system in any tooling and equipment organization should include

- Design and certification
- Capability assessment (over time)
- Mastering
- Gauge repeatability and reproducibility (gauge R&R)
- Operational definition
- Control
- Repair and recertification

Perhaps the most important of these is *gauge R&R*. It is a technique to measure inherent variation in measuring equipment and variation between gauge operators. It does reveal how much of the specification limit spread is used by measurement nonrepeatability. The gauge R&R establishes the following:

- *Gauge accuracy*, which is the difference between the observed average of measurements and the master value.
- *Repeatability*, which is the measure of the gauge to repeat a given measurement. Gauge repeatability is the variation in measurement obtained with one gauge when used several times by one operator while measuring the identical characteristics on the same parts.
- *Reproducibility*, which is the measure of the ability of people to repeat a given requirement. Gauge reproducibility is the variation in the average of the measurements made by different operators using the same gauge when measuring identical characteristics on the same parts.
- *Stability*, which is the total variation in the measurements obtained with a gauge on the same master parts when measuring a single characteristic over an extended time period. Sometimes *stability* is also referred to as *drift*.
- *Linearity*, which is the difference in the accuracy values through the expected operating range of the gauge.

It is very important to remember that, during gauge calibration, linearity and accuracy should be performed before assessing gauge R&R.

$$RR = (AV^2 + EV^2)^{1/2}$$

Gauge R&R quantifies measurement system variation (RR) attributable to equipment variation (EV) and appraiser variation (AV). Repeatability is equipment variation, and reducibility is appraiser variation. Measurement system variation may be expressed as a percentage of tolerance, such as

$$\% \text{ error} = 100\% \text{ RR/Tolerance}$$

Finally, in evaluating gauge R&R, the following are guidelines for acceptable percentage error:

- Under 10%: Gauge is *acceptable*.
- 10%–30%: Gauge *may be acceptable* depending on the importance of the application, cost of the gauge, cost of repairs, and so on.
- Over 30%: Gauge *needs improvement;* make every effort to identify and correct causes of variation, or use a different measurement system.*

Using Capability Analysis to Improve Machine Design

From personal experience over the last 30 years in the field of quality, I have observed that a large portion of the cost of poor quality is caused by manufacturing processes (machine and equipment) that are *incapable of producing parts to the specifications.* This brings us to the tooling and equipment industry because it provides the machinery, tooling, and fixturing. What is interesting, however, is that quite a few of the tooling and equipment suppliers will *justify* their delivery of unsuitable machines and/or equipment by claiming that "the specifications were misunderstood," "the material variation was greater than predicted," or "the machine was the first of its kind," and so on.

To avoid such a predicament, a tooling, equipment, and fixturing supplier must understand the concept of capability and what the requirements are for documenting such capability. Again, it is beyond the scope of this book to address the complete topic of capability; however, a short review of the topic follows.

A capability analysis evaluates the ability of the process to meet specifications. Stability, of course, is generally a prerequisite for calculating capability. There are several indices that can be used to quantify process capability.

---

* For more information, see DaimlerChrysler, Ford, and General Motors (2002).

All are calculated based on data collected from the process. Selection of a capability index depends in part on the type of data collected. Examples are the following:

Variable Data	Attribute Data
C_p	First-time capability (FTC; percentage good)
C_{pk}	Average percentage defective
	Average defects per sample
	Average defects per unit

C_p is sometimes referred to as the process potential because it considers only process *variation* (i.e., spread), not process *location*. It predicts the potential performance if the process was perfectly centered. C_{pk} accounts for both variation and location. The higher the C_p or C_{pk} number is, the better the process capability will be. For a perfectly centered process, the C_{pk} and C_p are equal.

In the tooling and equipment industry, because of limited samples, the designations P_p and P_{pk} are sometimes used instead of C_p and C_{pk} when the process spread is calculated using total sample deviation rather than standard deviation estimated from a control chart (R-bar/d2). This alternative calculation is also particularly useful prior to the start of full production when there are insufficient data to generate meaningful control charts. P_p and P_{pk} are referred to as *process performance*.

First-Time Capability (FTC) is the ratio of correctly processed items to the total number of items processed, usually expressed as a percentage. Both the FTC and the average percentage defective have the ability to be rolled up mathematically so that the performance of individual operations or zones can be combined to calculate total plant FTC, or average percentage defective. In contrast, C_p and C_{pk} values can be calculated only for a single process characteristic. For example, it is not valid to calculate a C_{pk} if the process is not stable, and it certainly is not valid to combine C_{pk}'s into an average C_{pk}.

Another way to calculate FTC is through *a graphical representation* using a normal probability paper. Usually, the graphic provides a way to present a visual picture of the spread found among a sample group of parts. A straight line or close to it is interpreted as the process being capable. If the line is not straight, then there is unusual variation in the process, and it must be removed through problem-solving techniques.

Once the basic understanding of capability has been communicated in the organization, then it is imperative that the *documentation* of such an activity be demonstrated. This demonstration must be done at least by identifying the following:

- The critical characteristics and required capability ratio
- The specified source for material
- The quantity and cycles to be run
- The person who will make the evaluation
- The source and type of gauging
- The measurement methods
- The process requirements

After the equipment is built, the supplier performs the capability study. Major items the supplier should report to the customer are contained in a "capability document," which should include a summary of the measurements and explanations of the results. The items addressed for explanations might include the following:

- Evidence of drift
- The shape of the process distribution
- Any adjustments, stoppages, or interruptions that occurred during the run
- For multiple-stream processes, identification of the fixture, hand, pallet, or cavity gauge variability and calibration
- The existence of statistical control for mean and variability process, if the characteristics satisfy the capability requirements
- If the process is not capable, recommendations for the type of corrective action needed to fix the machine.

Minimum recommended criteria for machine and process capability acceptance are

1. In the six-sigma world: a machine (short-run) capability C_{pk} of 1.67 and a process (long-run) capability P_{pk} of 1.25.
2. In the automotive world: a machine (short-run) capability P_{pk} of 1.67 and a process (long-run) capability C_{pk} of 1.33.*

---

* For more information, see DaimlerChrysler, Ford, and General Motors (2002), and Bothe (1997).

Using Design of Experiments (DOE) in Reliability Applications to Improve Machine Design

We can certainly use DOE in *passive observation* of the covariates in the tested components. We can also use DOE in *directed experimentation* as part of our reliability improvement. Covariates are usually called *factors* in the experimentation framework. However, when standard methods of experimental design are employed, two main technical problems arise in the reliability area.

The first problem is that *failure time data are rarely normally distributed*, so standard analysis tools that rely on symmetry (e.g., normal plots) do not work too well. This problem can be overcome by considering a transformation of the fail times to make them approximately normal: the log transformation is usually a good choice. The exact form of the fail time distribution is not important because we are looking for effects that improve reliability, rather than exact predictions of reliability itself.

The second problem is *censoring*, and this problem is a little bit trickier, but it can be dealt with by iteration as follows:

1. Choose a basic model to fit to the data.
2. Fit the model to the data, treating the censor times as failure times.
3. Using this model, make a conditional prediction for the unobserved fail times for each censored observation. The prediction is conditional because the actual failure time must be consistent with the censoring mechanism.
4. Replace censor times with the fail time predictions from step 3.
5. Go back to step 2.

Eventually, this process will converge—that is, the predictions for the fail times of the censorings will stop changing from one iteration to the next. If necessary, the process can be tried with several model choices for step 1. In fact, the algorithm of the five steps leads to the same results as maximum likelihood estimation.*

---

* For more information, see Stamatis (2003b); Montgomery (1991).

Using Hazard Analysis and Critical Control Points (HACCP) to Improve Machine Design

The basic HACCP is an analysis for generating information regarding hazards in general. The basis of HACCP is the following seven steps:

1. *Analyze hazards.* Potential hazards associated with a food machine and measures to control those hazards are identified.
2. *Identify critical control points.* These are points in a food's production.
3. *Establish preventive measures.* These are established with critical limits for each control point.
4. *Establish monitoring procedures.* These monitor the critical control points.
5. *Establish corrective actions.* These need to be taken when monitoring shows that a critical limit has not been met.
6. *Establish verification procedures.* These verify that the system is working properly.
7. *Establish effective record keeping.* These document the HACCP system.

These seven steps are very closely related with failure mode and effect analysis (FMEA) in the sense that they try to predict potential hazard. The difference is that, whereas FMEA focuses *first* on the severity of the failure, *then* on criticality, and *last* on detection, HACCP focuses first on hazards of the critical points and *then* on controls. There is a huge difference between the two. If one has to evaluate FMEA and HACCP, one will conclude that FMEA provides more useful information for improvement.

Using HALT and HASS Test Processes to Improve Machine Design

HALT (Highly Accelerated Life Test) and HASS (Highly Accelerated Stress Screens) are two types of accelerated test processes used to simulate aging in manufactured products. The HALT/HASS process was invented by Dr. Gregg Hobbs in the early 1980s. It has since been used with much success in various military and commercial applications. The HALT/HASS methods and tools are still in the development phase, and will continue to evolve as more companies embrace the concept of accelerated testing. Some companies, such as the Ford Motor Company, have adopted their own versions. Ford calls its versions FAST (Ford Accelerated Stress Test) and PASS (Production Accelerated Stress Screen).

FAST: Ford's Accelerated Stress Test

The objective of FAST is to discover the operational and destruct limits of a design, and to verify how close these limits are to the technological limits of the components and materials used in the design. FAST also verifies that the component/module is strong enough to meet the requirements of the customer and vehicle application. These requirements must be balanced with reasonable cost considerations.

As mentioned, the goal of accelerated testing is to simulate aging. If the stress–strength relationships are plotted, the design strength and field stress are distributed around means.

For example, let us assume that the stress and strength distributions are as shown in Figure 8.4. Anytime the tails of the two distributions overlap, there is an opportunity for the product to fail in the field. This area of overlap is called *interference*.

Many products, including some electronic products, have a tendency to grow weaker with age. This is reflected in a greater overlap of the curves, thus increasing the interference area. Accelerated testing attempts to simulate the aging process so that the limits of design strength are identified quickly and the necessary design modifications can be implemented.

FAST is a highly accelerated test designed to fail the target component or module. The goal of this process is to cause failure, discover the root cause, fix it, and retest it. This process continues until the "limit of technology" is reached and all the components of one technology (i.e., capacitors, diodes, resistors) fail. Once a design reaches its limit of technology, the tails of the stress–strength distribution should have minimal overlap. The FAST method uses step-stress techniques to discover the operating and destruct limits of the component or module design.

Before FAST is run on a product, the product development team should verify that

- The component/module meets specification requirements at minimum and maximum temperature.
- The vibration evaluation test (sine-sweep) is complete.
- Data are available for review by the reliability engineer.
- A copy of all schematics is available for review.

The product development team will provide the component/module monitoring equipment used during FAST, and the team will work with the reliability engineer to define what constitutes a "failure" during the test.

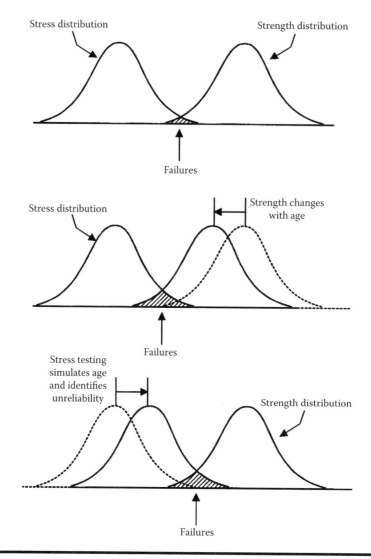

Figure 8.4 Accelerated stress testing—application stress versus design strength.

This method is used in the Preprototype and/or Prebookshelf phase of the development cycle as soon as the first parts are available. Let us look at an example.

Example

Suppose we want to discover the operating and destruct limits of a component/module design for minimum temperature. The unit is placed in a test chamber, stabilized at −40°C, then powered up to verify the operation. The unit is then unpowered, the temperature lowered to −45°C, and the unit is allowed to stabilize at that temperature. It is then powered on and verified.

This process is repeated as the temperature is lowered by 5° increments. At −70°C, the unit fails. The unit is warmed to −65°C to see if it recovers. Normally, it will recover. The temperature of −65°C is said to be its operational limit.

The test continues to determine the destruct limit. The limit is lowered to −75°C, stabilized, powered to see if it operates, then returned to −65°C to see if it recovers. If when this unit is taken down to −95°C and returned to −65°C, and does not recover, the minimum temperature destruct limit for this module is −95°C. The failed module is then analyzed to determine the root cause of the failure. The team must then determine if the failure mode is the limit of technology or if it is a design problem that can be fixed.

Experience has shown that 80% of the failures are design problems accelerated to failure using the FAST or similar accelerated stress test methods. The benefits of FAST include

- Easier system and subsystem validation due to
  - Elimination of component-/module-related failures
  - Verification of worst-case stress analysis and derating requirements
- A list of failure modes and corrections to be shared with the design team and incorporated into future designs
- Products that allow the manufacturing team to use PASS and eliminate the in-process "build and check" types of tests

The failure modes from FAST and PASS are used by the manufacturing team to ensure that they do not see any of these problems in their products.

PASS: Ford's Production Accelerated Stress Screen

On the other hand, PASS is incorporated into a process after the design has been first FASTed. The purpose of PASS is to take the process flaws created in the component/module from latent (invisible) to patent (visible). This is accomplished by severely stressing a component enough to make the flaws "visible" to the monitoring equipment. These flaws are called outliers, as shown in Figure 8.5, and they result from process variation, process changes, and different supplier sources.

The goal of PASS is to find the outliers that will assist in the determination of the root cause and the correction of the problem before the component reaches the customer. This process offers the opportunity to eliminate Module Conditioning and Burn-in.

In a sense, the objective of PASS is to precipitate all manufacturing defects in the component/module at the manufacturing facility, while still

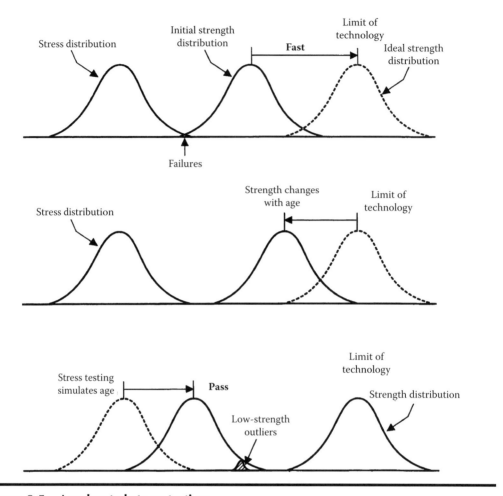

Figure 8.5 Accelerated stress testing.

leaving the product with substantially more fatigue life after screening than is required for survival in the normal customer environment.

PASS development is an iterative process that starts when the prepilot units become available in the Prepilot phase of the development cycle. The initial PASS screening test limits are the FAST operational limits, and will be adjusted accordingly as the components/modules fail and the root cause determinations indicate whether the failures are limits of technology or process problems. PASS also incorporates findings from PFMEA (Process Failure Mode and Effect Analysis) regarding possible "significant" process failure modes that must be detected if present.

When PASS development is complete, a Strength-of-PASS test is performed to verify that PASS has not removed too much useful life from the product. A sample of 12–24 components is run through 10–20 PASS cycles.

These samples are then tested using the Design Verification Life test. If the samples fail this test, the screen is too strong. PASS will be adjusted based on the root cause analysis and the Strength-of-PASS is rerun. By conducting a PASS analysis, in essence the benefits will include

- Accelerated manufacturing screens
- Reduced facility requirements
- Improved rate of return on tester costs

Using ROCOF Analysis to Improve Machine Design

By definition, ROCOF is the Rate of Change of Failure or Rate of Change of Occurrence of Failure. Another way of looking at ROCOF is as the Reliability Bathtub curve for real data. ROCOF analysis is useful in situations where you have a lot of data over time or miles, and so on. It is an excellent tool to be used in conjunction with warranty data.

A warning here is appropriate: Do *not* plot TGW (Things gone wrong)/1,000 data using a ROCOF chart because TGW/1,000 is not cumulative.

Figure 8.6 shows an example of a ROCOF analysis.

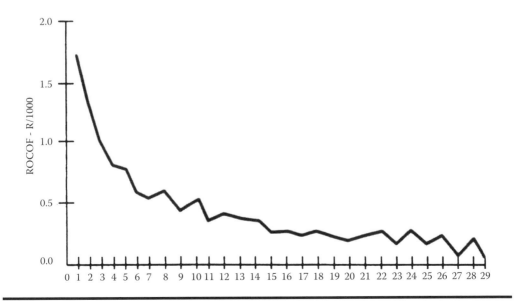

Figure 8.6   An example of a ROCOF plot.

Example

Figure 8.6 plots a car customer's concern that the "Vehicle Pulls Left While Braking." It was created by warranty data. The X-axis is "Months in Service (MIS)," and the Y-axis is the incremental repair rate, or "delta R/1,000."

In this example, it is important to note that the MIS is 29 months. However, the last four points are immature data. Possible repairs for this situation are

- Adjust the brake system.
- Burnish (remove) rust from braking surfaces (e.g., drums).
- Fix parking brake release mechanism to ensure full disengagement of the parking brake.

Where do the plotted values come from? They come from the Warranty Time-In-Service (TIS) Matrix, as shown in Table 8.4.

It is worth noting here that

- MIS = Months in service
- YTD = Year-to-date average (weighted by the divisors, the total number of parts, machines in the months of production)
- MOP = Month of production of the product, part, machine for the given model year
- An R/1,000 is essentially the number of repairs divided by the number of parts, machines that have achieved a certain number of months in service.

The most important matter to remember about warranty data is that it is always changing (maturing) until the last MOP has achieved the highest TIS value covered by the warranty policy (see Table 8.5).

Table 8.4 Warranty TIS Matrix

MIS	YTD	1st MOP	2nd MOP
0	0.19	0.55	0.33
1	1.90	2.73	1.90
2	3.26	3.69	3.37
3	4.27	4.78	4.32
4	5.33	5.33	5.41
REPAIRS	2,379	88	262
COST ($)	380,240	14,165	40,832
DIVISORS	213,586	7,317	21,088

Table 8.5 Key Characteristics of the TIS Matrix: Maturity

MIS	YTD	SEP	OCT	NOV	DEC
0	0.30	0.41	0.27	0.29	0.30
1	2.05	2.71	2.01	2.35	
2	4.32	5.47	4.06		
3	6.07	6.07			

Table 8.6 ROCOF Chart Values

Statistic Selected: R/1000				
Observation	Model Year	TIS Value	Delta	Weighted YTD
1	95	1	1.71	1.90
2	95	2	1.36	3.26
3	95	3	1.01	4.27
4	95	4	0.86	5.13
5	95	5	0.84	5.97
		X-axis ↑⎯⎯⎯⎯⎯	Y-axis ↑	
		Plotted values		

- During successive monthly reporting, the weighted YTD R/1,000 value for a given TIS can change up or down, as additional machines, parts, and repairs enter the system.
- A common practice is to heavily discount the last four points on the ROCOF chart for this reason unless ROCOF is performed on mature data.

Special note: Always remember that the ROCOF's purpose is the same as that of the reliability bathtub curve: to understand the life stages of a product or process (early, useful life, wearout) and take corrective action as soon as possible.

So what values make up the ROCOF chart? The values actually plotted are the TIS Value (on the X-axis) versus "delta" (on the Y-axis) columns. As shown in Table 8.6, also from warranty data, we may be able to plot the "weighted YTD" or "cumulative R/1,000" column from the TIS matrix.

Note: TIS and MIS are used interchangeably in the warranty system. For example, the weighted YTD R/1,000 through 1-MIS (which includes predelivery) is 1.90 R/1,000. The same value through 2-MIS is 3.26. The difference of these gives (3.26 − 1.90) = 1.36 R/1,000, the incremented or delta R/1,000 at 2-MIS and so on. The first value of 1.71 is the difference between the predelivery R/1,000, which is not shown, and the R/1,000 through 1-MIS which, we said previously, is 1.90 R/1,000.

Benefits of Using ROCOF Analysis

There are two key benefits to using ROCOF analysis. The first key benefit is that *you can visualize performance* relative to the "reliability bathtub" model. Just by looking at the plot, you can compare performance across different parts, machines, and so on. This helps to visualize the rate of change of repair rate: is it decreasing, constant, or increasing? Knowing this performance suggests certain types of failure modes and their corresponding remedial actions, as we have said. Also, with warranty coverage data, ROCOF charts usually exhibit early (decreasing rate) and useful life (constant rate) performance because coverage is generally 3 years or less. If evidence of wearout is detected, it should be promptly investigated. Finally, knowing the TIS value where performance changes from one bathtub region to another helps the engineer specify what failed hardware to collect, and it aids in calibrating the severity of development tests.

The second key benefit is that *you can compare performance*. With warranty data available, you can also plot two or more parts or machines on the same graph to compare performance across these items. Also, different parts and machine line graphs for the same system can also be useful to engineers with cross-knowledge line commodity responsibilities.

Cautions for Using ROCOF Plots with Warranty Data

Suppose an engineer is looking at electrical warranty across three machines—A, B, and C. Each had a different launch date such that A and C had 10 months in service, whereas B had only 8 months. To avoid an erroneous conclusion, the ROCOF plot should be interpreted only through 8 months in service.

It is also important to recognize that cost per unit (CPU) data can be ROCOF charted, and it may be useful to do so even though the reliability bathtub interpretation is not appropriate. CPU is really a compound metric comprised of the effect of R/1000 (frequency of repair) and CPR (cost per repair, a leveraging or weighting factor). So,

$$CPU = (R/1{,}000 \times CPR) \times 1{,}000$$

It is useful to develop a ROCOF chart using CPU data to see how the CPU trend changes with increasing TIS. For example, if the CPU dramatically rises at 20 months in service, it is usually because the mix of repairs has changed—a single new but costly repair could be appearing at 20 MIS. At this point, the manager of R&M needs to investigate promptly—even if the frequency of repair is low—in order to prevent incurring high warranty costs and customer dissatisfaction. Remember, who pays for these costly repairs after warranty coverage expires? The customer!

In addition to warranty interpretation, you need to know about the following cautions as they apply:

- *Statistical Control:* The time frame for all MOPs must be in-control. A midyear process shift can distort (and invalidate) a ROCOF analysis. The solution is to treat the "before" and the "after" process shift data as two separate populations and ROCOF chart each.
- *Coverage Limitations:* When products or machines begin to drop out of coverage because they have exceeded the warranty ceiling, the ROCOF plot will tail down. This does not mean the product is "curing itself."
- *Aggregates:* Aggregates containing commodities with widely varying TIS should be shortened to the length of the commodity with the lowest TIS to avoid distorted ROCOF charts.

As an alternative to ROCOF plots, it is possible to plot the Cumulative Hazard Plots, as shown in Figure 8.7.

Here are some factors to keep in mind when plotting a cumulative hazard plot instead of a ROCOF plot:

- If the number of failures is small, the "ROCOF" or "reliability bathtub" plot approach can be difficult to interpret.
- What is needed is a way of smoothing the data without losing the essential information.

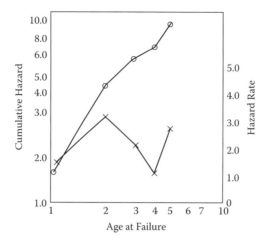

Figure 8.7 Alternative to ROCOF plots: the Cumulative Hazard plot.

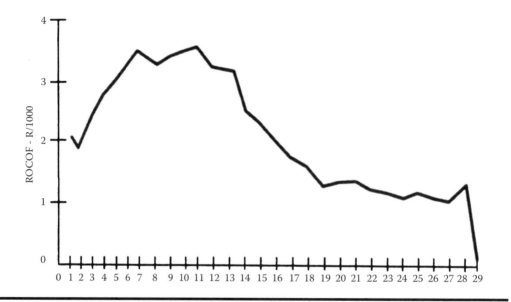

Figure 8.8 ROCOF plot—another example.

- Statistical packages such as Minitab can perform this smoothing.
- Another approach is to plot the cumulative failure rate (the weighted YTD column from the TIS matrix) on log paper.

Figure 8.8 shows another example of a ROCOF plot. Here, notice the last point at 29 MIS, which drops to zero. Generally, the last four data points in each MY represent very immature data and should be heavily discounted.

Next month, more parts/equipment and claims information will be processed by the warranty system and change the value of this point.

This graph really shows an infant mortality situation (decreasing failure rate) possibly transitioning to useful life (flat failure rate). However, there is some delay in the TIS that these early life failures begin manifesting themselves in warranty. (Note the data trend between 0 and 7 MIS). Possible reasons for this delay are

- Different people may have different thresholds for their tolerance of this particular problem—brakes.
- It may take a few months and/or miles to "break in" the rotor pads and linings before the "noise" becomes noticeable.
- People may wait to bring their concern for repair until after other concerns have surfaced.
- Be clear on one thing—the rising data trend from 0 to 7 MIS does not represent wearout!

Summary

This chapter covered some new and old methodologies on innovation techniques that focus on design improvements such as TRIZ, Pugh, GD&T, SPC, hazard analysis, and ROCOF. The next chapter continues the discussion with reliability growth and accelerated testing.

Chapter 9
Reliability Growth

The last chapter described some old and new methodologies relating to innovation techniques. This chapter continues that discussion; however, the focus is on reliability growth and accelerated testing.

Reliability growth analysis is the process of collecting, modeling, analyzing, and interpreting data from the reliability growth development test program (testing). In addition, reliability growth analysis can be done for data collected from the field (fielded systems). Fielded systems also include the ability to analyze the data of complex repairable systems. Depending on the metrics of interest and the method of data collection, different models can be used or developed to analyze the growth processes.

For effective reliability growth, the following factors may be considered:

- Management: decisions are made regarding the management strategy to correct problems or not correct problems and the effectiveness of the corrective actions.
- Testing: provides opportunities to identify the weaknesses and failure modes in the design and manufacturing process.
- Failure-mode root cause identification: funding, personnel, and procedures are provided to analyze, isolate, and identify the cause of failures.
- Corrective action effectiveness: design resources to implement corrective actions that are effective and support attainment of the reliability goals.
- Valid reliability assessments.

Calculating Reliability Growth

Reliability growth can be defined as the improvement as a result of identifying and eliminating machinery or equipment failure that is caused during machine test and operation. It is the result of an interactive design between the customer and the supplier. The essential components for developing a growth program are

- In-depth analysis of each failure, effects of the failure, and cause of the failure that occurs after installation
- Upgrading of equipment and people, thus eliminating the root cause failures
- Managing failures by redesign of equipment, maintenance, and/or training
- Upgrading the equipment when necessary
- Continuous improvement of the equipment

Reliability growth will continue even after the equipment is installed in the plant. However, it will take an effort on the part of the reliability and maintainability (R&M) teams and the supplier in order to develop a solid reliability growth program.

A reliability growth plot is an effective method that can be used to track R&M continuous improvement activities. In addition, it can be used to predict reliability growth of machinery from one machine to the next.

Reliability growth plotting is generally implemented during the last part of the Failure Investigation process associated with failure reporting, analysis, and corrective action system (FRACAS). The information developed for reliability growth should be sent back to the supplier and to internal engineering personnel in order to take corrective actions on new equipment designs.

Reliability growth is calculated based on information relating to the time between failure events and the number of failures observed during the recording period. The data obtained from this information is then plotted on log–log paper, and the slope of the line is calculated to determine if the machine's performance is improving or decreasing.

Let us walk through the process of calculating reliability growth.

Step 1: Collect Data on the Machine and Calculate the Cumulative Mean Time between Failures (MTBF) Value for the Machine

Table 9.1 illustrates the difference between two machines relating to reliability growth information.

Table 9.1 Two Machines Relating to Reliability Growth

Machine A			Machine B		
Failures	Time	MTBF	Failures	Time	MTBF
1	30	30	1	50	50
1	40	35	2	40	30
2	90	40	3	30	20
1	90	50	1	20	20

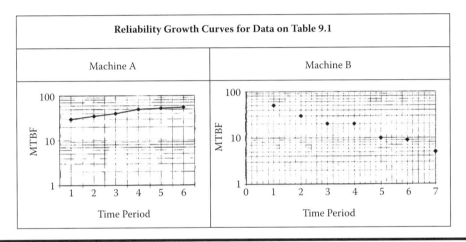

Figure 9.1 Reliability growth for machines A and B.

Step 2: Take the Data and Plot the Data on Log–Log Paper

Figure 9.1 offers an example of how to take data and plot them.

Notice that Machine A climbs to a certain level and remains constant over several periods of time. The increasing slope of the line indicates reliability growth. The flat portion of the line indicates that the machine has achieved its inherent level of MTBF and cannot get any better. It is the job of the plant's maintenance department to maintain this level of performance. Machine B is declining, which means that reliability is also declining.

Step 3: Calculate the Slope

When the reliability plot is moving in a positive direction, a statement regarding equipment performance can be determined by measuring the slope of the line. Once the slope of the line is determined, the Duane Model can be applied to determine the overall effectiveness of the R&M program.

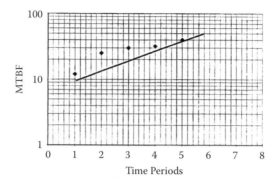

Figure 9.2 Reliability growth plot with slope.

Table 9.2 A Typical Slope Interpretation

β	Reliability Growth Rate
0.4–0.6	Eliminating failures takes top priority. Immediate analysis and corrective action for all failures.
0.3–0.4	Priority attention to reliability improvement. Normal (typical stresses) environment utilization. Well-managed analysis and corrective action for important failure modes.
0.2–0.3	Routine attention to reliability improvement. Corrective action taken for important failure modes.
0–0.2	No priority given to reliability improvement. Failure data not analyzed. Corrective action taken for important failure modes, but with low priority.

(The Duane Model is one that is used as a basis for predicting the reliability growth that could be expected in an equipment development program.) From the slope measurement, the determination is made by using the Duane Model information supplied in the table. The beta is the slope measurement factor, as shown in Figure 9.2; then, Table 9.2 shows a typical slope interpretation.

How the Equipment Supplier Can Help Improve Machine Reliability

During installation and operation of the equipment, the supplier should provide to the customer recommendations on how the reliability of the equipment can be improved. This action can be accomplished by visiting, on a regular basis, the equipment on the plant floor and developing interaction with the customer's R&M teams to conduct root cause analysis on the equipment.

The participation of the supplier at this point would greatly enhance the reliability growth of the equipment on the floor and future equipment the supplier will be providing. The results of this growth can also affect the mean time to repair (MTTR) numbers associated with the equipment. Any new improvements should be logged and included as part of new designs for the equipment.

The Customer's Responsibility in Improving Machine Reliability

The customer's responsibility is to keep accurate failure history logs on the equipment. This log should identify

- Classification of the failure
- Time of the failure
- Root cause of the failure
- Corrective action for the failure

How to Implement a Reliability Growth Program

It may sound very difficult and seem like a Herculean feat to implement a reliability growth initiative in your organization, but in reality, it is very simple. It needs true commitment from top management and a vision for continual improvement in R&M. The following three approaches can help:

1. Set the overall reliability goals for the equipment. These goals can be divided into short- and long-term goals.
2. Review the overall equipment effectiveness (OEE) calculations for the equipment. Set new goals for availability, performance, and quality for the short and long term.
3. Establish a universal maintenance tagging system for identification of failures and their corrective actions.

Initiate an R&M Feedback Process

Feedback of performance (machinery) data, along with solid data collection, validation, and uniform calculation, serves as the foundation for continuous improvement activities. Without performance data feedback, continuous improvement activities would most likely fail or not meet the expectations of the R&M team. Therefore, an effective R&M feedback process needs to be an integral part of the R&M program.

Create a Failure Data Feedback Process Flowchart

The following process outlines the strategy that should be implemented when machinery and/or components experience a failure. This process is an important precursor to the performance data feedback plan, which establishes specific feedback channels:

1. Identify the failed machine.
2. Fill the failure tag as appropriate.
3. Enter appropriate information into the database.
4. Ship appropriate information to the supplier.
5. Evaluate the root cause.
6. Conduct a machine failure mode and effects analysis (MFMEA) (refer back to Chapter 6 for details), a preventive maintenance (PM) plan, and/or a FRACAS analysis (refer back to Chapter 4 for details).

Implement a Performance Data Feedback Plan

Feedback of the data is needed in order to proceed with continuous improvement activities and may be from several sources. In order to support a concise data feedback process that fulfills continuous improvement objectives, a comprehensive performance data feedback plan needs to be developed and implemented. The key component of a performance data feedback plan is the Performance Data Feedback Responsibilities Matrix (see Table 9.3).

Table 9.3 also shows the responsibilities of the following groups:

- Customer's manufacturing team representatives
- Customer's R&M team
- Machinery and equipment OEMs (Original Equipment Manufacturers)
- Component manufacturers and suppliers

The matrix, in general, identifies the key feedback channels that should be in place. It is important to note that the groups identified in the foregoing list, without exception, are responsible for securing, validating, uniformly

Table 9.3 Performance Data Feedback Responsibilities Matrix

Performance Data/Activity	OEM		Component supplier		Mtg.	R&M
	Feedback to Ford Prior to Acceptance of Machinery	Post Job 1 and Installation	Feedback to Ford Prior to Acceptance of Contract or P.A.	Post Job 1 and Installation	Post Job 1 and Installation	Post Job 1 and Installation
Performance metric predictions						
Reliability predictions			X			
MTBE prediction	X					
MTBF prediction			X			
MTTR prediction	X					
Testing data						
Life testing data			X			
Environmental testing data			X			
Ancillary data						
Bill of material	X					
Quantitative field performance data						
MTBE achieved						X
MTBF achieved			X			X
MTTR achieved						X
Availability						X
Starve and block trend						X
Wellness trend						X

(Continued)

Table 9.3 Performance Data Feedback Responsibilities Matrix (Continued)

Performance Data/Activity	OEM		Component supplier		Mtg.	R&M
	Feedback to Ford Prior to Acceptance of Machinery	Post Job 1 and Installation	Feedback to Ford Prior to Acceptance of Contract or P.A.	Post Job 1 and Installation	Post Job 1 and Installation	Post Job 1 and Installation
Warranty rate			X			
Field audit reports		X	X			
Qualitative field performance data						
Fault code reporting	X					X
Summary						
MMS data capturing					X	
Accuracy						
Service report	X	X				
Summary						
Service alert bulletins	X	X	X	X		
TGR/TGW list	X	X	X	X	X	X
FRACAS						
Component failure tag	X				X	
Root cause failure	X	X	X	X		
Analysis						
Failure/corrective action						
Reports	X	X	X	X		
Component problem log						
Reports						X

calculating, and distributing (communicating) their own performance data. This, in turn, will enhance and contribute to the success of the performance data feedback process.

A customer and its manufacturers/suppliers will realize the following benefits when an effective data feedback plan is in place (see Table 9.4):

R&M Information Systems

Many different systems have been developed to alert other areas within a customer's system. For example, in the automotive industry, these systems are shown in a high level on Table 9.5.

Preparing Reports on Things Gone Right/ Things Gone Wrong (TGR/TGW)

TGR/TGW reports are key elements in the R&M feedback process; Table 9.6 shows an example of such a report. Most important, they help facilitate R&M continuous improvement activities by providing the R&M team with historical R&M data that can be referenced when new machines are being considered for acquisition. A typical report may provide the R&M team with

- A way to track failure areas of a particular machine
- A list of failures that are specific to a particular machine

Table 9.4 Benefits of a Performance Data Feedback Plan

Customer	Supplier
Improved machinery performance	Clear understanding of their product's performance
Reduced life-cycle cost	Insight into the reduction of warranty costs through improved reliability
Baseline definitions for supplier selection and continuous improvement activities	Industry competitive edge based on clear data
Best practice definitions in M&E design More responsive suppliers	Baseline definition for design standardization and continuous improvement
Global productivity and R&M success	Improved customer satisfaction

Table 9.5 R&M Information System Example from an Automotive Organization

Organization	R&M Information Systems
High level i.e., vehicle office	Generally, the organization will have an internal system that contains a listing of plant personnel, identified as the Early Equipment Listing. When a problem or failure is observed in the plant, a distribution can be made to all of the responsible individuals—even at the plant level—to query about the failure and any solutions that may reduce the impact of the failure mode. Ask your plant personnel or the acquisition engineer about this listing.
Second level i.e., engine	In today's world of high technology, it is not unusual to find organizations with Intranet access to reliability and maintainability information. The site is generally created to serve as an R&M information exchange system. It may include historical as well as surrogate data in the form of • Reliability growth curves • Reliability alerts • Completed R&M • Design reviews • 8Ds • TGR/TGW (i.e., "Things Gone Right/Things Gone Wrong": see the next section for more information) • FMEAs for machines and components
Third level i.e., powertrain	Suppliers and points of contacts may also be listed. At this level, the organization may develop a database to collect and report information relating to the failure tag. In addition, it may also be plugged into the Plant Floor Information Systems data to capture overall performance of the equipment or machine. The information captured allows for the development of component reliability growth curves.

- An action item and follow-up list
- A failure history of a particular machine prior to acquisition
- A record of R&M improvements over a specific period of time
- A compilation of R&M data that can be used in an LCC (i.e., life-cycle costing) analysis

The OEM is responsible for completing TGR/TGW reports. Ultimately, the OEM is able to hand the customer a book of TGR/TGW reports that lays out the R&M history of the machine that the customer is targeting to acquire.

Table 9.6 A Typical TGR/TGW Report

	TGR/TGW							Plant:		
	Machinery Installed: Robots to Line 50									
Action Item #	Date Recorded	TGR	TGW	TGR or TGW	Cause	Effect	Corrective Action Taken	Corrective Action Agent	Date Completed	Est. Hours to Complete
1	3/15/98	X		Cells 2 and 3 floor cleaned of hydraulic fluids by 12:15	Previous experience with bilsings		Request cleaner at 11:00	Cleaner	3/15/98	0.5
2	3/19/98		X	SE locator bracket on cell 3 could not be installed	Bilsing locator in the way		Wait until shutdown to install it	Millwright, welder	3/25/98	1.5
3	3/31/98	X		Cut the deacon block and mount plate that the bracket is welded to (#3 SE)		Save time when we do install bracket on SE corner		Millwright, welder, carpenter	4/5/98	2.5
4	4/10/98		X	Wheel on cart #2 is missing taper lock		Delay installation of brackets	Replace taper lock	Millwright	4/10/98	1.5
5	4/12/98		X	19-Pin sine cable was disconnected from press control box	Miscommunication		Rewire	Electrician	4/14/98	4.0

(Continued)

Table 9.6 A Typical TGR/TGW Report (Continued)

TGR/TGW

Machinery Installed: Robots to Line 50 | Plant:

Action Item #	Date Recorded	TGR	TGW	TGR or TGW	Cause	Effect	Corrective Action Taken	Corrective Action Agent	Date Completed	Est. Hours to Complete
6	4/18/98		X	The bolts to mount the robot plate to the CPI cart are too short	Bolts misordered		Plant provides correct-length bolts		4/18/98	1.5
7	4/28/98		X	Through holes in robot plate do not line up with CPI cart holes	CPI carts are not uniform in their design		Burn away portion of cart to allow clearance	Welder	4/28/98	1.0

Reliability

In the last section, we discussed reliability growth in which a reliability goal (or goals) is set and should be achieved during the development testing program, with the necessary allocation or reallocation of resources. It was emphasized that planning and evaluation are essential factors in a growth process program through well-thought-out structured planning and assessment techniques.

A reliability growth program differs from a conventional reliability program in that there is a more objectively developed growth standard against which assessment techniques are compared. Reliability requirements, on the other hand, are often set based on a mission probability that a system will operate in a particular environment for a stated period of time without failure. This notion of reliability is concerned with any unscheduled maintenance actions affecting a defined mission objective. In this section, we are going to address reliability from the classical point of view.

Reliability is the probability that machinery/equipment can perform continuously, without failure, for a specified interval of time when operating under stated conditions. Increased reliability implies less failure of the machinery and, consequently, less downtime and loss of production.

Reliability modeling is used to define the structure of the equipment or machinery under investigation. It allows the team to review where the "weak link" of the equipment is located, which is causing equipment downtime due to failures. There are three types of models that can be used to evaluate the design of the equipment:

1. Reliability allocation models
2. Series models
3. Parallel models

Let us take a closer look at each type.

Using a Reliability Allocation Model to Evaluate Equipment Design

Figure 9.3 illustrates the effect of reliability numbers for system, subsystem, and component levels for an engine after 10 years of operation in the field. Notice that the engine has an overall reliability of 70%, but at the component level, the reliability of support must be 99.999% after 10 years.

ENGINE	ROCKER LEVER & CAM FOLLOWER ASSEMBLY (97.20)		
RELIABILITY	LUBE SYSTEM (97.35)		
GOAL	EXTERNAL SYSTEM (98.60)		
0.70	CYLINDER BLOCK (94.0)	CRANK (99.65)	
		PISTON (99.5)	
	CYLINDER HEAD (96.25)	BLOCK CYLINDER (99.55)	
		BLOCK-MISC. (98.05)	
	DRIVE SYSTEM (98.25)	VIBRATION DANGER (99.9)	
		CAMSHAFT (99.5)	O-RING (99.86)
	FUEL SYSTEM (92.75)	FLYWHEEL (99.65)	GASKET (99.86)
	COOLING SYSTEM (96.10)	MAIN BEARING (99.88)	CAMSHAFT (99.8)
		LINEAR (99.75)	BUSHING (99.9)
	ACCESSORIES (96.70)	CONNECTING ROD (99.8)	BOLT (99.999)
	TURBO SYSTEM (98.15)	PISTON RING (99.0)	THRUST BEARING (99.999)
	MISCELLANEOUS (99.70)	MISCELLANEOUS (99.79)	COVER PLATE (99.98)
			KEY (99.999)
			SUPPORT (99.999)

Figure 9.3 Reliability allocation.

Using a Series Model to Evaluate Equipment Design

The series model is defined as a complex system of independent units connected together (or interrelated) in such a way that the entire system will fail if any one of the units fails. For example, Figure 9.4 shows how MTBF can be calculated.

Step 1: Convert the MTBF values to failure rates

$$\text{Failure rate (FR)} = \frac{1}{\text{MTBF}}; FR1 = 1/30 = 0.0333; FR2 = 1/60 = 0.0167;$$

$$FR3 = 1/45 = 0.0222$$

Step 2: Add the failure rates

$$\text{Failure rate of series} = FR_1 + FR_2 + FR_3 = 0.0333 + 0.0167 + 0.0222 = 0.0722$$

Step 3: Convert the failure rate total to MTBF value

$$\text{MTBF} = \frac{1}{FR} = 1/0.0722 = 13.85 \text{ h}$$

The series model can be used to calculate the following information for developing an overall allocation study. For example, Table 9.7 was developed for a Rear Axle Tester/Balancer. The general formula for a series model is

$$R(t)_{Total} = R_1(t) \times R_2(t) \times R_3(t) \ldots$$

where R(t) = reliability of a product at time (t).

Figure 9.4 Serial model.

Table 9.7 An Example of Series Model Calculations

Component	MTBF hours	FR	JPH	QTY	MTTR minutes	Avail	R1 shift
Roller	80,000	1.25×10^{-5}	30	18	30.0	0.9998875	0.9982
Prox Switch	90,175	1.10×10^{-5}	30	15	7.41	0.9999795	0.9986
Valve	20,000	5.0×10^{-5}	60	5	6.50	0.999972	0.9980
Cylinder	12,357	8.09×10^{-5}	60	1	9.72	0.9999869	0.9993
Rodless Cylinder	40,000	2.5×10^{-5}	30	2	28.0	0.99997	0.9996
Clamp	20,000	5.0×10^{-5}	30	6	45.0	0.99979	0.9976
Safety Guard Switch	20,000	5.0×10^{-5}	30	11	6.00	0.99994	0.9956
Pin	25,000	4.0×10^{-5}	30	4	4.72	0.99998	0.9987
Ball Bushing	40,000	2.5×10^{-5}	60	8	90.0	0.99970	0.9984
Clevis Connection	40,000	2.5×10^{-5}	60	1	45.0	0.99998	0.9998
Total	2710.0	36.9×10^{-5}			24.16	0.9992	0.9840

Using a Parallel Model to Evaluate Equipment Design

A parallel reliability model is a complex set of interrelated components connected together so that a redundant or standby system is available when a failure occurs, as shown in Figure 9.5.

Parallel models represent equipment with redundant, backup systems. For example, a system with four welding arms performs a series of welds using only two welding arms, while the remaining two arms are backups. Then, if one of the operating arms fails, one of the backup arms can begin welding in its place. Therefore, the equipment remains operational, resulting in greatly improved reliability.

However, backup systems may substantially increase the purchase price of the equipment; therefore, they are usually only deployed on crucial subsystems. For example, in Figure 9.5, we have a parallel system, and we are interested in the system MTBF.

Calculating the MTBF for the System MTBF is

$$\text{System MTBF} = 80 + 80 - 1/[1/80 + 1/80] = 120 \text{ MTBF}$$

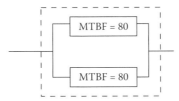

Figure 9.5 Parallel model.

The general equation for calculating parallel reliability is

$$R(t)_{total} = 1 - [(1 - R_1(t)) \times (1 - R_2(t)) \times (1 - R_3(t)) \ldots]$$

where R(t) = reliability of a product at time (t).

Reasons for Performing Reliability Tests

The purpose of performing a reliability test is to answer the question "Does the item meet or exceed the specified minimum reliability requirement?" If you recall, the warranty requirement (R/1000) was given as part of the initial set of requirements. It was converted to a reliability requirement, the criteria for performance were established, and the requirement among the parts of the system was allocated. Reliability testing then is used to

- Determine whether the system conforms to the specified, quantitative reliability requirements.
- Evaluate the system's expected performance in the warranty period and its compliance with Useful Life targets set by the customer.
- Compare performance of the system to the goal that was established earlier.
- Monitor and validate reliability growth.
- Determine design actions based on the outcomes of the test.

In addition to its other uses, the outcomes of reliability testing are used as a basis for design qualification and acceptance. Reliability testing should be a natural extension of analytical reliability models so that test results will clarify and verify the predicted results in the "customer" environment.

A reliability test is effectively a "sampling" test in that it involves a sample of objects selected from a "population." From the sample data, some statements are made about the population parameters. In reliability testing, as in any sampling test,

- The "sample" is assumed to be representative of the population.
- The characteristics of the sample (e.g., the sample mean) are assumed to be an estimate of the true value of the population characteristics (e.g., the population mean).

A key factor in reliability test planning is choosing the proper sample size. Most of the activity in determining sample size is involved with either:

- Achieving the desired confidence that the test results give the correct information
- Reducing the risk that the test results will give the wrong information

When to Conduct Reliability Testing

Before the hardware is available, simulation and analysis should be used to find design weaknesses. Reliability testing should begin as soon as hardware is available for testing. Ideally, much of the reliability testing will occur "on the bench" with the testing of individual components. There is good reason for this: the effect of failure on schedule and cost increases progressively with the program timeline. The later that failure and corrective action occurs in the process, the more it costs to correct, and the less time there is to make the correction.

Here are some key points to remember regarding test planning:

- Develop the reliability test plan early in the design phase.
- Update the plan as requirements are added.
- Run the formal reliability testing according to the predetermined procedure. This is to ensure that results are not contaminated by development testing or procedural issues.
- Develop the test plan in order to get the maximum information with the fewest resources possible. Here you may want to consider performing a design of experiments (DOE).
- Increase test efficiency by understanding stress/strength and acceleration factor relationships. This may require accelerated testing, such as

FAST (Ford Accelerated Stress Test, described in Chapter 8), which will increase the information gained from a test program.
- Make sure your test plan shows the relationship between development testing and reliability testing. Although all data add to the overall knowledge about a system, other functional development testing is an opportunity to gain insight into the reliability performance of your product.

Note: A "control sample" should be maintained as a reference throughout the reliability testing process. Control samples should not be subjected to any stresses other than normal parametric and functional testing.

Setting Your Objectives for Reliability Testing

Always have in mind the objectives described in the following paragraphs.

Test with regard to production intent. Make sure the sample that is tested is representative of the system that the customer will receive. Of course, consider that these elements may change, or that they are not known. However, use the same production intent to the extent known at the time of the test plan. This means that the test unit is representative of the final product in all areas, including
  – Materials (e.g., metals, fasteners, weight)
  – Processes (e.g., machining, casting, heat treat)
  – Procedures (e.g., assembly, service, repair)
Determine performance parameters before testing is started. It is often more important in reliability evaluations to monitor percentage change in a parameter rather than performance to specification.
Duplicate/simulate the full range of customer stresses and environments. This includes testing to the 95th percentile.
Quantify failures as they relate to the system being tested. A failure results when a system does not perform to customer expectations even if there is no actual broken part. Remember, customer requirements include the specifications and requirements of internal customers and regulatory agencies as well as the ultimate purchaser.
Test structure to identify hardware interface issues. These relate to the system being tested.

Classical Reliability Terms

In addition, keep in mind that there are some basic terms associated with understanding and predicting reliability. Some of these terms have already been introduced in previous chapters; however, we are repeating them in the rest of this chapter to emphasize that these items are very important and are associated with understanding and predicting reliability.

Calculating MTBF

MTBF is the average time between failure occurrences. The equation would be the operating time of equipment divided by the total number of failures.

Determining MTBF Point Estimates and Confidence Intervals

When calculating MTBF, the value is known as a *point estimate*, which is defined as a single value that is computed from a set of sample measurements and used to estimate the value of a specific population. All data collected to calculate the MTBF values are point estimates.

When data are collected to calculate the MTBF values, it becomes important to understand the level of confidence in which the MTBF value will fall. This is known as the *confidence interval* for MTBF. Confidence intervals allow a band to be put around the MTBF data, which adds more meaning to the measurement. A confidence interval of 90% means that 90% of the calculated data from the sample will contain the true MTBF value, while 10% will fall outside this value. A 95% confidence interval means that 5% of the calculated data will fall outside the true MTBF mean value.

As the number of failures increases, the size of the interval decreases. The interval remains constant for more than 30 failures. MTBF can be expressed in terms of the following equation:

$$\text{MTBF} = [\text{Total operating time}/N], \text{ or}$$

$$= 1/\text{Failure rate}$$

where
    Operating time = Total time scheduled to operate
    N = Total number of failures observed during the operating period

Example 1

If machinery is operating for 400 h and there are eight failures, then the MTBF is

$$\text{MTBF} = [\text{Total operating time}/N] = 400/8 = 50 \text{ h}$$

Note: Where applicable, the time (T) component of MTBF can be changed to C (cycles). The steps taken to calculate MCBF are identical to those used to calculate MTBF.

Example 2

What is the MTBF of a pump that has an operating time of 500 h and experiences two replaced seals, one motor replacement, one replaced coupling, one PM meeting, and two scheduled filter cleanings during this time? Perform the following appropriate calculations. If necessary, refer to the previous example.

$$\text{MTBF} = [\text{Total operating time}/N] = 500/4 = 125 \text{ h}$$

The pump has a mean time between failure occurrence rate of 125 h. This means that, on average, the water pump will operate within its performance parameters for 125 h before experiencing a failure mode.

Calculating Equipment Failure Rates

Failure rates estimate the number of failures in a given period of time, events, cycles, or number of parts. It is the probability of failure within a unit of time. Failure rate can be expressed in terms of the following equation:

$$\text{Failure rate} = 1/\text{MTBF}$$

Example 1

The failure rate of a water pump that experiences one failure within an operating time period of 2000 h is

$$\text{Failure rate} = 1/\text{MTBF} = 1/2000 = 0.0005 \text{ failures per hour}$$

This means that there is a 0.0005 probability that a failure will occur with every hour of operation.

Example 2

What is the failure rate of a pump that experiences one failure within an operating time period of 2500 h? Perform the following appropriate calculations. If necessary, refer to the previous example.

$$\text{Failure rate} = 1/\text{MTBF} = 1/2500 = 0.0004 \text{ failures per hour}$$

Calculating Failure Rates vis-à-vis the Total Population of Equipment Components

Because the failure rate is the rate of failure per period of time, this value can be calculated over the population of components under the failure rate measurement. For example, the failure rate of a limit switch has been calculated at 0.00005 (MTBF ≥ 20,000) failures per hour. The plant has 1000 of the same type of limit switches located throughout the plant. Therefore, the failure rate for the population of limit switches can be calculated as

$$\text{Failure rate} = 1/\text{MTBF} = 1/20000 = 0.00005$$

$$\text{Population failure rate} = \text{Failure rate} \times \text{population}$$

$$= 0.00005 \times 1000$$

$$= 0.05 \text{ failure per hour}$$

This indicates that five limit switches will fail in the plant every 100 h of normal operation.

Measuring Equipment Maintainability

Maintainability is a characteristic of design, installation, and operation; it is usually expressed as the probability that machinery/equipment can be retained in, or restored to, a specified operable condition within a specified interval of time when maintenance is performed in accordance with prescribed procedures.

Maintainability is defined in terms of mean time to repair (MTTR). The Society of Automotive Engineers (SAE) defines MTTR as *the average time to restore machinery or equipment to specified conditions*. All machines/equipment eventually fail and, when they do, the cost and time of repairing them must be the lowest possible. One way to accomplish this is to identify the equipment's "lowest replaceable unit" (LRU). For example, if the equipment's drive fails, it might be quicker, and thus cheaper, to replace the drive rather than the specific failed component. That is why it is very important to know what the repair is and how long it lasts. Generally, repair time consists of the following six basic elements, and it can be shown graphically as in Figure 9.6:

1. Report the failure.
2. Locate the resource.

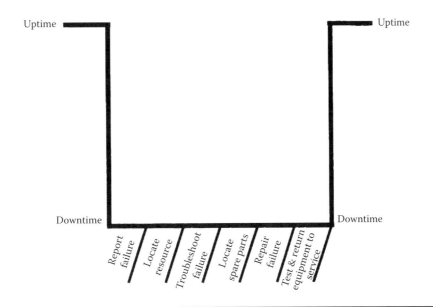

Figure 9.6   A pictorial view of MTTR.

3. Troubleshoot the failure.
4. Locate spare parts.
5. Repair the failure.
6. Test and return equipment to service.

It is important to note that the MTTR calculation is based on repairing one failure, and one failure only. The longer each failure takes to repair, the more the equipment's cost of ownership goes up. Additionally, MTTR directly affects uptime, uptime percent, and capacity. MTTR can be expressed in terms of the following equation:

$$MTTR = \frac{\sum t}{N}$$

where
  $\sum ta$ = Summation of repair time t
  N = Total number of repairs

Example 1

A pump operates for 300 h. During that period, there were four failure events recorded. The total repair time was 5 h. The MTTR is

$$\text{MTTR} = \frac{\sum t}{N} = \frac{5}{4} = 1.25 \, \text{h}$$

Example 2

A pump operated for 450 h. During that period, there were six failure events recorded. The total repair time was 10 h. What is the MTTR for this pump? Perform the appropriate calculations below. If necessary, refer to the previous example.

$$\text{MTTR} = \frac{\sum t}{N} = \frac{10}{6} = 1.67 \, \text{h}$$

Five Tests for Defining Failures

There are many ways to define and quantify failures. The next sections identify the most common.

1. Sudden-Death Testing

Sudden-death testing allows you to obtain test data quickly and reduces the number of test fixtures required. It can be used on a sample as large as 40 or more, or as small as 15. Sudden-death testing reduces testing time in cases where the lower quartile (lower 25%) of a life distribution is considerably lower than the upper quartile (upper 25%).

The philosophy involved in sudden-death testing is to test small groups of samples to a first failure only and use the data to determine the Weibull distribution of the component. Only common failure mechanisms can be used for each Weibull distribution. Care must be taken to determine the true root cause of all failures. Failure must be related to the stresses applied during the test.

1. Choose a sample size that can be divided into three or more groups with the same number of items in each group. Divide the sample into three or more groups of equal size, and treat each group as if it were an individual assembly.
2. Test all items concurrently in each group until there is a first failure in that group. Testing is then stopped on the remaining units in that group as soon as the first unit fails (hence the name "sudden death").

3. Record the time to first failure in each group.
4. Rank the times to failure in an ascending order.
5. Assign median ranks to each failure based on the sample size equal to the number of groups. Median rank charts (see Appendix K) are used for this purpose.
6. Plot the times to failure versus median ranks on Weibull paper.
7. Draw the best fit line. (Eye the line or use the regression model.) This line represents the sudden-death line.
8. Determine the life at which 50% of the first failures are likely to occur (B_{50} life) by drawing a horizontal line from the 50% level to the sudden-death line. Drop a vertical line from this point down.
9. Find the median rank for the first failure when the sample size is equal to the number of items in each subgroup. Again, refer to the median rank charts. Draw a horizontal line from this point until it intersects the vertical line drawn in the previous step.
10. Draw a line parallel to the sudden-death line passing through the intersection point from Step 9. This line is called the population line and represents the Weibull distribution of the population.

Sudden-death testing is a good method to use to determine the failure distribution of the component.

Example 1

The example finishes when the accelerated testing begins. Please note that the example includes the tables.
Assume you have a sample of 40 parts from the same population run (production) available for testing purposes. The parts are divided into five groups of eight parts as shown:

   Group 1 --------------→ 1 2 3 4 5 6 7 8
   Group 2 --------------→ 1 2 3 4 5 6 7 8
   Group 3 --------------→ 1 2 3 4 5 6 7 8
   Group 4 --------------→ 1 2 3 4 5 6 7 8
   Group 5 --------------→ 1 2 3 4 5 6 7 8

All parts are put on test in each group simultaneously. The test proceeds until any one part in each group fails. At that time, testing stops on all parts in that group. In the test, we experience the following first failures in each group:

   Group 1 --------------→ Part #3 fails at 120 h
   Group 2 --------------→ Part #4 fails at 65 h

Group 3 --------------→ Part #1 fails at 155 h
Group 4 --------------→ Part #5 fails at 300 h
Group 5 --------------→ Part #7 fails at 200 h

Failure data are arranged in ascending hours to failure, and their median ranks are determined based on a sample size of N = 5. (There are five failures, one in each of five groups.) Table 9.8 illustrates the data.

The median rank percentage for each failure is derived from the median rank tables. The data appropriate for sample sizes of 1 through 10 are shown in Table 9.9 (for a more detailed table see Appendix K.2).

If the life hours and median ranks of the five failures are plotted on Weibull paper, the resulting line is called the sudden-death line.

- The sudden-death line represents the cumulative distribution that would result if five assemblies failed, but it actually represents five measures of the first failure in eight of the population.
- The median life point on the sudden-death line (the point at which 50% of the failures occur) will correspond to the median rank for the first failure in a sample of eight, which is 8.30%.
- The population line is drawn parallel to the sudden-death line through a point plotted at 8.30% and at the median life to first failure, as determined here.

This estimate of the population's minimum life is just as reliable as the one that would have been obtained if all 40 parts were tested to failure.

2. Accelerated Testing

Accelerated testing is another approach that may be used to reduce the total test time required. It requires stressing the product to levels that are more severe than normal. The results that are obtained at the accelerated stress levels are compared to those at the design stress or normal operating

Table 9.8 Failure Data

Failure order number	Life hours	Median ranks, %
1	65	12.95
2	120	31.38
3	155	50.00
4	200	68.62
5	300	87.06

Table 9.9 Partial Median Rank Table

Failure #	Sample size									
	1	2	3	4	5	6	7	8	9	10
1	50.000	29.289	20.630	15.910	12.945	10.910	9.428	8.300	7.412	6.697
2		70.711	50.000	38.573	31.381	26.445	22.849	20.113	17.962	16.226
3			79.370	61.427	50.000	42.141	36.412	32.052	28.624	25.857
4				84.090	68.619	57.859	50.000	44.015	39.308	35.510
5					87.055	73.555	63.588	55.984	50.000	45.169
6						89.090	77.151	67.948	60.691	54.831
7							90.572	79.887	71.376	64.490
8								91.700	82.038	74.142
9									92.587	83.774
10										93.303

conditions. We will look at examples of this comparison in this section. We use accelerated testing to

- Generate failures, especially in components that have long life under normal conditions
- Obtain information that relates to life under normal conditions
- Determine design/technology limits of the hardware

Accelerated testing is accomplished by reducing cycle time, such as by *overstressing* or by *compressing* cycle time by reducing or eliminating idle time in the normal operating cycle. There are some pitfalls in using accelerated testing:

- Accelerated testing can cause failure modes that are not representative.
- If there is little correlation to "real" use (such as aging, thermal cycling, and corrosion), then it will be difficult to determine how accelerated testing affects these types of failure modes.

There are many methods that can be used for accelerated testing. However, before we discuss some of the more basic methods, keep in mind that, very often, some parts are subjected to multiple stresses and combinations of stresses. These stresses and combinations should be identified very early in the design phase. When accelerated tests are run, assure that all the stresses are represented in the test environment and that the product is exposed to every stress.

Furthermore, the data from accelerated tests are interpreted and analyzed using different models. The model that is used depends on the product, the testing method, and the accelerating variables. The models give the product life or performance as a function of the accelerating stress.

Constant-Stress Testing

In constant-stress testing, each test unit is run at constant high stress until it fails or its performance degrades. Several different constant-stress conditions are usually employed, and a number of test units are tested at each condition. In Figure 9.7, a constant-stress test is shown that uses three stress levels. Failed test units are indicated by the stars. The arrows show unfailed test units. At its lowest level of stress, five components failed and four survived without failure out of the nine units tested. At its highest level of stress, all four components that were tested failed.

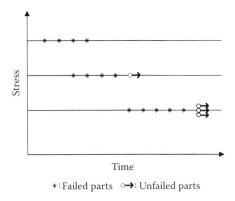

Figure 9.7 Constant-stress testing.

It should be noted here that some products run constant stress, and this type of test represents the actual use of those products. Constant stress is most helpful for simple components, and it will usually provide greater accuracy in estimating time to failure. However, keep in mind that, in systems and assemblies, acceleration factors often differ for different types of components.

Step-Stress Testing

In step-stress testing, the item is tested initially at a normal, constant stress for a specified period of time. Then the stress is increased to a higher level for a specified period of time. Increases continue in a stepped fashion. The graphic in Figure 9.8 illustrates this concept, with the stars indicating the points of failure.

The main advantage of step-stress testing is that it quickly yields failure because increasing stress ensures that failures occur. A disadvantage is that failure modes that occur at high stress may differ from those at normal-use conditions.

Quick failures do not guarantee more accurate estimates of life or reliability. A constant-stress test with a few failures usually yields greater accuracy in estimating the actual time to failure than a shorter step-stress test; however, we may need to do both to correlate the results so that the results of the shorter test can be used to predict the life of the component.

Note: Failures must be related to the stress conditions to be valid. Other test discrepancies should be noted, repaired, and the testing continued.

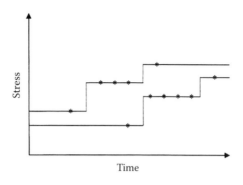

Figure 9.8 Step-stress testing.

Progressive-Stress Testing

Progressive-stress testing is step-stress testing carried to the extreme. In this test, the stress on a test unit is continuously increased rather than being increased in steps. Usually, the accelerating variable is increased linearly with time. Several different rates of increase are used, and a number of test units are tested at each rate of increase. Under a low rate of increase of stress, failure is prolonged, and the component fails at lower stress because of the natural aging effects or cumulative effects of the stress on it. Figure 9.9 illustrates these points. Progressive-stress testing has some of the same advantages and disadvantages as step-stress testing.

There are two points to keep in mind as you analyze accelerated test data:

- Units run at a constant high stress tend to have shorter life than units run at a constant low stress. You can see this phenomenon in Figure 9.9.
- Distribution plots show the cumulative percentage of the samples that fail as a function of time. In Figure 9.10, the plotted points are the individual failure times. The smooth curve is the estimate of the actual cumulative percentage failing as a function of time.

There are two models that deal specifically with accelerated tests, however, which are more appropriate for component-level testing: the inverse power law model (which is a special kind of mathematical relationship between two quantities) and the Arrhenius model (which is used to describe the kinetics of an activated process). Let us look at each individually.

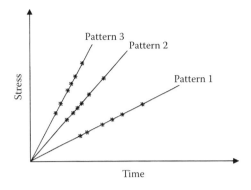

Figure 9.9 Progressive-stress testing.

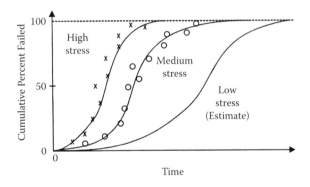

Figure 9.10 Cumulative percentage of samples failed as function of time at test stress.

Using the Inverse Power Law Model

The inverse power law model applies to many failure mechanisms as well as to many systems and components. This model assumes that, at any stress, the time to failure is Weibull distributed. The Weibull shape parameter β has the same value for all the stress levels. Always remember that the Weibull scale parameter θ is an inverse power function of the stress.

The model assumes that the life at rated stress divided by the life at accelerated stress is equal to accelerated stress divided by rated stress, raised to the power n:

$$\frac{\text{Life at rated stress}}{\text{Life at accelerated stress}} = \left(\frac{\text{Accelerated stress}}{\text{Rated stress}}\right)^n$$

where
n = Acceleration factor determined from the slope of the S-N diagram on the log–log scale

Using the stated equation, we can say that

$$\theta_u = \theta_s = \left(\frac{\text{Accelerated stress}}{\text{Rated stress}}\right)^n$$

where
θ_u = Life at the rated usage stress level
θ_s = Life at the accelerated stress level

Example

Let us assume we tested 15 incandescent lamps at 36 V until all items in the sample failed. A second sample of 15 lamps was tested at 20 V. Using this data, we will determine the characteristic life at each test voltage and use this information to determine the characteristic life of the device when operated at 5 V.

From the accelerated test data, we have

$$\theta_{20\,\text{volts}} = 11.7\,\text{h}; \text{ and } \theta_{36\,\text{volts}} = 2.3\,\text{h}$$

Because we know these two factors, now we can calculate the acceleration factor, n, using the relationship

$$\frac{\text{Life at rated stress}}{\text{Life at accelerated stress}} = \left(\frac{\text{Accelerated stress}}{\text{Rated stress}}\right)^n$$

Substituting, it becomes:

$$\frac{\theta_{20\,V}}{\theta_{36\,V}} = \left(\frac{36\,V}{20\,V}\right)^n$$

$$\frac{11.7\,h}{2.3\,h} = \left(\frac{36\,V}{20\,V}\right)^n$$

Therefore, n = 2.767

Now using this information, we can determine any characteristic life. In this example, we will use the value of 5 V:

$$\theta_u = \theta_s = \left(\frac{\text{Accelerated stress}}{\text{Rated stress}}\right)^n$$

$$\theta_{5V} = \theta_{36V} \left(\frac{\text{Accelerated stress}}{\text{Rated stress}}\right)^n = 2.3\left(\frac{36}{5}\right)^{2.767} = 542\,h$$

Therefore, the characteristic life at 5 V is 542 h, as shown in Figure 9.11.

Note: Not all electronic parts or assemblies will follow the inverse power law model; therefore, its applicability must usually be verified experimentally before use.

Using the Arrhenius Model

An alternative to the inverse power law model is the Arrhenius relationship for reaction rate, which is often used to account for the effect of temperature on electrical/electronic components. The Arrhenius relationship is as follows:

$$\text{Reaction rate} = A\,\exp\left(\frac{-E_a}{k_B T}\right)$$

where
 A = Normalizing constant
 T = Ambient temperature in degrees Kelvin
 K_B = Boltzmann's constant = 8.63×10^{-5} eV/°K
 E_a = Activation energy type constant (unique for each failure mechanism)

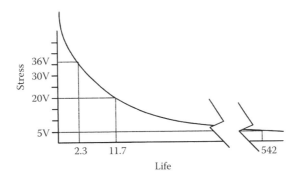

Figure 9.11 Inverse power law model.

In those situations where it can be shown that the failure mechanism rate follows the Arrhenius rate with temperature, the following Acceleration Factor (AF) can be developed:

$$\text{Rate}_{use} = A \exp\left(\frac{-E_a}{k_B T_{use}}\right)$$

$$\text{Rate}_{accelerated} = A \exp\left(\frac{-E_a}{k_B T_{accelerated}}\right)$$

$$\text{Acceleration factor} = AF = \frac{\text{Rate}_a}{\text{Rate}_u} = \frac{A \exp\left(\frac{-E_a}{K_B T_a}\right)}{A \exp\left(\frac{-E_a}{K_B T_u}\right)}$$

$$AF = \exp\left[\frac{-E_a}{K_B}\left(\frac{1}{T_a} - \frac{1}{T_u}\right)\right] = \exp\left[\frac{-E_a}{K_B}\left(\frac{1}{T_u} - \frac{1}{T_a}\right)\right]$$

where
T_a = Acceleration test temperature in degrees Kelvin
T_u = Actual use temperature in degrees Kelvin

Example

Assume that we have a device that has an activation energy of 0.5 and a characteristic life of 2,750 h at an accelerated operating temperature of 150°C. We want to find the characteristic life at an expected use of 85°C.
The conversion factor for Celsius to Kelvin is: °K = °C + 273.
Therefore: T_a = 150 + 273 = 423°K; T_u = 85 + 273 = 358°K; E_a = 0.5
Therefore, by substituting the values in the equation, we have

$$AF = \exp\left[\frac{0.5}{8.63 \times 10^{-5}}\left(\frac{1}{358} - \frac{1}{423}\right)\right]$$

AF = exp [2.49] = 12. The acceleration factor is 12.

To determine life at 85°C, multiply the acceleration factor by the characteristic life at the accelerated test level of 150°C.
Therefore, the characteristic life at 85°C = (12) (2750 h) = 33,000.

Characteristics of a Reliability Demonstration Test

There are eight characteristics that are important in reliability demonstration testing. These are

1. **Specified reliability, Rs**: This value is sometimes known as "customer reliability." Although it is usually represented as the probability of success (i.e., 0.98), other measures may be used, such as a specified MTBF.
2. **Confidence level of the demonstration test**: While a customer desires certain reliability, he wants the demonstration test to prove reliability at a given confidence level. A demonstration test with a 90% confidence level is said to "demonstrate with 90% confidence that the specified reliability requirement is achieved."
3. **Consumer's risk, β (beta risk or Type II error)**: Any demonstration test runs the risk of accepting bad product or rejecting good product. From the consumer's point of view, the risk is greatest if bad product is accepted. Therefore, he wants to minimize that risk. The consumer's risk is the risk that a test can accept a product that actually fails to meet the reliability requirement. Consumer's risk can be expressed as β = 1 – confidence level
4. **Probability distribution**: This is the distribution that is used for the number of failures or for time to failure. These are generally expressed as normal, exponential, or Weibull.
5. **Sampling scheme**: Here we address the issue of sampling. What kind of sampling should we have, and how should it be conducted?
6. **Number of test failures to allow**: Here we address our specifications and the relationship of the failures to our reliability expectations.
7. **Producer's risk, (α—alpha risk or Type I error)**: From the producer's standpoint, the risk is greatest if the test rejects good product. Producer's risk is the risk that the test will reject a product that actually meets the reliability requirement.
8. **Design reliability, Rd**: This is the reliability that is required in order to meet the producer's risk (α) requirement at the particular sample size chosen for the test. Small test sample sizes will require high design reliability in order to meet the producer's risk objective. As the sample size increases, the design reliability requirement will become smaller in order to meet the producer's risk objective.

In discussing the characteristics of a reliability demonstration test, we also must be familiar with the Operating Characteristic Curve (OCC) because the relationship between the probability of acceptance and population reliability becomes very obvious, and it can be very useful in showing the relationship between the probability of acceptance and MTBF or failure rate.

A specific OC curve will apply to each test situation and will depend on the number of pieces tested and the number of failures allowed. Figure 9.12 illustrates an OC curve using reliability as a measure. As you can see, when the population reliability is equal to R_s, the probability of acceptance is equal to β. When the population reliability is equal to R_d, the probability of acceptance is equal to $1 - \alpha$.

Types of Reliability Demonstration Tests

Demonstration tests are classified according to the method of assessing reliability. There are four major types of tests:

1. Attributes tests
2. Variables tests
3. Fixed-sample tests
4. Sequential tests

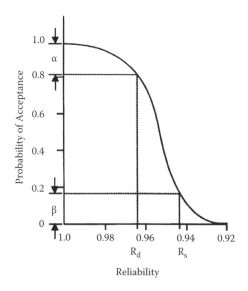

Figure 9.12 Operating characteristic curve (OCC).

Each is described in a bit more detail in the following sections. Note also that the four test types are not mutually exclusive; you can have fixed-sample or sequential-attributes tests as well as fixed-sample or sequential-variables tests.

Attributes Tests

If the components being tested are merely being classified as acceptable or unacceptable, the demonstration test is an attributes test. These tests

- May be performed even if a probability distribution of the time to failure is not known.
- May be performed if a probability distribution such as normal, exponential, or Weibull is assumed by dichotomizing the life distribution into acceptable and unacceptable time to failure.
- Are usually simpler and cheaper to perform.
- Usually require larger sample sizes to achieve the same confidence or risks as variables tests.

Variables Tests

Variables tests are used when more information is required than whether the unit passed or failed; for example, "What was the time to failure?" The test is a variables test if the life of the items under test is

- Recorded in time units
- Assumed to have a specific probability distribution such as normal, exponential, or Weibull

Fixed-Sample Tests

When the required reliability and the test confidence/risk are known, statistical theory will dictate the precise number of items that must be tested if a fixed-sample size is desired.

Sequential Tests

A sequential test may be used when the units are tested one at a time and the conclusion to accept or reject is reached after an indeterminate number of observations. In a sequential test,

- The "average" number of samples required to reach a conclusion will usually be lower than in a fixed-sample test. This is especially true if the population reliability is very good or very poor.
- The required sample size is unknown at the beginning of the test and can be substantially larger than the fixed-sample test in certain cases.
- The test time required is much longer because samples are tested one at a time (in series) rather than all at the same time (in parallel), as in fixed-sample tests.

Test Methods

Attributes tests can be used when

The accept/reject criterion is a go/no-go situation.
The probability distribution of times to failure is unknown.
Variables tests are found to be too expensive.

When to Use Each Type of Attributes Tests

We will look at five types of attributes tests in this section:

1. The fixed-sample test using the hypergeometric distribution for very small populations
2. The fixed-sample test using the binomial distribution for large populations
3. The fixed-sample test using the Poisson distribution for large populations
4. Success testing (when no failures are allowed) for large populations
5. A sequential, binomial test

Small Populations—Fixed-Sample Test Using the Hypergeometric Distribution

When testing items from a small population, and the accept/reject decision is based on attribute data, the hypergeometric distribution is applicable for test planning. The condition for use is the definition of successfully passing the test. This is determined by an item surviving the test. The parameter to be evaluated is population reliability. The estimation of the parameter is based on a fixed-sample size and testing without repair. The method to use is

- Define the criteria for success/failure.
- Define the acceptance reliability, R_s.
- Specify the confidence level or the corresponding consumer's risk, β.
- Specify, if desired, producer's risk, α. (Producer's risk can be used to calculate the design reliability target, R_d, needed in order to meet the requirements.)

The process consists of a trial-and-error solution of the hypergeometric equation until the conditions for the probability of acceptance are met. The equation that is used is

$$pr(x \leq f) = \sum_{x=0}^{f} \frac{\binom{N(1-R)}{x}\binom{NR}{n-x}}{\binom{N}{n}}$$

where
  $pr(x \leq f)$ = Probability of acceptance
  f = Maximum number of failures to be allowed
  x = Observed failures in sample
  R = Reliability of population
  N = Population size
  n = Sample size

$$\binom{N}{n} = \frac{N!}{n!(N-n)!}$$

Example

We are the supplier of an airbag system that causes the airbag to deploy in an automobile crash. Airbags that do not inflate properly are considered failures. Our total population size, N, is 10 airbag systems. Our specified reliability required is $R_s = 0.80$. How many systems do we need to test from this population, with zero failures allowed, to demonstrate the reliability with a 60% confidence level? (Producer's risk, α, will not be considered in this example.) By trial and error, we will solve for the probability of acceptance for various sample sizes to find the condition that

$$\text{Pr}(0) \leq \beta \text{ when } R = R_s$$

where

$$\beta = 1 - 0.6 = 0.4$$

The first trial will be n = 1

$$Pr(0) = \frac{\binom{10(1-0.8)}{0}\binom{10(0.8)}{1}}{\binom{10}{1}} = \frac{\left(\frac{2!}{0!(2-0)!}\right)\left(\frac{8!}{1!(8-1)!}\right)}{\frac{10!}{1!(10-1)!}} = 0.8$$

Because 0.8 is greater than 0.4, we must test more than one sample.
For n = 2,

$$Pr(0) = \frac{\binom{2}{0}\binom{8}{2}}{\binom{10}{2}} = 0.62$$

Because 0.62 is greater than 0.4, we must test more than two samples.
For n = 3,

$$Pr(0) = \frac{\binom{2}{0}\binom{8}{3}}{\binom{10}{3}} = 0.47$$

Because 0.47 is greater than 0.4, we must test more than three samples.
For n = 4,

$$Pr(0) = \frac{\binom{2}{0}\binom{8}{4}}{\binom{10}{4}} = 0.33$$

Because 0.33 < 0.4, we must test four airbag systems with zero failures to demonstrate a reliability of 0.80 at a 60% confidence level.

Large Population—Fixed-Sample Test Using the Binomial Distribution

When testing parts from a large population, and the accept/reject decision is based on attribute data, binomial distribution can be used. Note that, for a large N (one in which the sample size will be less than 10% of the population), binomial distribution is a good approximation for hypergeometric distribution. The binomial attribute demonstration test is probably the most versatile for use in the automotive industry (however, not unique to that industry) on vehicle components for several reasons:

- The population is large.
- The time-to-failure distribution for the parts is probably unknown.
- Pass/fail criteria is usually appropriate.

Like hypergeometric distribution, the procedure begins by the identification of

- Specified reliability, R_s
- Confidence level or consumer's risk, β
- Producer's risk, α (if desired)

The process consists of a trial-and-error solution to the binomial equation until the conditions for the probability of acceptance are met. The equation that is used is

$$\Pr(x \leq f) = \sum_{x=0}^{f} \binom{n}{x} (1-R)^x (R)^{n-x}$$

where
  $\Pr(x \leq f$ = Probability of acceptance
  f = Maximum number of failures to be allowed
  x = Observed failures in sample
  R = Reliability of population
  n = Sample size

Example

Let us assume we have an antilock brake system that is considered a success when it performs its function. If the product fails to perform the function, it is a failure. The specified reliability, R_s, requirement is 0.90. The customer would like this reliability demonstrated with a confidence of 85%. Because our population is quite large and we have go/no-go, accept/reject criteria, this binomial fixed-sample reliability demonstration test method is appropriate. If we have test facilities to check 40 systems, how many failures can we allow and still meet the customer requirement?

From the information given, we can determine the customer's risk, β:

$$\beta = 1 - \text{Confidence level} = 1 - 0.85 = 0.15$$

We can determine the maximum number of failures allowed by solving for

$$\Pr(x \leq f) \leq \beta \text{ or } \Pr(x \leq f) \leq 0.15$$

When we solve the equation for $\Pr(x \leq f)$, we find that for

$$\text{Zero failures allowed } (f = 0), \Pr(x \leq 0) = \binom{40}{0}(0.1)^0 (0.9)^{40} = 0.01478$$

$$\text{One failure allowed } (f = 1), \Pr(x \leq 1) = 0.01478 + 0.06569 = 0.08047$$

$$\text{Two failures allowed } (f = 2), \Pr(x \leq 2) = 0.01478 + 0.06569 + 0.14233 = 0.22280$$

Because $\Pr(x \leq 2) = 0.22280$ is greater than 0.15, we can see that one failure will be the maximum that we can allow because $\Pr(x \leq 1) \leq \beta$.

Our test condition will be: Test 40 pieces and allow a maximum of one failure to demonstrate that the population reliability is equal to or greater than 0.90. Because $\Pr(x \leq 1) = 0.08047$, we can say that, if we have the maximum of one failure, we are 92% confident that the population reliability is equal to or greater than 0.90 because $1 - 0.08047 = 0.91953 \approx 92\%$. In order to determine the design reliability, R_d, required, we must determine a producer's risk with which we can live. If we choose y α = 0.20, the probability of acceptance will be $1 - 0.20 = 0.80$. We can determine R_d by solving the equation for $\Pr(x \leq f) = 0.80$ for $R = R_d$.

By trial and error, we can find that $R_d = 0.98$ is required. This means that, to be able to meet the producer's risk of 0.20 when testing 40 pieces and allowing one failure, the population reliability (or design reliability) should be at least 0.98. Due to the trial-and-error method of solving the binomial probability equation, a computer program is a good alternative because it can make the required calculations in a short amount of time. One such program, "Binomial Test Plan," is available from E. Harold Vannoy, P.E. Another alternative is the Binomial Distribution

Nomograph developed for comparing reliability and confidence levels by Montgomery (1996). Such a nomograph is helpful for test plan development.

Example

If we use the conditions for the previous example where we have a reliability requirement of 0.90, test 40 pieces, and allow one failure, we can use the nomograph to determine the confidence level by drawing a line from 0.90 on the reliability scale through the point where the one-failure-allowed line intersects the 40-piece sample-size line. This line can be seen to cross the confidence level scale at 92%, which agrees with the previous calculations. Using the nomograph, we can easily determine the sample size required for any confidence level and reliability requirement by

- Drawing a line between the two points
- Reading the sample size from the intersection of the drawn line and the line for the number of failures allowed and sample size

For the previous example (R_s = 0.90 and confidence level = 0.85), we can see that, for one failure allowed, the sample size should be 33 pieces, as illustrated on the nomograph.

Large Population—Fixed-Sample Test Using the Poisson Distribution

The Poisson distribution can be used as an approximation of both the hypergeometric and the binomial distributions if

- The population, N, is large compared to the sample size, n.
- The fractional defective in the population is small ($R_{population}$ > 0.9).

The process consists of a trial-and-error solution using the following equation or Poisson tables, R_s, R_d, and various sample sizes until the conditions of α and β are satisfied.

$$\Pr(x \le f) = \sum_{x=0}^{f} \frac{\lambda_{Poi}^{x} e^{-\lambda_{Poi}}}{x!}$$

where
 $\Pr(x \le f)$ = Probability of acceptance
 f = Maximum number of failures to be allowed

x = Observed failures in sample
$\lambda_{poi} = (n)(1 - R)$
R = Reliability of population
n = Sample size

Additional information on this method is available in the references and selected bibliography sections of this chapter.

Success Testing

Success testing is a special case of binomial attributes testing for large populations where no failures are allowed. It is the simplest method for demonstrating a required reliability level at a specified confidence level. In this test case, n items are subjected to a test for the specified time of interest, and the specified reliability and confidence levels are demonstrated if no failures occur. The method uses the following relationship:

$$R = (1 - C)^{1/n} = (\beta)^{1/n}$$

where
C = Confidence level
R = Reliability required
n = Number of units tested
β = Consumer's risk

The necessary sample size to demonstrate the required reliability at a given confidence level is

$$n = \frac{\ln(1 - C)}{\ln R}$$

For example, if a specified reliability (R_s) of 0.90 is to be demonstrated with 90% confidence, we have

$$n = \frac{\ln 0.10}{\ln 0.90} = 22$$

Thus, we must place 22 items on test and allow no failures.

Sequential Test Plan for the Binomial Distribution

The sequential test is a hypothesis testing method in which a decision is made after each sample is tested. When sufficient information is gathered, the testing is discontinued. In this type of testing, sample size is not fixed in advance but depends upon the observations. Sequential tests should not be used when the exact time or cost of the test must be known beforehand or is specified. This type of test plan may be useful when the

- Accept/reject criterion for the parts on test is based on attributes.
- Exact test time available and sample size to be used are not known or specified.

The test procedure consists of testing parts one at a time and classifying the tested parts as good or defective. After each part is tested, calculations are made based on the test data generated to that point. The decision is made as to whether the test has been passed or failed, or if another observation should be made. A sequential test will result in a smaller average number of parts tested when the population tested has a reliability close to either the specified or design reliability. The method to use is

- Determine R_s, R_d, α, β
- Calculate the accept/reject decision points using

$$\frac{\beta}{1-\alpha} \quad \text{and} \quad \frac{1-\beta}{\alpha}.$$

As each part is tested, classify it as either a failure or success. Evaluate the following expression for the binomial distribution

$$L = \left(\frac{1-R_s}{1-R_d}\right)^f \left(\frac{R_s}{R_d}\right)^s$$

where
  f = total number of failures
  s = total number of successes

$$\text{If } L > \frac{1-\beta}{\alpha}, \text{ the test is failed.}$$

$$\text{If } L < \frac{\beta}{1-\alpha}, \text{ the test is passed.}$$

if $\dfrac{\beta}{1-\alpha} \leq L \leq \dfrac{1-\beta}{\alpha}$, the test should be continued.

Example

Assume that you have set up a test where (a) Samples that show specified reliability, $R_s = 0.90$, are to be accepted with $\beta = 0.10$; and (b) Samples that show design reliability, $R_d = 0.96$, are to be accepted with $\alpha = 0.05$.

The decision points can be calculated as follows:

$$\dfrac{\beta}{1-\alpha} = \dfrac{0.10}{1-0.05} = 0.105$$

and

$$\dfrac{1-\beta}{\alpha} = \dfrac{1-0.10}{0.05} = 18$$

To evaluate success and failures we use the equation

$$L = \left(\dfrac{1-R_s}{1-R_d}\right)^f \left(\dfrac{R_s}{R_d}\right)^s, \text{ which becomes } L = \left(\dfrac{0.10}{0.04}\right)^f \left(\dfrac{0.90}{0.96}\right)^s$$

In this example, if L is greater than 18, reject the samples. If L is less than 0.105, accept the samples. If L is between 0.105 and 18, continue testing.

Graphical Solution: A graphical solution for critical values of f and s is possible by solving the following equations:

$$\text{Ln}\dfrac{1-\beta}{\alpha} = (f)\ln\left(\dfrac{1-R_s}{1-R_d}\right)^f + (s)\ln\left(\dfrac{R_s}{R_d}\right)^s$$

and

$$\ln\dfrac{\beta}{1-\alpha} = (f)\ln\left(\dfrac{1-R_s}{1-R_d}\right)^f + (s)\ln\left(\dfrac{R_s}{R_d}\right)^s$$

using the values from the example,

$$\ln\dfrac{1-\beta}{\alpha} = (f)\ln\left(\dfrac{1-R_s}{1-R_d}\right)^f + (s)\ln\left(\dfrac{R_s}{R_d}\right)^s = \ln 18 = (f)\ln 2.5 + (s)\ln 0.9375$$

By substituting values for f or s, the equation will generate the decision points shown in Table 9.10 (plan a)

Table 9.10a and b Values for f and s in a Sequential Test Plan

Plan a		Plan b	
f	s	f	s
0	−44.79	−2.46	0
3.15	0	−1.76	10
3.86	10	0	34.92
10	97.19	10	176.90

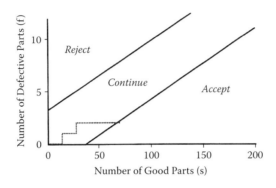

Figure 9.13 Graphical solution of sequential binomial test.

These points can be plotted on graph paper to form a decision line between "continue testing" and "reject" (see Figure 9.13).

In the same way, the equation

$$\ln \frac{\beta}{1-\alpha} = (f)\ln\left(\frac{1-R_s}{1-R_d}\right)^f + (s)\ln\left(\frac{R_s}{R_d}\right)^s$$

becomes ln0.105 = (f)ln 2.5 + (s)ln 09375

Substituting values for f or s in this equation will generate the points shown in Table 9.10 (plan b).

These points can also be plotted on graph paper to form the decision line between continue testing and accept. The resulting graph is shown in Figure 9.13. As each good part is observed, a horizontal line is drawn on the graph. Each defective part is recorded as a vertical line. By comparing the test results to the decision lines, the appropriate action (accept, reject, continue) can be taken.

Table 9.11 Values for f and s in a Sequential Test Plan

Plan a		Plan b	
f	s	f	s
0	−44.79	−2.46	0
3.15	0	−1.76	10
3.86	10	0	34.92
10	97.19	10	176.90

Failure-Truncated Test Plans—Fixed-Sample Test Using the Exponential Distribution

This test plan is used to demonstrate the life characteristics of items whose failure times are exponentially distributed and when the test will be terminated after a preassigned number of failures. The method to use is as follows.

First, obtain the specified reliability (R_s), failure rate (λ_s) or MTBF (θ_s), and test confidence. Remember that, for the exponential distribution,

$$R_s = e^{-\lambda_s t} = e^{-t/\theta_s}$$

Then, solve the following equation for various sample sizes and allowable failures:

$$\theta \geq \left(\frac{2 \sum_{i=1}^{n} t_i}{\chi^2_{\beta, 2f}} \right)$$

where
 θ = MTBF demonstrated
 t_i = Hours of testing for unit i
 f = Number of failures

$\chi^2_{\beta, 2f}$ = The β percentage point of the chi-square distribution for 2f degrees of freedom
β = 1 − confidence level

Example

What MTBF is demonstrated at a 90% confidence level when 40 units are placed on test, and the test is stopped after the second failure that occurred at 950 h? The first failure occurred at 900 h.

From the given information, we know that

$$n = 40; \beta = 1 - 0.90 = 0.10$$

and

$$\sum_{i=1}^{n} t_i = 900 + 39(950) = 37{,}950$$

From the χ^2 table we can determine the value of $\chi^2_{0.10, 4} = 7.779$. Refer to the table in Appendix A. III.

Therefore,

$$\theta \geq \left(\frac{2 \sum_{i=1}^{n} t_i}{\chi^2_{\beta, 2f}} \right) = \left(\frac{(2)(37{,}950)}{\chi^2_{0.1, 4}} \right) = \left(\frac{(2)(37{,}950)}{7.779} \right) = 9{,}757 \, h$$

We can say that we are 90% confident that the MTBF demonstrated is equal to or greater than 9757 h.

Time-Truncated Test Plans—Fixed-Sample Test Using the Exponential Distribution

This type of test plan is used when

- A demonstration test is constrained by time or schedule.
- Testing is by variables.
- Distribution of failure times is known to be exponential.

The method to use will be the same as with the failure-truncated test.

$$\theta \geq \left(\frac{2 \sum_{i=1}^{n} t_i}{\chi^2_{\beta, 2(f+1)}} \right)$$

where
 θ = MTBF demonstrated
 t_i = Hours of testing for unit i
 f = Number of failures
 $\chi^2_{\beta, 2(f+1)}$ = The β percentage point of the chi-square distribution for 2(f + 1) degrees of freedom
 β = 1 − confidence level

For the time-truncated test, the test is stopped at a specific time and the number of observed failures (f) is determined. Because the time at which the next failure would have occurred after the test was stopped is unknown, it will be assumed to occur in the next instant after the test is stopped. This is the reason that the number 1 is added to the number of failures in the degrees of freedom for χ^2.

Example

How many units must be checked on a 2000 h test if zero failures are allowed and θ_s = 32,258? A 75% confidence level is required.
From the information, we know that

$$= 1 - 0.75 = 0.25$$

$$2(f + 1) = 2(0 + 1) = 2$$

Therefore,

$$\theta \geq \left(\frac{2 \sum_{i=1}^{n} t_i}{\chi^2_{0.25, 2}} \right) = \theta \geq \left(\frac{2 \sum_{i=1}^{n} t_i}{2.772} \right) = 32,258$$

By rearranging this equation, we see that

$$\sum_{i=1}^{n} t_i = \frac{(2.772)(32,258)}{2} = 44,709.59$$

Because no failures are allowed, all units must complete the 2,000 h test and

$$44,709.59 = (n)(2,000)$$

Solving for n,

$$N = \frac{44,709.59}{2,000} = 22.35 \text{ or } 23 \text{ units}$$

We can say that, if we place 23 units on test for 2,000 h and have no failures, we can be 75% confident that the MTBF is equal to or greater than 32,258 h.

Note: This assumes that the test environment duplicates the use environment such that 1 h on test is equal to 1 h of actual use.

(Failure-truncated and time-truncated demonstration test plans for the exponential distribution can also be designed in terms of θ_s, θ_d, α, and p by using methods covered in the sources listed in the References and Additional Readings section at the end of the book.)

Weibull and Normal Distributions

Fixed-sample tests using the Weibull distribution and for the normal distribution have also been developed. If you are interested in pursuing the tests for either of these distributions, see Chapter 15, which covers these in more detail.

Sequential Test Plans

Sequential test plans can also be used for variables demonstration tests. The sequential test leads to a shorter average number of part hours of test exposure if the population MTBF is near θ_s or θ_d (i.e., close to the specified or design MTBF).

Exponential Distribution Sequential Test Plans

This test plan can be used when

- The demonstration test is based upon time-to-failure data.
- The underlying probability distribution is exponential.

The method to be used for the exponential distribution is to

- Identify θ_s, θ_d α, and β.
- Calculate accept/reject decision points $\beta/1 - \alpha$ and $1 - \beta/\alpha$.

Evaluate the following expression for the exponential distribution:

$$L = \frac{\theta_d}{\theta_s} \exp\left[-\left(\frac{1}{\theta_s} - \frac{1}{\theta_d}\right)\right] \sum_{i=1}^{n} t_i$$

where
t_i = time to failure of the i-th unit tested.
N = Number tested

If $L > \frac{1-\beta}{\alpha}$, the test is failed.

If $L < \frac{\beta}{1-\alpha}$, the test is passed.

If $\frac{\beta}{1-\alpha} \leq L \leq \frac{1-\beta}{\alpha}$, the test should be continued.

A graphical solution can also be used by plotting decision lines as shown in Figure 9.14:

$$nb - h_1 \text{ and } nb + h_2$$

where
n = Number tested
$b = \frac{1}{D} \ln \frac{\theta_d}{\theta_s}$
$h_1 = \frac{1}{D} \ln \frac{1-\beta}{\alpha}$
$h_2 = \frac{1}{D} \ln \frac{1-\alpha}{\beta}$
$D = \frac{1}{\theta_s} - \frac{1}{\theta_d}$

The conclusion is based on the notion that t_i equals time to failure for the i-th item and

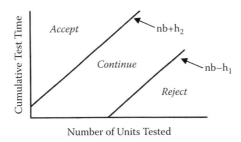

Figure 9.14 Sequential testing graph (a).

If $\Sigma t_i < nb - h_1$, the test is failed.
If $\Sigma t_i \geq nb + h_2$, the test is passed.
If $nb - h_1 \leq \Sigma t_i < nb + h_2$, continue the test.

The graph on Figure 9.14 shows these relationships.

Example:

Assume you are interested in testing a new product to see whether it meets a specified MTBF of 500 h, with a consumer's risk of 0.10. Further, specify a design MTBF of 1,000 h for a producer's risk of 0.05. Run tests to determine whether the product meets the criteria.

First, determine D based on the known criteria:

$$D = \frac{1}{\theta_s} - \frac{1}{\theta_d} = \frac{1}{500} - \frac{1}{1000} = 0.002 - 0.001 = 0.001$$

Next solve for h_2:

$$h2 = \frac{1}{D} \ln \frac{1-\alpha}{\beta} = \frac{1}{0.001} \ln \frac{1-0.05}{0.10} = 2251.2918 \approx 2251$$

Then, solve for h_1:

$$h1 = \frac{1}{D} \ln \frac{1-\beta}{\alpha} = \frac{1}{0.001} \ln \frac{1-0.10}{0.05} = 2890.3718 \approx 2890$$

Finally, solve for b:

$$b = \frac{1}{D} \ln \frac{\theta_d}{\theta_s} = \frac{1}{0.001} \ln \frac{1000}{500} = 693.1472 \approx 693$$

Table 9.12 Results of Sequential Failures

Failure #	Time to Failure	Cumulative Test Time, $\sum t_i$
1	400	400
2	500	900
3	700	1600
4	800	2400
5	1000	3400
6	1200	4600
7	1500	6100
8	2000	8100

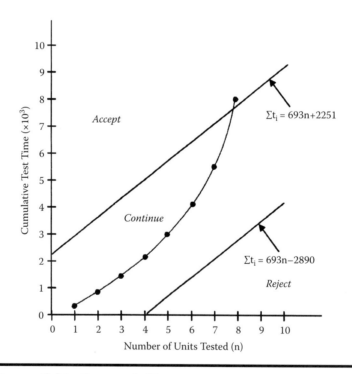

Figure 9.15 Sequential testing graph (b).

Using these results, we can determine at which points we can make a decision. Therefore,

If $\sum t_i < 693n - 2{,}890$, the test is failed.
If $\sum t_i \geq 693n + 2{,}251$, the test is passed.
If $693n - 2{,}890 \leq \sum t_i < 693n + 2{,}251$, continue the test.

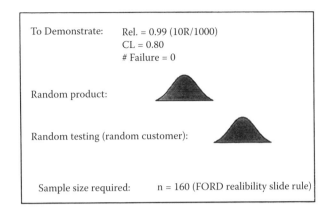

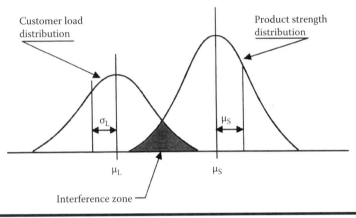

Figure 9.16 Interference (tail) testing.

Table 9.13 Sample Sizes at a Point Stress

Point Stress at Given Percentile	Required Sample Size
50th	3212
84th (1σ from μ stress)	145
90th	71
95th	31
99th	9

Now, assuming that the tests give the values in Table 9.12, we can plot the points and the decision lines as illustrated in Figure 9.15.

Now we can see that, when n = 8, the cumulative test time point is above the upper decision line. We can therefore assume that the product meets the design requirement of 1,000 h MTBF.

Interference (Tail) Testing

Interference demonstration testing can sometimes be used when the stress and strength distributions are accurately known. If a random sample of the population is obtained, it can be tested at a point stress that corresponds to a specific percentile of the stress distribution. By knowing the stress and strength distributions, the required reliability, the desired confidence level, and the number of allowable failures, it is possible to determine the sample size required.

Example

Assume that the stress and strength distributions are known to be normal distributions. Further assume that σ stress = σ strength (see Figure 9.16). How many pieces must be tested with zero failures allowed in order to demonstrate a reliability of 0.99 with an 80% confidence level?

Under the conditions given, it can be shown that the following sample sizes, shown in Table 9.13, are required when testing at a point stress corresponding to the given percentile.

These sample sizes would compare to a required sample size of 161 pieces, which can be determined using the success (binomial) testing methods given earlier. As you can see, the required sample size can be greatly reduced if the stress and strength distributions are accurately known. Note that the example deals with only one stress and strength distribution. Under normal circumstances, several stresses must be considered individually and combined in order to determine the demonstrated reliability.

Summary

This chapter discussed the reliability growth and accelerated testing approaches for understanding and improving machines. The next chapter introduces the concepts and importance of availability, failure, failure rate, life-cycle costs (LCC), and the requirements of each phase of LCC. Also, design reviews will be discussed.

Chapter 10

Maintenance Issues and Concerns

Chapter 9 concluded the discussion on improving OEE using reliability growth and accelerating testing. This chapter introduces maintenance of equipment and machines as they are operating.

All machines eventually fail, and when they do, the cost and time of repairing them must be the lowest possible. One way to accomplish this is to identify a machine's "lowest replaceable unit" (LRU). For example, if a machine's drive fails, it might be quicker, and thus cheaper, to replace the drive rather than the specific failed component.

To address optimizing the LRU and minimizing the failures of machinery, we use the term *maintainability*. Maintainability is a characteristic of design, installation, and operation, usually expressed as the probability that a machine can be retained in, or restored to, specified operable condition within a specified interval of time when maintenance is performed in accordance with prescribed procedures.

Benefits of Improved Equipment Maintainability

Why should we look at maintainability? Because maintainability provides information about the vital characteristics of manufacturing machinery and equipment that enable suppliers and customers to be world-class competitors. Efficient production planning depends on a process that yields high-quality parts at a specific rate without interruption. Predictable equipment R&M (i.e., reliability and maintainability) of the manufacturing machinery and equipment is a key ingredient in maintaining production efficiency and the effective deployment of *just-in-time* (JIT) principles. Improved equipment R&M leads to lower total life-cycle costs that are necessary to maintain the competitive edge. Improved equipment R&M also results in improved availability, while highly available production machinery produces consistently high-quality products at lower costs and higher output levels.

It is also important here to mention that maintainability is defined in terms of the mean time to repair (MTTR). MTTR addresses single repairs (one failure) and only one repair at the time.

The calculation for MTTR is:

$$MTTR = \frac{\sum t}{N}$$

where
 $\sum t$ = summation of repair time
 N = total number of repairs

The longer the repair takes, the more the equipment's cost goes up. Therefore, successful equipment R&M deployment has a positive effect on *reducing equipment downtime.* A cursory summary of user and supplier benefits is shown on Table 10.1.

Key Concepts Pertaining to Maintainability

Some of the major concepts that relate to maintainability are described in the following sections.

Calculating the Availability of Machinery/Equipment

Availability is a measure of the degree to which machinery/equipment is in an operable and committal state at any point in time. Specifically, it is the

Table 10.1 A Cursory Summary of Benefits of Improved Equipment Maintainability for Users and Suppliers

User Benefits	Supplier Benefits
Higher machinery and equipment availability	Reduced warranty costs
Unscheduled downtime reduced/eliminated	Reduced build costs
Reduced maintenance costs	Reduced design costs
Stabilized work schedule	Improved user relations
Improved JIT performance capability	Higher user satisfaction
Improved profitability	Improved status in the marketplace
Increased employee satisfaction	A competitive edge in the marketplace
Lower overall cost production	Increased employee satisfaction
More consistent part/product quality	Increased understanding of product applications
Less need for in-process inventory to cover downtime	Increased sales volume
Lower equipment LCC	

percent of time machinery and/or equipment will be operable when needed. Availability can be expressed in terms of the following equation:

$$Availability = \frac{MTBF}{MTBF + MTTR}$$

where
  Mean time between failure (MTBF) = Average time between failure occurrence.
  Mean time to repair (MTTR) = Average time to restore machinery or equipment to specific conditions

Example

Find the availability of a piece of equipment that has an MTBF of 100 h and a MTTR of 20 h.

Availability = MTBF/(MTBF + MTTR) = 100/(100 + 20) = 100/120 = 0.833 or 83.3%

Notice that, if the MTTR increases, the availability of the equipment will be reduced. If the MTBF increases, the availability of the equipment will also be increased.

Addressing Failures of Equipment or Machinery

A failure is an event where machinery/equipment is not available to produce parts at specified conditions when scheduled or is not capable of producing parts or scheduled operations to specifications. For every failure, an action is required.

Failures may be classified into several different areas such as

- *Catastrophic failures* occur randomly. They are chance failures that will cause a sudden failure without warning. These types of failures are the most costly and will cause emergency shutdown of the equipment at the most inappropriate times.
- *Wear-out failures* occur slowly, and frequently signal when failure is imminent. This type of slow degradation of quality is often observed before the failure occurs, and maintenance actions can be taken to replace the worn-out components.

Failures can be grouped into three time periods over the useful life of the equipment:

1. Infant mortality
2. Useful life
3. Wearout

These areas are described in the following and can be represented by the graph identified as the bathtub curve.

Failures during the Infant Mortality Period

During the infant mortality time period of the equipment's life cycle, the failures that occur are associated with the equipment's *manufacturing and installation*. This area of the bathtub curve is typically identified as *the burn-in period* for the equipment. Failures typically seen during this period of time are

- Poor welds
- Cold solder joints

- Nicks, voids, cracks
- Incorrect part positioning
- Poor workmanship
- Contamination
- Substandard materials
- Improper installation

For example, consider a newly installed grinder that runs 1 h before a belt breaks. Upon examining the belt, it is discovered to be nicked as a result of supplier assembly error. A replacement belt is then installed, and the grinder continues to run with no subsequent belt failures. This represents "Infant Mortality" and is a failure occurring in the first bathtub curve region that will not repeat itself once it is corrected.

Often, this early phase of an equipment's life cycle is referred to as "working the bugs out!" A high number of failures resulting from these bugs creates an aggressive infant mortality curve and is indicative of low-quality supplier-assembly work. Conversely, a low number of failures during this time creates a less aggressive, almost flat infant mortality curve and is indicative of high supplier quality.

Failures during the Useful Life Period

During the useful lifetime period of the equipment's life cycle, the failures that occur are associated with the *design* of the equipment. Failures that are typically seen in this time period are identified as

- Low safety factors
- Higher stress on the equipment
- Abuse of equipment
- Operator errors
- Misapplication of equipment
- Maintenance procedures and practices

During this phase, failures are due to normal wear, and if they are properly tracked, they become very predictable.

For example, a bearing may have a rated life of 2,000 h. However, because of its high-temperature operating environment, failures may occur every 1,200 h. Therefore, bearing replacements may be scheduled every 1,100 h, or bearings more suitable for high-temperature operations should be obtained.

It is this region of the bathtub curve that is useful to maintenance when developing their preventive maintenance program and failure tracking system.

Failures during the Wearout Period

The wearout period is defined as the area where the failures on the equipment will begin to increase over time. At this point, the equipment must be repaired or rebuilt in order for it to remain in service. Typical failures that occur during this period of time are

- Fatigue
- Shrinking or cracking
- Hardening of rubber components
- Frictional wear
- Chemical changes

Failures occur more frequently as components reach the end of their useful life. At this time, a decision must be made to decommission or rebuild the equipment. There are no hard-and-fast rules as to which to choose. However, the Equipment R&M tools can help evaluate the alternative costs, thereby allowing formulation of the most efficient solution.

It is obvious from this short discussion here that reliability plays a major role in maintainability. In fact, reliability can be measured against a time reference point. This point is identified as the *operating time* or the *mission time* for the machine or equipment. Mission times are generally specified by the engineering organization during the design cycle of the equipment. It must be remembered that the reliability calculation only applies to the flat part of the bathtub curve.

Reliability can be calculated based on this information:

$$R(t) = 1 - F(t)$$

where
 $R(t)$ = reliability at time t
 $F(t)$ = failure at time t
This means that reliability can never be 100%, and it can also be shown as

$$R(t) = e^{(-t/MTBF)}$$

where
- R(t) = reliability at time t
- t = Operation time or mission time of the equipment
- MTBF = Mean time between failure

Example 1

A drill press is designed to operate for 100 h. The MTBF for this drill press is also rated at 100 h. What is the probability the drill press will not fail during the mission time?

$$R(t) = e^{(-t/\text{MTBF})}$$

$$R100 = e^{(-100/100)} = 0.37$$

This means that the drill press will have a 37% chance of not breaking down during the 100 h mission time. This also means that the drill press will have a 63% chance of breaking down during the 100 h mission time.

Example 2

The same drill press from Example 1 has an increased MTBF of 300 h. What is the probability that the drill will not fail during the mission time?

$$R(t) = e^{(-t/\text{MTBF})}$$

$$R100 = e^{(-100/300)} = 0.72$$

This means that the drill press now has a 72% probability of not breaking down during the 100 h mission time. The probability of breakdown is only 28%. This reduction in probability of breakdown means that the drill press will be available to produce more parts during its operational time.

Determining Failure Rates

Failure rates are defined as *the number of failures per gross operating period in terms of time, events, cycles, or number of parts*. An equation can be used to define failure rates as follows:

$$\lambda = \frac{n}{\sum t}$$

where
- λ = Failure rate
- n = Number of failures

Σ = summation sign

t = time to failure, cycles to failure for a single unit

Example

A machine is operated for 2000 cycles. During that period, 10 failures were observed and recorded. What is the failure rate for this equipment?

$$\lambda = \frac{n}{\sum t} = \frac{10}{2,000} = 0.005$$

Ensuring Safety in Using Equipment

The priority concern in all maintainability issues is *safety first*. Throughout the Equipment R&M implementation phases, it becomes very important to understand that safety is the top concern with equipment manufacturing. The benefits of the improved design must *not* be allowed to compromise the ability of the manufacturing machinery and equipment to operate safely and be maintained without risk to the personnel. Passive safety features should always be major criteria for good designs.

Keeping an Eye on Life-Cycle Costs (LCC)

LCC refers to the total costs of a system during its life cycle. It is the sum of nonrecurring costs plus support costs. Typically, the support costs associated with the life cycle can account for more than 50% of the total LCC. This value can be reduced by identifying failures and taking corrective actions early in the design phase. These actions taken early in the life cycle of the equipment can contribute to overall reduction in LCC factors associated with the equipment. To optimize the recognition of the appropriate R&M, an appropriate and successful strategy must be developed. (LCC is covered in more detail in Chapter 14.)

Successful Implementation of Equipment R&M

Successful implementation of equipment R&M depends on thorough communication between the user and the supplier. The communication must begin at the project conception phase and continue throughout the equipment life cycle. This communication link will ensure that problems are identified, root causes are determined, and corrective actions are put in place.

Attainment of reasonable levels of equipment R&M seldom occurs by chance. It requires planning, goal definition, a design philosophy, analysis, assessment, and feedback for continuous improvement. Management must recognize the value of the equipment R&M program and commit the resources to obtain the goal. Without such commitment, the probability of attaining the goals of equipment R&M are low. Successful attainment of equipment R&M in terms of quantitative and qualitative objectives requires a team effort involving all functions of the business. A simplified way of visualizing this process is the five-phase approach. Although this process may be shown as a linear one, in reality, it may be overlapping in several phases:

- Phase 1: Concept
- Phase 2: Development and design
- Phase 3: Build and install
- Phase 4: Operation and support
- Phase 5: Conversion and/or decommission

The following sections of this chapter offer an overview of each phase; Chapters 11–14 describe the specific requirements of each phase.

Phase 1: Concept—Establishing System Requirements for Equipment R&M

The concept phase is research and limited design or development. The action taken here will typically result in a proposal. During this phase, the customer and the supplier must work together to establish system requirements. Team members might include, but not be limited to,

- Machine operators
- Maintenance personnel
- Engineering personnel
- Supplier personnel (if the supplier is selected)

Although nothing is definite or "locked-in" at this phase, when it is completed, project leaders should have established general requirements and identified available options. If the supplier is selected, then its key personnel should be involved in the concept phase. Also, if appropriate, concurrent engineering is initiated.

The responsibilities of the *customer* (i.e., *user*) may be to specify all of the following:

- Mean time between failure (MTBF)/Mean cycles between failure (MCBF)/Mean time to failure (MTTF)
- The expected MTTR
- How the machinery will be used
- The expected duty cycle
- Type of machinery environments the machine will be operating in
- Continuous improvement ideas

The responsibilities of the *supplier* may be to understand the quantitative and qualitative requirements as specified by the user.

Phase 2: Developing and Designing for Equipment R&M

The first goal of the design/development phase is finalizing an equipment performance specification and establishing concurrent engineering. Relying on input from plant production and supplier personnel, the team formalizes the specification to ensure that the machine's capabilities satisfy the plant's requirements. It is important to note during this phase that the equipment specification is evaluated in terms of "Life-cycle costs," which determine the "total cost of ownership." *Savings realized by a low purchase price can quickly be wiped out in the maintenance costs of a poorly designed machine.*

In addition, all the issues from the concept phase are incorporated, and safety, ergonomics, accessibility, and other maintainability issues are designed into the system. Equipment R&M allocation requirements are also formalized, as well as component configuration and suppliers. Of course, their selection is based on the predictive Equipment R&M statistics they provide.

By conducting an in-depth design review, the customer and the equipment supplier gain greater assurance that the machine will meet its equipment R&M goals, resulting in higher profits and enhanced reputations for both.

The responsibilities of the *customer user* may be to specify R&M program requirements. Here the user may review and modify the specifications, as well as make sure that they are appropriate and applicable to the intended use of the equipment.

The responsibilities of the *supplier* may be as follows:

- Establish design margins
- Incorporate maintainability design concepts
- Conduct reliability analysis and prediction
- Perform EFMEA (equipment failure mode and effect analysis)
- Conduct design reviews.

Phase 3: Building and Installing for Equipment R&M

In this phase, the supplier begins subassembly and component procurement with a clear understanding of the equipment R&M requirements. However, it is inevitable that problems may be encountered and design changes required. Therefore, it is crucial that concurrent engineering be deployed and the team maintain clear channels of communication during this phase. As changes are implemented, each should be evaluated using equipment R&M guidelines (as agreed between the user and the supplier). Manufacturing process variables affecting equipment R&M should be identified and targeted for control.

The following is a list of milestones occurring during the Build and Install phase:

- Develop maintenance procedures
- Develop training materials
- Conduct training classes
- Execute machine RQT (Reliability Qualification Tests)
- Establish equipment R&M database
- Ship equipment
- Install equipment
- Identify early failures (e.g., infant mortality or design flaws)

The responsibilities of the *customer* (i.e., *user*) may be to

- Monitor the building of the equipment
- Monitor equipment testing
- Monitor the installation process

The responsibilities of the *supplier* may be to

- Provide machinery parts if needed
- Perform tolerance studies
- Conduct stress analysis
- Perform reliability testing
- Collect data during testing
- Identify root cause analysis

Phase 4: Operation and Support of Equipment R&M

In this phase, the equipment is installed in its location and fully operational. However, if equipment R&M capabilities are to be fully deployed, data collection and feedback are essential. Therefore, collection mechanisms should be in place and agreed on by both parties. This allows for the most efficient continued equipment R&M growth and continuous improvement while providing the framework for an effective preventive maintenance program. For an equipment R&M initiative to be successful, the manufacturing machinery, equipment, and component suppliers must have access to maintenance records and equipment R&M databases.

The responsibilities of the *customer* (*user*) may be to implement the data collection system. The responsibilities for the *supplier* may be to

- Reliability growth
- Maintainability improvement
- Failure analysis and reporting
- Corrective action
- Data exchange

Phase 5: Converting and/or Decommissioning Equipment

This phase represents the end of the equipment's expected life and is validated by an aggressive increase in the number of "wear-out failures" or a need to modernize or convert the machine to remain competitive. When a machine reaches this phase, there are four possible options:

1. Decommission and replace with a new machine.
2. Rebuild it to a "good-as-new" condition.
3. Modernize it.
4. Convert it.

Option #1: Decommission and Replace with a New Machine

This option provides an opportunity to maximize on the benefits of equipment R&M. If a failure reporting and analysis system was used on the old equipment to track performance, the team has an excellent profile to begin its new equipment specification. However, if no previous equipment R&M information is available, the activities associated with the concept, development/design, and build and install phases (i.e., Phases 1–3) should still be conducted.

Option #2: Rebuild the Equipment to a "Good-as-New" Condition

If this option is chosen, previous information is essential for effective equipment R&M deployment. Without understanding the machine's performance profile, many of the inferior aspects of the machine would no doubt remain unchanged. If there is no previous performance profile available, Phases 1 and 2 should be deployed to ensure the highest reliability improvement possible.

Option #3: Modernize the Equipment

Quite often, modernizing is done when updated subsystems are available at economical rates. By replacing subsystems such as feeders, stackers, or drill heads, performance can be greatly enhanced. However, you should give particular attention to how the subsystems interface with each other and the machine. If this option is selected, you should conduct the first four phases of equipment R&M to maximize your return on your investment.

We have talked about the R&M being a team activity. As such, team members are appropriately assigned roles and responsibilities. In addition, one of equipment R&M's most powerful attributes is concurrent engineering. As the multifunctional team works together through the various equipment R&M phases, responsibilities are delegated to individual team members. However, because of the various roles assumed by the customer and the supplier, these responsibilities vary.

The following is a typical list of the *customer's* responsibilities:

1. Give the supplier realistic goals to achieve. The goals should be attainable and, at the same time, allow the customer's expectations for achieving continual improvement.
2. Assume that the supplier is competent. If the supplier has achieved its "Quality certification" status, it has demonstrated some of the

preliminary customer's requirement as a quality vendor. It is not engineering's responsibility to reevaluate them.

3. Establish critical project review checkpoints. Then follow through to see that established goals are reached. (In the automotive world, this is established through Advanced Product Quality Planning (APQP) and Production Part Approval Process (PPAP). The reader is encouraged to see the approach that Ford Motor Company takes in its PPAP requirements. (See Appendix H.)
4. Ensure that all possible failure modes are considered.
5. When failures are not eliminated, strive to establish maintainability. For example, develop preventive maintenance (PM) procedures that are simple in nature and quick to perform.
6. Ensure that the root causes of anticipated problems are understood and recommended actions are in place to improve reliability.
7. Provide the factory/user's perspective. This improves the design and the customer's relationship with the supplier.
8. Clarify requirements. No matter how detailed the specification, the customer must be willing to clarify or refine the equipment's requirements. In fact, in some cases, the customer must change the design specifications due to unreasonable requests.
9. Address critical design issues. If the equipment's design appears to be flawed, it is the customer's responsibility to raise the specific issues with the supplier.

Finally, the following is a list of the *supplier's* R&M responsibilities:

1. Accomplish the basic equipment design based on the customer's request.
2. *Always* alert the customer to any possible specification errors or conflicts.
3. Obtain the necessary resources to deploy equipment R&M.
4. Enlist customer manufacturing personnel for team participation (if necessary).
5. Train supplier personnel in equipment R&M awareness, equipment failure mode and effect analysis (FMEA), fault tree analysis, and failure-reporting methods.
6. Secure quality components from quality suppliers.
7. Maintain agreed-upon schedule dates to ensure timely equipment delivery.

Now that we have addressed some of the issues of LCC on a per-phase basis, let us look further into the exact requirements of each phase, beginning with Phase 1 in Chapter 11.

This phase is very similar to modernizing the machine with one fundamental difference. Whereas both phases update the machine for better productivity, the modernization focuses on improvement of the machine to produce a better existing product. On the other hand, conversion focuses on improvement as well, but this improvement is for a new product. In other words, conversion uses updates to the same machine with specific improvements with the purpose of producing something completely different than the original intent.

In relation to conversion one may be interested in figuring out the mean time to convert (MTTC). That is the average time it takes to convert a machine from producing one style of parts to another. Conversions are common and represent idle time for the machine. Therefore, conversions impact uptime, uptime % and capacity. The MTTC generally is reported on per week basis (Conversions/Week). If this number is greater than zero, there may be an opportunity to improve the uptime %. The conversions should be evaluated to see if they could:

- Be done off-shift
- Be done during lunch and breaks
- Be simplified, thereby reducing complexity of design and consequently reduce time requirements
- Eliminated by having duplicate machines

Summary

This chapter provided an overview of maintenance requirements, breaking them down for each of the LCC components. The next chapter discusses the first phase in detail.

Chapter 11

Requirements of Phase 1 of Implementing Equipment R&M: The Concept

Chapter 10 introduced the life-cycle costs (LCC), with all its phases. This chapter discusses the details of Phase 1 of the LCC.

Since the years of the Industrial Revolution, industry has continually sought to improve equipment performance and reliability. However, as we have moved into the 21st century, documenting equipment performance results and specifications remains challenging. Only through painstaking detail and industrial publishing practices were the most basic information made available.

With the advent of the computer age, more sophisticated data collection capabilities became possible. As suppliers found better ways to track machine performance, design and maintenance deficiencies became clearer, and suppliers could focus on improvement as they never had before.

Overview of Phase 1 Functions: Establishing Specs for Reliability and Maintainability (R&M)

As mentioned in Chapter 10, the focus of Phase 1 is to establish the specifications. But what does that mean? This chapter elaborates: simply put, this phase requires a way to:

1. *Determine factory requirements*: This is a simple process of asking the right questions of the personnel and translating their data into relevant factory experience.
2. *Investigate the supplier and machine capability*: Simultaneously with the first step, sift through the information regarding the equipment under investigation. This should result in understanding machine performance in order to ascertain the equipment's real capabilities.
3. *Utilize models to determine fit and areas for leverage (typically through historical data)*: Bring together the requirements of the factory with the capability of the supplier. Here is where the maximum leverage can be accomplished on the equipment under investigation. If historical data do not exist, you may use surrogate data until such time that you can generate data from the machine.
4. *Document performance*: Prepare a listing of performance requirements that have been developed for the equipment. These performance parameters should include the following information:

- Safety issues
- Electrical or power consumption
- Pneumatic requirements
- Environmental issues
- Other equipment performance requirements

In addition to the requirements just mentioned, in the 21st century, we also have to address environmental specifications: It is important to develop a set of environmental specifications for the equipment. This includes the various conditions that the equipment will be subject to while in the plant. These factors will play into the amount of failure and repair time needed for equipment on the plant floor. Environmental factors could include

- Temperature
- Humidity
- Electrical power
- Mechanical shock
- Pressure or vacuum
- Pneumatic pressure
- Corrosive material
- Immersion or splash

Table 11.1 A Typical Specification Matrix

	R&M Specification Matrix			
	Phase 1	Phase 2	Phase 3	Phase 4
Specification	Concept	Design/ Development	Build/ Install	Operation/ Support
R&M Program				
Supplier R&M Program Plan	X			
Quality Control	X			
Life-Cycle Costing (LCC) Model Development and R&M Optimization	X			
Periodic R&M Reporting Requirements	X			
R&M Design Reviews	X			
Formal R&M Evidence Book	X			
R&M Program Review, 6 months after Job #1				X
R&M Engineering				
Field History Review		X		
Failure Mode and Effects Analysis (FMEA)		X		
Electrical Stress Analysis and Derating		X		
Mechanical Stress Analysis and Derating		X		
Thermal Analysis		X		
Applications Engineering		X		
R&M Design Review Guidelines		X		
R&M Testing and Assessment		X		
Maintainability Matrix		X		

(Continued)

Table 11.1 A Typical Specification Matrix (*Continued*)

	R&M Specification Matrix			
	Phase 1	Phase 2	Phase 3	Phase 4
Specification	Concept	Design/ Development	Build/ Install	Operation/ Support
R&M Continuous Improvement Activities				
Failure Reporting, Analysis, and Corrective Action System (FRACAS)			X	
Machinery and Equipment Performance Data Feedback Plan				X
R&M Continuous Improvement Activities				X
Ford Total Productive Maintenance Activities (FTPM)		X		
Things Gone Right/Things Gone Wrong (TGR/TGW)		X		

- Hydraulic pressure
- Electrical noise—electromagnetic fields
- Relative humidity
- Ultraviolet radiation
- Vibration
- Utility services
- Contamination and their sources

To help both the customer and the supplier in the specification definition, a matrix may be developed. During each of the phases of the equipment's life cycle, it is important to document the performance specifications of the equipment. A matrix can be developed that will assist the R&M team in making sure that all areas of the R&M strategy are addressed. The matrix in Table 11.1 was developed using the Society of Automotive Engineers (SAE) guidelines.

Developing R&M Equipment Specifications

Developing a good equipment specification follows a simple Six-step process, summarized in Table 11.2; each step is described in more detail in the following sections.

Step 1: Determining Factory Requirements

The first step in developing a good equipment specification is to identify the plant's needed capacity. This is accomplished by first meeting with the necessary plant personnel to identify the required production volumes.

Example

If a transmission plant produces 575,000 transmissions a year with each containing six pinions, then the plant needs a machine capable of providing 3,450,000 pinions per year.

$$575,000 \times 6 = 3,450,000$$

Table 11.2 A Typical R&M Equipment Specification

	Capacity	Overall Equipment Effectiveness (OEE)	Specific Performance Specifications
Factory	Determine yearly production requirements production weeks/year and hours/week		
Operation	Determine required capacity and number of machines	Determine current equipment availability, performance, quality rates	
Machine	Determine weekly, hourly capacity rates per machine	Calculate the required availability % to meet capacity demands	Determine specific machine performance goals that meet LCC and capacity

If four (4) machines were used, then the capacity per machine would be

$$\frac{3,450,000}{4} = 862,500/\text{machine}.$$

Since there are 48 weeks of production this means that

$$\frac{862,500}{48} = 18,000/\text{week}/\text{machine}$$

Step 2: Develop Current Equipment Baseline

The next step is to develop the current baseline of the equipment on the floor. This step is accomplished by using the overall equipment effectiveness (OEE) value of the current equipment and then developing a set of OEE numbers for the equipment under investigation.

Step 3: Utilize the Information from Steps 1 and 2 to Determine Fit and Areas of Leverage

By bringing together the requirements of the factory with the baseline capability, the leverage of the equipment under investigation can be maximized.

Step 4: Document Performance Requirements

A list of performance requirements should be prepared for the equipment. Typical items should be

- Safety issues
- Electrical or power consumption
- Environmental issues
- Pneumatic requirements
- Other requirements as needed

Step 5: Develop Environmental Specifications

In this step, it is important to develop a set of environmental specifications for the equipment. This means generating a list of various conditions that the equipment will be subjected to while in the operation mode at the customer's facility. These items on the list will play a major role in the amount of failures and repair times that will be needed to be accounted for on the plant floor. Typical environmental factors should include

- Temperature
- Humidity
- Electrical noise
- Vibration
- Corrosive material
- Electromagnetic fields
- Others as needed

Step 6: Analyze Performance

You can either use manual tools and equations for analyzing the data or use software packages. There are many such packages available in the market. However, for this discussion, we reference the RAMGT software program. The main menu definitions are described in the following list.

Capacity analysis: The ability of a machine to produce the requirements of the customer in a given time.

Utilization %: Percentage of time that the machine is utilized.

Uptime %: Percentage of time that the machine is available for operation.

JPH: The total number of *jobs per hour* (JPH) needed to achieve the required capacity from the equipment to meet the overall factory requirements. Measurement can also be provided in the total number of cycles performed by the equipment under investigation.

MTBF (mean time between failures): It is defined as the average time between failure occurrences. The equation would be the sum of the operating time of a machine divided by the total number of failures:

$$\text{MTBF} = \frac{\sum t}{N}$$

Special note: You may notice that the formula is similar to the mean time to repair (MTTR). However, the Σt in the case of the MTTR is for total repair time, whereas in the case of the MTBF, it is for total operating time. Similarly, the N in the case of MTTR is for total number of repairs, whereas in MTBF it is for total number of failures.

Example

If a machine operates for 400 h and breaks down eight times, the MTBF is 50 h. (MTBF = 400/8 = 50 h).

MTTR (mean time to repair): The average time to restore machinery or equipment to its specified conditions. It is important to note that the MTTR calculation is based on repairing one failure—and one failure only. The longer each failure takes to repair, the more the equipment's cost of ownership goes up. Additionally, MTTR directly affects uptime, uptime %, and capacity. The formula for MTTR is

$$\text{MTTR} = \frac{\sum t}{N} = \frac{\text{Total Repair Time}}{\text{Number of failures}}$$

Example

A machine operates for 400 h. During that period, there were eight failures recorded. The total repair time was 4 h. What is the MTTR for this machine?

$$\text{MTTR} = \frac{\sum t}{N} = \frac{\text{Total Repair Time}}{\text{Number of failures}} = \frac{4}{8} = 0.5 \text{ h}$$

Response: Response is the average time it takes for maintenance personnel to respond to a failure. Response time is affected by factors such as maintenance procedures, staffing, or equipment location. Additionally, response time can directly affect uptime, uptime %, and capacity.

MTTC (mean time to convert): The average time it takes to convert a machine from producing one style of parts to another. Because conversions represent idle time for the machine, they directly impact uptime, uptime %, and capacity.

Scheduled downtime (DT): The amount of time in a week that the machine is scheduled to be down for items such as preventive maintenance or quality team meetings. However, it does not include downtime resulting from equipment failures or material shortages, because these are not scheduled; they are unscheduled. It is also important to note that scheduled DT directly affects uptime, uptime %, and machine capacity.

% Starved: This represents the amount of time the machine is available to run product, but because of upstream process interruptions, no product is available. Consequently, no parts are produced. This directly impacts the machine's % utilization and capacity.

MTBA (mean time between assists): The average time between assists. For example: If a machine runs 8 h and is assisted four times by the operator, then the MTBA = 8/4 = 2 h.

Because an assist represents lost production capability, it directly impacts capacity, uptime hours, and % uptime.

MTTA (mean time to assist): The average time required by the operator to assist the machine. For example, if the machine requires three assists in a 6 h period of time and the total time assisting is 8 min, then the MTTA = 8/3 = 2.66 min.

1st Run %: The percentage of product which, when processed through the equipment for the first time, results in a satisfactory product. It is inversely proportional to the scrap rate. Therefore, if a machine processes 100 pieces and two are rejected, the 1st run % = [(100 − 2)/100] • 100 = 98%

Hours/week: The number of hours per week the equipment is scheduled to run. Any change in it proportionately changes capacity. Additionally, the more a machine is run in a given time period, the lower the apportioned cost per piece will be. Consequently, higher operating hours/week tend to decrease per-piece costs.

Conversions per week (conv/week): The number of scheduled conversions per week. This number is multiplied by the MTTC to calculate the downtime generated by the conversions. Therefore, an increase in the number of conversions decreases the uptime %, uptime, and capacity, while increasing the equipment's labor costs.

Breaks?->: The equipment's capability to run unattended. The selection is made by entering Yes (1) or No (0). If the machine can run unattended through breaks, then Yes (1) is selected. If the machine cannot run unattended through breaks, then No (0) is selected. It is important to note that when Yes (1) is selected, the uptime hours are calculated based on an available 8 h per shift. If, however, No (0) is selected, then the uptime hours are calculated based on an available 7 h per shift.

Total downtime: The total time the machine is not operating.

Uptime hours: The total time that the machine is available for operation.

Determine Equipment R&M Goals

It is the customer's responsibility to specify the R&M goals for new equipment. However, as already mentioned, the supplier should be prepared to address how the R&M qualitative and quantitative R&M goals will be met.

During the concept phase of the equipment R&M strategy, the goals should address the areas described in the following sections.

Determining Goals for Acceptable Failure Requirements for the Equipment

These should be listed as

- MTBF
- Mean cycles between failure (MCBF) for repairable equipment
- MTTF (mean time to failure)

Determining Goals for Equipment Usage

Equipment usage is also very important to establishing the overall R&M specifications. The usage should address the following:

- Cycle times of the equipment
- Preventive maintenance schedules
- Inspection
- Various support activities for the equipment

Determining Maintainability Requirements

The maintainability requirements are inherent factors that are designed into the equipment or machinery. Both the customer and the supplier must recognize how the equipment will be used and the maintenance impacts. The machinery user specifies maintainability requirements in terms of mean time to repair or mean time to replace (MTTR).

Setting Goals for Documenting Failure Definition

Failure definition is reported as all downtime events associated with the equipment. This needs to be documented by the customer to ensure that all failures are addressed and actions are in place to remove the failures from the equipment.

Setting Goals for Environmental Considerations

The customer must detail all the environmental considerations and usage for the equipment. This information should include (but not be limited to) heat, shock, humidity, vibration, dust, and contamination.

Improving Equipment Performance

The equipment or machinery performance specifications can be improved by using the RAMGT software package. The screens shown in the following figures allow various equipment scenarios to be developed and identify areas where improvements can be made. These improvements are only suggestions, and some of the improvements may be too costly for implementation or may void some type of safety regulation on the equipment or machinery. These parameters can be evaluated through several commercial software packages; in our case, we use the RAMGT software package.

Evaluating ISO-Capacity

The ISO-capacity screen evaluates how well the current performance parameters will be able to meet the new capacity requirements for the new equipment. The data will indicate the JPH and required uptime to meet the new capacity requirements. Two types of screens can be viewed to evaluate the data. The first is the ISO-capacity screen, shown in Figure 11.1, and the second is the ISO-capacity graph, shown in Figure 11.2.

Calculating Available Production Time

After the relationship between JPH and the available production hours becomes close enough to achieve the needed weekly capacity, the specification must be analyzed to ensure that each of its parameters represents the most efficient value. This is best accomplished by constructing a time chart to represent the available time and how it is allocated among the following hours:

- Utilization
- Unmanned
- Scheduled downtime
- Failures
- % Starved/blocked
- Assists
- Conversions

Using the RAMGT software, a time chart can be created and accessed under the "Goal Setting" menu, as shown in Figure 11.3.

ISO-Capacity Model

RANGE/GOAL GRAPH PRINT PAGE HELP QUIT

Change Capacity Goal & JPH Range

Capacity Analysis–Gear Hob			WEEKLY CAP GOAL:		24000
UTIL %	62%		START RANGE:		180
UPTIME %	76%		INTERVAL:		10
CAPACITY	15063				
JPH	212			UPTIME	
			JPH	HOURS	PERCENT
MTBF	10	HOURS			
MTTR	1	HOURS	180	169.6	141%
RESPONSE	0.25	HOURS	190	160.7	134%
MTTC	0	MINUTES	200	152.6	127%
SCHED DT	10	HOURS	210	145.4	121%
% STARVED	5%		220	138.8	116%
MTBA	0.5	HOURS	230	132.7	111%
MTTA	2	MINUTES	240	127.2	106%
1ST RUN %	95.000%		250	122.1	102%
HRS/WEEK	120		260	117.4	98%
CONV/WEEK	0		270	113.1	94%
BREAKS?–>	0	YES(1) NO(0)	280	109.0	91%
Total DT=	28.3	23.61%	290	105.3	88%
UptimeHrs	91.7		300	101.8	85%
		NUM CMD			

Figure 11.1 ISO-capacity screen.

Calculating Mean Time between Failures Parameters

The MTBF parameter can be displayed on the MTBF screen. This screen will illustrate the relationship between MTBF, capacity, and uptime %, as shown in Figure 11.4.

Examining the MTBF versus uptime graph reveals that an MTBF of 2.5 h results in an uptime % of 52, whereas an MTBF of 5 h yields an uptime % of 68, which is an increase of 15%. However, an increase to an MTBF of 10 (twice the improvement) results in an uptime of 75%, which is an increase of only 7.5%.

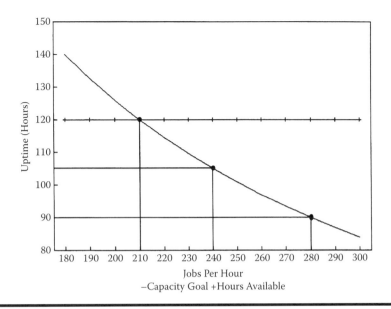

Figure 11.2 ISO-capacity graph.

Calculating Mean Time to Repair or Replace

To further understand the impact of failures on the specification's capacity, you must consider the MTTR and response times:

- The lower the MTBF value, the greater its impact of MTTR.
- The higher the MTBF value, the lower its impact of MTTR.

RAMGT can demonstrate the relationship between MTTR, capability, and capacity, as shown in Figure 11.5.

Therefore, with this graph, the impact of repair time on uptime % can be seen and evaluated immediately. It is also important to note that response time has the same relationship on uptime % and therefore can be evaluated using the same graph.

Calculating Scheduled Downtime

Scheduled downtime is defined as the elapsed time that equipment is down for scheduled maintenance or because it is turned off for other reasons. It also could be caused by one or more of the following factors:

- Preventive maintenance
- Quality team meetings scheduled at the start of the shift
- Production practices structured to eliminate scheduled downtime

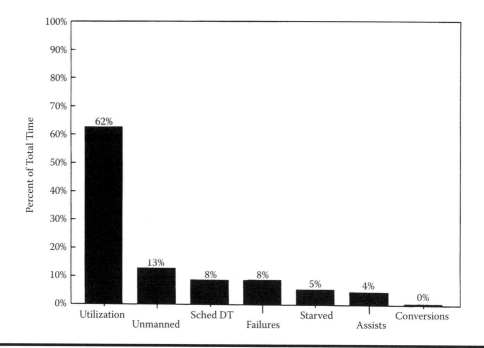

Figure 11.3 Time chart using the RAMGT software.

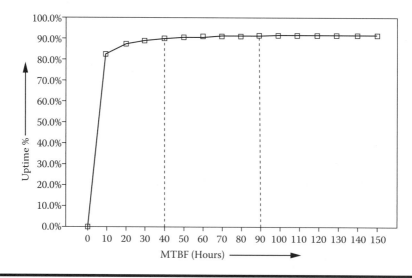

Figure 11.4 Uptime versus MTBF.

Using the RAMGT software, the "WHAT IF?" menu can be used to illustrate the relationship between scheduled downtime and uptime %. When first viewing the screen, notice under the "RANGE" command the input for the "START RANGE" and "INTERVAL." These entries define the "*X*" and "*Y*" axes of a graph for later viewing. Therefore, enter "0" for the "START RANGE"

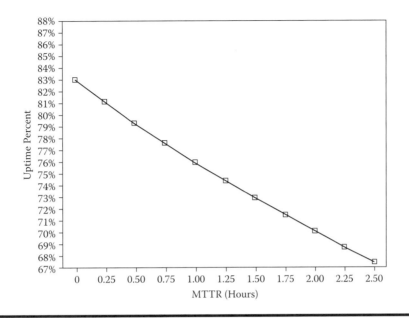

Figure 11.5 Uptime versus MTTR graph.

and "1" for the "INTERVAL." The following data columns illustrate the relationship between scheduled downtime and the uptime %. For many, a graph better represents this relationship (see Figures 11.6 and 11.7).

Therefore, under the "GRAPH" command, the linear relationship between the scheduled downtime hours and uptime % is illustrated in Figures 11.5–11.7.

Calculating the % Starved Time

This represents the amount of time the machine is up and available to produce product, but no product is available to run: therefore, the machine is starved. Because this relationship is directly proportional to the uptime %, no specific graph under the "WHAT IF?" menu is needed. Instead, it is sufficient to understand that for every 1% starved that is eliminated, 1% in uptime % is gained.

Calculating the Number of Assists

Assists are considered failures because they show the time that is lost because of the frequency of assists (mean time between assists) and the

Sensitivity to Sched Down Time

RANGE CHANGE PARAMETERS GRAPH PRINT QUIT

Change the Range for Scheduled Down Time

Capacity Analysis–Gear Hob				START RANGE:	0
UTIL %	62%			INTERVAL:	1
UPTIME %	76%		SCHED DT	CAPACITY	UPTIME %
CAPACITY	15063				
			0	16741	83%
JPH	212	HOURS	1	16574	83%
MTBF	10		2	16406	82%
MTTR	1	HOURS	3	16238	81%
RESPONSE	0.25	HOURS	4	16070	81%
MTTC	0	MINUTES	5	15902	80%
SCHED DT	10	HOURS	6	15734	79%
% STARVED	5%		7	15567	78%
MTBA	0.5	HOURS	8	15399	78%
MTTA	2	MINUTES	9	15231	77%
1ST RUN %	95.000%		10	15063	76%
HRS/WEEK	120		11	14895	76%
CONV/WEEK	0		12	14727	75%
BREAKS?–>	0	YES(1) NO(0)	13	14560	74%
Total DT=	28.3	23.61%	14	14392	74%
UptimeHrs	91.7		15	14224	73%
		NUM CMD			

Figure 11.6 RAMGT screen.

length of time required for each assist (mean time to assist). For example, in the specification example, a stoppage occurs every 30 min, requiring 2 min to resolve and return to production. This results in a lost production time of 4%. If, however, the rate of assists can be reduced to once every hour, the lost production time is reduced to 2%. If the length of time required to return the equipment to service can be reduced from 2 min to 1 min, then the lost production time is reduced once again to 1%.

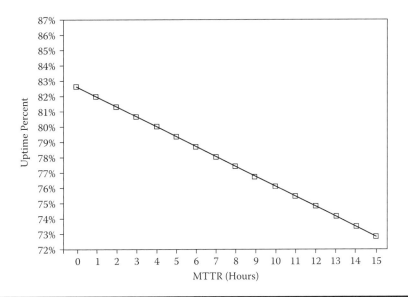

Figure 11.7   Scheduled downtime graph.

Evaluating Equipment Conversions

In the specification example, there were no conversions; therefore, the lost time due to conversions is zero. This number is compiled by the MTTC (mean time to convert) and the number of conversions per week (conv./week). If this number is greater than zero, look for an opportunity to improve the uptime %. The conversions should be evaluated to see if they could be changed in any of the following ways:

- Could they be done off-shift?
- Could they be done during lunch and breaks?
- Could they be simplified, thereby reducing time requirements?
- Could they be eliminated by having duplicate machines?

Calculating the 1st Run %

Another parameter to consider as an improvement opportunity is the 1st run %. As stated earlier, this is the inverse percentage of the scrap rate. If 1% of the product is scrapped, then the 1st run % is 99%. Any improvement in the number relates linearly to the capacity number.

For example, if 1st run % increases by 1%, so does capacity. This can also be viewed in graphical form to better understand the relationship each change will make, as illustrated in Figure 11.8. I call it a figure because it is a screen from the computer. You are right in that it can also be viewed as a table; it's your choice. Using the RAMGT software, this particular graph can be accessed under the "WHAT IF?" menu, as shown in Figure 11.9.

Sensitivity to 1st Run %

RANGE CHANGE PARAMETERS GRAPH PRINT QUIT

Change the Range for 1st Run % Capability

Capacity Analysis–Gear Hob					START RANGE:	100.0%
UTIL %	62%				INTERVAL:	0.5%
UPTIME %	76%					
CAPACITY	15063					
				1ST RUN %	CAPACITY	
JPH	212					
MTBF	10	HOURS		100.0%	15856	
MTTR	1	HOURS		99.5%	15777	
RESPONSE	0.25	HOURS		99.0%	15697	
MTTC	0	MINUTES		98.5%	15618	
SCHED DT	10	HOURS		98.0%	15539	
% STARVED	5%			97.5%	15459	
MTBA	0.5	HOURS		97.0%	15380	
MTTA	2	MINUTES		96.5%	15301	
1ST RUN %	95.000%			96.0%	15222	
HRS/WEEK	120			95.5%	15142	
CONV/WEEK	0			95.0%	15063	
BREAKS?–>	0	YES(1) NO(0)		94.5%	14984	
Total DT=	28.3	23.61%		94.0%	14904	
UptimeHrs	91.7			93.5%	14825	
		NUM CMD				

Figure 11.8 1st Run menu screen.

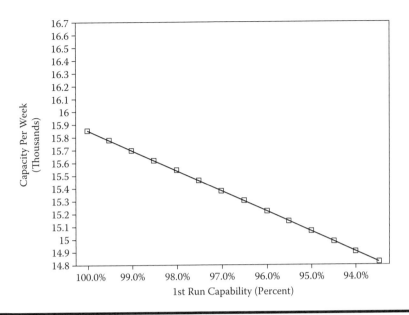

Figure 11.9 1st Run graph.

Summary

This chapter discussed in detail Phase 1 of the LCC model—specifically, the best way to develop an efficient specification using the R&M specification matrix, and to systematically review the time chart's greatest indicated leverage areas, thus revealing where the most efficient changes need to occur. The best leverage is obtained early when establishing the specifications. Chapter 12 discusses Phase 2.

Chapter 12

Requirements of Phase 2 of Implementing Equipment R&M: Development and Design

Chapter 11 discussed the requirements of Phase 1 and the specifications of generating the data for the reliability and maintainability (R&M) matrix. This chapter addresses Phase 2.

The design and development phase focuses on the equipment design and verification of the capabilities of the evolving design to meet the R&M requirements specified in Phase 1, the Concept Phase, described in Chapter 11. The output from the design and development phase of the R&M process should be as follows:

1. *Preliminary design reviews*: Validate the capability of the evolving design to meet all technical requirements.
2. *Critical design reviews*: Verify that the documented design and related analysis are complete and accurate.

Techniques for Designing Equipment Maintainability

There are several techniques available to improve a machine's maintainability during the design phase. However, their individual deployment may vary depending on the application. Therefore, the following sections describe techniques to consider when improving a machine's maintainability characteristics.

Minimize Maintenance Requirements

The ultimate goal is to *design out* the need for maintenance. For example, if the machine uses bearings that require lubrication every 500 operating hours, a sealed, prelubricating bearing providing 10,000 operating hours results in significantly improved maintainability and the equipment's overall availability.

Minimize Maintenance Handling Requirements

During a machine's design stage, many aspects of its configuration are determined. Therefore, review is necessary to ensure the maintenance handling requirements are reduced to the minimum.

For example, a machine may encounter a sensor failure requiring 15 min to replace the sensor. However, if this sensor is located behind a drive assembly taking 3 h to remove and replace, an additional 3 h of maintenance is encountered. Frequent removal of subassemblies to perform maintenance increases the likelihood of additional equipment failure and subcomponent damage.

Design for Interfacing Parts Reliability

As the machine's configuration is determined through design, the tolerances and alignments of interfacing parts are determined. Therefore, close care must be exercised to ensure that part alignments are correctly specified and that frequent replacement of high-wear components does not accelerate any mating parts' failure rates.

For example, a large grinder transmitting power using five drive belts may require more frequent drive belt replacement. Therefore, the machine may be designed with self-aligning pulleys and quick-release guards that are accessible to maintenance personnel. This allows quick, frequent belt replacements while not damaging other machine parts.

Fault Tolerance Design

Fault tolerance design provides a machine with backup "features" that allow continued operation if a component fails.

For example, a backup computer is placed in the system. This computer can be thrown into service within moments of a failure, allowing the system to remain in production, while the main computer is under repair. This computer can also be used during regularly scheduled maintenance on the primary system. This method is costly, but it will increase the overall reliability and availability of the equipment or machinery under investigation.

Maintenance Tools and Equipment

Quite often when machines are designed, special tools may be required to perform routine maintenance. This is usually the result of the configuration established during the design. Therefore, if your machine requires a special tool, be sure to obtain it when the machine is delivered—*not* when it fails, and one must then be obtained from the supplier.

Remove and Replace

As Equipment R&M deployment actively influences machine builders to modify machines complementing today's factories, greater emphasis is placed on defining how repairs can most efficiently be performed. One technique is to define in advance the level of repair maintenance people should perform. Then, review the repairs associated with various potential failures to determine the most efficient solution. This is further defined by identifying a machine's "lowest replaceable unit" (LRU), for example, if the machine has a programmable logic controller (PLC). One possible failure may be a blown output module. Once diagnosed, the output module could be removed, and a new output module can be reinstalled.

The failed module can be returned to the original equipment manufacturer (OEM) for repair or repaired in-house if the technical capabilities are present. Removal and replacement provides for quick and easy maintenance. This also means that a supply of modules must be kept in stock to ensure mean time to repair (MTTR) values are kept to the lowest levels. Increasing MTTR will reduce availability and cause reduced production.

Strive for Interchangeability and Standardization of Equipment

Because of the large number of machines purchased by a typical customer, a tremendous potential savings is associated with standardizing components when possible and striving to obtain interchangeable parts.

For example, if all electrical motors are provided by one manufacturer, the spare parts inventory is reduced because multiple, same-size motors are not necessary. If one machine is out of service, its electrical motors can be reused on other production equipment. Therefore, every effort should be made to consider standardization opportunities.

Keep in Mind Working Environment Considerations

When machines are installed in various locations by the customer, the plant's environmental aspects must be reviewed to determine their effects on reliability.

For example, if the machine is installed in a glass plant, its reliability must be evaluated when considering high temperatures, the presence of sand, and around-the-clock operation. If these conditions are ignored, there is a much greater probability of frequent machine failure.

Customer's Equipment Design Responsibilities

Phase 2 also places great emphasis on the cooperation between the customer and the equipment supplier. The responsibilities of each party must be fully understood for successful completion of this phase. The importance of this understanding may be shown with an illustration adopted from www.projectcartoon.com and modified by the author (see Figure 12.1).

Typically, customers purchase large numbers of machines—some small, some large, some very expensive, and some off-the-shelf, at a very reasonable pricing. However, the amount of volume or price is not as important as bearing in mind that the most reliable results are achieved when concerned individuals fulfill their specific responsibilities. When purchasing new

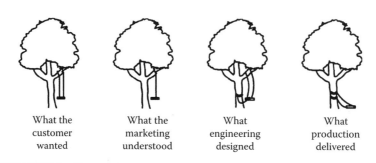

Figure 12.1 Customer and engineering requirements.

machines, the OEM personnel are the customers. Therefore, keep in mind the following factors:

- Give the supplier the proper goals. They should be reachable while allowing the customer to move toward world-class manufacturer status. This is best accomplished by establishing an efficient, concise specification. More specifically, focus on the following goals:
  - Reduced complexity of the system
  - Reusability of the parts or equipment
  - Carryover of the parts or equipment
- Reliability of the parts, machines, equipment, and system

In other words, make sure that the issue of the design for reliability is on the front of the discussion. That means that the discussion should consider at least the five following steps:

- Step 1. Design for maintainability.
  - Step 2. Perform functional analyses to determine failure modes as well as their consequences, severity, and methods of early detection.
  - Step 3. Analyze components with potential failures, and determine their failure models.
  - Step 4. Determine maintenance tasks, their frequency, and their effectiveness.
  - Step 5. Define and optimize the maintenance implementation plan.

- Assume the supplier is competent. If specific, in-depth analysis needs to be done on an aspect of the equipment, assume the supplier is capable and responsible for carrying it out.
- Establish preliminary design reviews to evaluate the capabilities of the evolving design.
- Establish critical design review checkpoints during the concept, design, and testing phases.
- Ensure that all possible failure modes are considered.
- When failures are not eliminated, push for easy maintainability. Preventive maintenance (PM) procedures should also be considered when designing for maintainability.
- Make sure that the root causes for existing problems are understood and corrected.
- Provide the factory user's perspective. This improves the design as well as your relationship with the rest of the factory.

- Clarify requirements. No matter how detailed the specification, always be prepared to clarify, sanitize, and refine the requirements.
- Establish the requirements for an equipment failure mode and effect analysis (EFMEA) and fault tree analyses (FTAs) on significant failure mode causes before you award a contract to a supplier.

Supplier Equipment Design Responsibilities

During this phase of the R&M development, the supplier has the following responsibilities:

- Select parts to meet overall R&M specifications.
- Complete tolerance stacking studies to ensure that electrical and mechanical tolerancing does not cause equipment failure.
- Conduct stress analysis on components that have shown high failure rates in the past.
- Develop a reliability qualification testing plan for the equipment in order to validate R&M specifications.
- Develop a data collection plan for the reliability testing phase of the equipment.
- Develop a plan that will enable the collection of reliability data at the customer's location.
- Develop and implement the EFMEA and the FTA.

Conducting Design Reviews

A design review is a formalized, documented, and systematic management process through which both the machinery supplier and the customer review all technical aspects of the evolving design, including R&M. This process typically involves the review of:

- Drawings
- Sketches
- Engineers' notebooks
- Analysis results
- Test documentation
- Mock-ups
- Assemblies
- Hardware

- Software
- Models/simulations

In order to be effective, the design review must be multiphased and keyed to the various phases of the design process. However, it must be recognized up front that *not* all design reviews are applicable to each machine. The supplier and the user should establish which reviews are required for a particular machine and when they are to be scheduled.

Designing Equipment Reliability

When conducting the equipment design phase, it is important to consider how the equipment's components and subsystem's performance support the machine's ultimate reliability goals. This is easily accomplished by:

1. Dividing the machine into subsystems at the LRU level. Usually, it is at this level that components are in series, and the failure of one component results in the failure of the machine.
2. Construct machine reliability models by apportioning mean time between failure (MTBF) and reliability requirements to individual subsystems. Also, remember to assess how quality and safety requirements are translated in each subsystem.
3. Prioritize according to judgment and experience. Begin with the focus on areas representing the largest risk, such as new technology or untested subsystems.

Constructing a machine's reliability model is accomplished through reliability modeling, in one of two ways: creating parallel models or creating series models.

Creating Parallel Reliability Models

Parallel reliability models are defined as a complex set of interrelated components connected together so that a redundant or stand-by system is available when a failure occurs. Parallel models represent machines with redundant backup systems.

For example, a machine with four welding arms performs a series of welds using only two welding arms, while the remaining two arms are backups. If one of the operating arms fails, one of the backup arms can begin welding in its place. The whole machine remains operational, resulting in greatly improved reliability. However, backup systems may substantially increase the purchase price of the machine; therefore, they are usually only deployed on crucial subsystems.

To calculate the MTBF or reliability values on a parallel system, use the following formulas:

$$\text{System } MTBF = MTBF + MTBF - \frac{1}{1/\text{MTBF} + 1/MTBF}$$

$$\text{System reliability} = (1 - R_1)(1 - R_2)$$

Example

A system has two subsystems in parallel with the following:

Subsystem 1 has *MTBF* = 80 and reliability of 0.90.
Subsystem 2 has *MTBF* = 80 and reliability of 0.90.

Substituting these values into the formulas, we get

System *MTBF* = 80 + 80 − [1/(1/80 + 1/80)] = 120 *MTBF*
System reliability = 1 − (1 − 0.90)(1 − 0.90) = 0.99

As seen in the preceding formulas, adding a parallel system improves an 80 h MTBF to a 120 h MTBF or a 90% reliable system into a 99% reliable system.

Special note: It is always important to remember that both MTBF and reliability represent a significant opportunity to improve the reliability of critical subsystems. Adding a backup (redundant) system increases the reliability exponentially!

Creating Series Reliability Models

Series reliability is defined as a complex system of independent units connected together or interrelated in such a way that the entire system will fail if any one of the units fails. When developing the modeling method, the developers should be aware of the type of reliability models that are being developed.

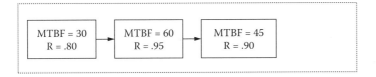

Figure 12.2 A series parallel system.

A static series model can be developed, which means that the failure of one component will not cause the premature failure of other components in the system. Series systems do not offer the reliability of backup systems. However, they are more economical to deploy.

In Figure 12.2, the model represents three series subsystems; we can calculate both the series model reliability and the MTBF as follows:

$$\text{System MTBF} = 1/[1/30 + 1/60 + 1/45] = 14 \text{ MTBF}$$

$$\text{System reliability} = (0.80)(0.95)(0.90) = 0.684$$

Here, it is important to remember when establishing machine reliability goals that each successive level of subsystems must have an exponentially larger MTBF.

A dynamic model means that the failure of one component will cause the failure of other components in the system.

Allocation of Reliability Goals

A reliability growth plot is an effective method that can be used to track R&M continuous improvement activities. In addition, it can be used to predict reliability growth of machinery from one machine to the next. Reliability growth plotting is generally implemented during the last part of the failure investigation process associated with failure reporting, analysis, and corrective action system (FRACAS). The information developed for reliability growth should be sent back to the supplier and to internal engineering personnel in order to take corrective actions on new equipment designs.

Recall that Figure 9.3 (in Chapter 9) illustrates the effect of reliability numbers for system, subsystem, and component levels for an engine after 10 years of operation in the field. Note that the engine has an overall reliability of 70%, but at the component level, the reliability of support must be 99.999% after 10 years.

Apportioning the Reliability Model

The RAMGT software tool allows for the development of a reliability model using the apportioning method. This method will equally divide the MTBF between all of the subsystems of the equipment under investigation.

Identifying the Subsystem Tree Model

The following breakdown of equipment can be used to determine MTBF at the subsystem level. In this example, the system MTBF has been identified as 90 h. The next step is to identify the subsystem level of MTBF.

```
SYSTEM
(MTBF = 90 HOURS )
Sub-Systems
LOADING (MTBF = ?)
   POSITIONING/LOCATING (MTBF = ?)
      PROCESSING (MTBF = ?)
UNLOADING (MTBF = ?)
```

Conducting Equipment Failure Mode and Effect Analysis (EFMEA)

An EFMEA is a systematic approach that applies a tabular method to aid the thought process used by engineers to identify equipment's potential failure modes and their effects. Of course, in today's environment, as machine builders design new equipment, a variety of techniques are available. Although many are useful, it has been found that EFMEAs, when used in conjunction with FTAs on significant failure modes, are the most powerful combination to improve a machine's reliability through design. Generally, the equipment supplier is responsible for conducting the initial EFMEA, with the customer's personnel assisting as team members, as and if needed. The need for an EFMEA is primarily to

- Identity potential failure modes that may adversely affect safety.
- Identify potential design deficiencies before releasing machinery to production.

For example, an EFMEA identifies all of the machine's potential failure modes and their first-level causes. In other words, the FMEA identifies the breadth of problems involving "all potential failures" of new designs. On the other hand, the FTA is deployed to analyze the root cause of significant failures, establish the probabilities of each cause, as well as illustrate how each of the causes is related. That is, the FTA goes to the depth on one individual failure: "what are all possible causes for one failure?"

The purposes of an EFMEA are to

- Identify potential failure modes and rate the severity of their effects.
- Rank-order potential design and process deficiencies.
- Help engineers focus on eliminating equipment design and process concerns, and help prevent problems from occurring.

Key benefits of EFMEAs are to

- Improve the quality, reliability, and safety of the customer's equipment.
- Improve the customer's image and competitiveness.
- Help increase customer satisfaction.
- Reduce equipment development, timing, and cost.
- Document and track actions taken to reduce risk.
- Identify and improve the operator's safety risk.

Who Prepares an EFMEA?

The team approach to preparing EFMEAs is strongly recommended, with the equipment engineer leading the team.

The supplier's equipment design engineer is expected to involve representatives from all affected activities. Team members should include the following personnel/departments:

- Purchasing
- Skilled trades
- Reliability engineers
- Operators
- Testing
- Quality engineers
- Industrial engineers
- Supervisors

In addition, the most responsible customer engineer involved with the equipment should also participate. Team members may change as the equipment matures through the design, build, and test phase. Again, the EFMEA process is intended to be conducted as a team, with both the suppliers and customers.

Who Updates an EFMEA?

The supplier's design engineer is responsible for keeping the EFMEA up to date. Suppliers keep their own copy of the EFMEA up to date.

When Is an EFMEA Started?

An EFMEA should be started at any of the following times:

- When new systems, equipment, and processes are being designed.
- When existing equipment or processes are changed.
- When carryover equipment or processes are used in new application or new environments.
- After completing a problem-solving diagnostic model study (to prevent recurrence of a problem).
- When the equipment concept is formulated.

When Is an EFMEA Updated?

An EFMEA should be updated as follows:

- Whenever a change is being considered to a machine's design, application, environment, material construction, or operational process.
- Whenever there are changes to a new machine's project timeline.
- Whenever a new machine's design is modified.
- Whenever a discovery is made regarding a change in the failure mode of a new machine.

When Is an EFMEA Completed?

An EFMEA is a living document and must be updated whenever significant changes occur in the equipment's design or application. An EFMEA is considered complete when the equipment is installed, has passed its reliability test, and the plant staff have signed off.

When Can an EFMEA Be Discarded?

The record retention requirements for FMEAs are generally developed by the customer and are communicated to the supplier. If the design is proprietary to the supplier, then the EFMEA is developed by the supplier. Generally, it is held until one year after the design is discontinued. The retention period should be part of the organization's document control and should be found in their procedures. All FMEAs are controlled documents and, as such, should have traceability numbers for reference.

Using an EFMEA Form to Analyze Equipment Failures

The EFMEA form is the main body of the analysis: each heading is numbered, representing a corresponding location on the illustration in Figure 12.3.

The first eight lines of the EFMEA form comprise the header; lines 9–21 comprise the body of the form. Each line and its intended functions are described in the following sections.

Line (1) FMEA Number

The FMEA Number is the space provided for the FMEA number. Each FMEA should have a unique number to aid in tracking the document and its information.

Line (2) Equipment Name

The Equipment Name space illustrates the equipment's name. For example: "ABC Boring Machine."

Line (3) Design Responsibility

Design Responsibility identifies the supplier as well as the department and group that is designing the machine. For example: "ABC Tool Company."

Line (4) Prepared By

Prepared By should contain the name, telephone number, and company of the engineer responsible for preparing the EFMEA. For example: "Stacey Wang, (313) 734–4567, ABC Machine Tool Company."

328 ■ *The OEE Primer*

FMEA Number _____ (1)_____				Prepared By _____ (4)_____					FMEA Date _____ (7)_____			Page _____ of _____			
Equipment Name _____ (2)_____				Model _____ (5)_____					Core Team _____ (8)_____						
Design Responsibility _____ (3)_____				Review Date _____ (6)_____											

Sub-System Name 9a / 9b / Function & Performance Requirements	10 Potential Failure Mode	11 Potential Effect(s) of Failure	12 Severity	13 Class	14 Potential Causes of Failure	15 Occurrence	16 Current Design Controls and Equipment Controls	17 Detection	RPN	19 Recommended Action(s)	20 Responsibility & Target Completion Date	21 Actions Taken	Action Results				
													Severity	Occurrence	Detection	RPN 22

Figure 12.3 A typical EFMEA form.

Line (5) Model Line

The Model Line contains the equipment's model number if applicable. For example: AG4685Y92.

Line (6) Review Date

Review Date is the initial date the EFMEA is started. For example: 2008-09-01 (year-month-day). The date should fall within the design and development phase of the equipment's life-cycle process.

Line (7) EFMEA Date

EFMEA Date is the initial date the EFMEA is completed and the latest revision date. For example: 2008-09-01 (year-month-day).

Line (8) Core Team

Core Team lists the individuals and departments who participated in the EFMEA. For example: Core Team.

- T. Stamatis–Customer–Manufacturing engineer
- J. Roberson–Customer–Manufacturing engineer
- S. Wang–Supplier–Design engineer
- S. Stamatis–Supplier–Service engineer

In addition, it is recommended that all team members' names, departments, telephone numbers, and addresses be included on a distribution list.

Line (9A) Subsystem Name

With this item, the main body of the EFMEA begins. Subsystem Name is the information used to classify the analyzed machine's subsystems. For example, as a machine is analyzed, it becomes clear that it consists of several subsystems that are better analyzed individually. Therefore, a hierarchy of the machine is quickly formulated and then transferred to the column listing all of the subsystems in the appropriate order. A typical hierarchy may be in terms of:

- System level: Generic Machine
- Subsystem level: electrical; mechanical; controls

- Module/assembly level: frame mounting; drives; material handling; fixture tools
- Component level

It is also important to note that the "Subsystem Name" column and the "Function and Performance Requirements" column have the same location. To separate these two types of information, it is helpful to consider them as two separate columns. For example, list all of the "Subsystem Information" aligned against the left column margin and all of the "Function and Performance Requirements" information against the right column margin. This technique prevents confusing the two different types of information.

Line (9B) Subsystem Function Performance Requirements

Subsystem Function relates directly to the "Subsystem Name" information and lists all of the subsystem's associated functions and the design goal of that system. This column's information corresponds to each of the machine's identified subsystems. In addition, using the following four recommended steps to fill out this column simplifies subsystem function identification.

- *Step 1: Brainstorm*: Brainstorming is the process of considering every possible function associated with the subsystem. It is accomplished by the team verbalizing the possibilities, while a recorder places them on a flip chart. During this process, the machine's wants, needs, and requirements are considered. Environmental, production, and tolerance requirements must also be considered.
- *Step 2: Evaluate*: Once the brainstorming is complete, the list must be evaluated. It is at this time that each potential function's validity is reviewed. Any functions determined to be inappropriate are removed at this time.
- *Step 3: Convert to Verb–Noun Format:* After the subsystem's functions are established, each one is reviewed to ensure that it is expressed in a verb–noun format. For example, if the subsystem is a drive and its function is to advance a spindle, then the verb–noun expression is "Advance Spindle." Expressing functions in a verb–noun format ensures that the component (noun), as well as its action (verb), is always identified.
- *Step 4: Establish measurement*: When the functions are identified and expressed in a verb–noun format, the final step is establishing its measurement. For example, if a spindle advances, it is important to establish

how far it advances. If the requirement is 2.3 in., then 2.2 in. is not fulfilling the functional requirement. It is also important to establish ± tolerances. For example, 2.3 in. ± 0.005 in. Ultimately, this information is used to establish when the function is met or a failure is encountered.

Line (10) Failure Modes

Potential failure mode is defined as the manner in which machinery or equipment could potentially fail to meet the design goal. Potential failure modes are identified as those that are observed by the operator of the equipment. The failure modes are typically identified as the inverse of the function of the equipment, subsystem, or component under investigation. There are two types of approaches:

- Functional—Relates to loss of function
- Hardware—When detailed part designs are available

Failures for this example are typically used with static models. This means that the failure will have no effect on the failure of other subsystems or components under investigation. The R&M team should investigate all possible failure modes during the design stage. This will reduce the number of failures that may be seen during start-up, the debug phase, and the useful life of the equipment.

Line (11) Potential Effects

Potential Effects of the failure represent the potential consequences of the failure mode. For example, if the failure mode is the drill head advancing greater than 2.3 in., the effect may be to break the drill head's tooling. Therefore, this is listed on the EFMEA form. To assist in defining all of the possible effects, each of the following is considered, and then entered onto the form:

- Breakdowns
- Reduced cycle time
- Tooling
- Setup and adjustments
- Start-up losses
- Idling and minor stoppages

- Defective parts
- Safety
- Government regulation

Line (12) Severity

Severity represents the seriousness of the effects listed in column 11 and consists of three components:

- Safety (primarily operator safety)
- Equipment downtime
- Product scrap

Each effect is assigned a ranking between 1 and 10, as illustrated in the "Severity Evaluation Criteria." After the evaluation is complete, the rankings are reviewed, and the highest ranking is entered in the Severity column. Only the highest ranking is entered because it represents the most serious effect that can occur, if the failure mode occurs.

Line (13) Class

The "Class" column is not currently used on the EFMEA form. Typically, class ratings will be related to safety issues and those failure modes that have a severity rating of 9 or 10. If this appears on the EFMEA, then action needs to be taken to correct this problem. The 9 or 10 is generally reserved for Operator Safety. The 9 is something that happens with warning, and a 10 is something that happens without warning. The symbol for Operator Safety is OS.

It must be emphasized that this is treated as a critical item, but it is not critical in the traditional sense. The classical interpretation of a critical item is something that affects the product characteristic for the customer. In contrast, the OS is critical *only* for the safety of the operator and has nothing to do with the customer.

Line (14) Potential Causes

The "Potential Causes" column of the EFMEA form represents the design deficiency or process variation that results in the failure mode. It is recommended that all of the first-level failure mode causes be identified and listed in this column. This is easily accomplished by brainstorming each failure mode by considering the following:

- What could cause the subsystem to fail in this manner?
- What circumstances could cause the subsystem to fail to perform its function?
- What can cause the subsystem to fail to deliver its intended function?

It is helpful to first identify the causes associated with the highest-risk failure modes. Then review the following:

- Historical test reports
- Warranty data reports
- Concern reports
- Product recalls
- Field reports
- Surrogate EFMEAs

For failure modes whose effects have a severity ranking of 9 or 10, identify the root causes of the failure mode. Root causes are the underlying reasons a first-level cause occurs that can be actionable—in other words, something can be done about them in the foreseeable future with the resources available. To help identify root causes, the following common techniques may be considered:

- Problem-solving diagnostics
- Cause-and-effect diagrams
- Fault tree analysis (FTA)

Line (15) Occurrence

The "Occurrence" column contains the rating corresponding to the likelihood a particular failure mode will occur within a specific time period. However, because the "Current Design Controls and Equipment Controls" in use directly affect the occurrence of a cause, it is sometimes helpful to first complete column (16). It is also helpful to consider other sources of cause frequency information:

- Service history
- Warranty data
- Maintenance experience with similar parts

After an occurrence rating for each cause is established, it is entered in the Occurrence column. Unlike the Severity column, where only the highest effect's severity rating is entered, all of the occurrence ratings are entered. If there are four causes, then there are four occurrence ratings.

Line (16) Current Design Controls

The "Current Design Controls and Equipment Controls" column represents a listing of all the design and equipment controls. These controls are used to detect or prevent the failure mode or its first-level cause. Design controls are deployed during the design stage to establish activities ensuring a reliable design, while equipment controls are used during the process (manufacturing) phase to establish techniques ensuring reliable performance. To further distinguish design controls from equipment controls, it is helpful to include them in a table as follows:

- Design Controls
  - Design Review (D, which stands for detection controls)
  - Simulation Studies (D)
  - RQT (reliability qualification test) Testing (D)
  - Safety Margins (D)

- Equipment Controls
  - Limit Switch (P, which stands for prevention controls)
  - Laser Sensor (P)
  - Preventive Maintenance (P)
- Oil Light (P)

Line (17) Detection

"Detection" is the column containing the ranking of the current design or equipment control. This ranking indicates the likelihood that the identified controls can detect or prevent the failure mode. This ranking is identified after completing the "Current Design Controls and Equipment Controls" column. Each identified control is evaluated using the "Detection Evaluation Criteria." This criteria represents the control's likelihood of detecting or preventing the failure mode.

Line (18) Risk Priority Number

The "Risk Priority Number (RPN)" column contains the RPN(s) for an identified failure mode. It is the product of Severity (S) × Occurrence (O) × Detection (D):

$$RPN = (S) \times (O) \times (D)$$

It is important to note that because the occurrence rating for each cause is entered into column 15, there is an RPN resulting from each of the failure mode's causes. The calculated RPNs only represent the ranking of a machine's potential failure modes. Therefore, they establish the priority for equipment design improvements or operational changes.

Special note: Whereas the RPN is the traditional way of determining the priority for action, the better way is to react (take action) against a severity or operator safety rating of 9 or 10 first, then against the criticality S × O, and then detection.

In addition, when evaluating RPNs, it is important to note their constitution. For example, an RPN of 27 may appear to be low. However, if its Severity is 9, Occurrence is 3, and Detection is 1, it may present a higher risk than another failure mode with an RPN of 27, resulting from a Severity of 3, Occurrence of 3, and Detection of 3. Therefore, it is recommended that an RPN's Severity rating be considered first, and its Severity and Occurrence rating combination (generally known as criticality) be considered second. Finally, the Detection rating should be considered. A typical FMEA layout and RPN calculation are shown in Figure 12.4.

Line (19) Recommended Actions

The "Recommended Actions" column contains the actions recommended to reduce the corresponding failure mode's Severity, Occurrence, and Detection, in that sequence. In general, it is recommended that failure-mode recommended actions be formulated, based on the following priorities:

1. Failure modes with a severity rating of 9 or 10 before the engineering release, thus eliminating any safety concerns.
2. Failure modes with a high Severity/Occurrence rating combination, determined by team consensus.
3. Failure modes with high RPN resulting from high Severity/Occurrence/Detection ratings, determined by team consensus.

Function (F)	Failure Mode (FM)	Effects (E)	Severity (S)	Class	Cause (C)	Occurrence (O)	Design Control (D)	Detection	S x O x D	RPN
F_1	FM_1	$E_1=8$ $E_2=6$ $E_3=3$ $E_4=7$	8	YS	$C_1=10$	10	$D_1=7$ = $D_2=3$	3	8 x 10 x 3	240
				—	$C_2=2$	2	$D_1=8$ $D_2=5$ $P_1=2$	5	8 x 2 x 5	80
				YS	$C_3=8$	8	$D_1=6$ $D_2=4$	4	8 x 8 x 4	256

Figure 12.4 FMEA layout and RPN calculation.

FM$_2$	$E_1=2$ $E_2=10$ $E_3=5$ $E_4=8$	10	YC	$C_1=4$	4	$D_1=7$ $D_2=5$	5	10 × 4 × 5	200
			YC	$C_2=3$	3	$D_1=8$ $D_2=6$ $D_3=3$	3	10 × 3 × 3	90
			YC	$C_3=2$	2	$D_1=9$ $D_2=4$	4	10 × 2 × 4	80
FM$_3$ FM$_4$									
F$_2$	FM$_1$ · · FM$_n$								

Note: Severity is of the effect; Cause is of the failure—Never of the effect; Every cause has a class, O, D, and RPN; Priority is based on Severity, criticality (S × O) and, finally, Detection; RPN = Highest severity × all causes × lowest detection, Prevention methods do not count. Only the effect of the control counts, which of course is the question of "How well is the control effective in catching the root cause?"

Figure 12.4 (Continued) FMEA layout and RPN calculation.

The intent of any recommended action is to reduce the Severity, Occurrence, and Detection rankings, in that order. This is because the most efficient way to address a failure is to design it out of the equipment by reducing Severity or Occurrence. Engineers should strive to design equipment that is robust and does not experience failure modes. This is better than depending on process procedures and operator skill (Detection) to prevent failure modes.

For example, reducing or eliminating Severity ratings of 9 or 10 will potentially remove any safety or governmental regulation violation concerns and may eliminate the failure mode entirely.

Reducing occurrence to 1 greatly improves reliability by reducing the failure occurrences. However, establishing elaborate controls or complex operating procedures does little to reduce the failure mode's Severity. Although such practices may improve reliability, they also tend to increase a machine's life cycle. Therefore, they should only be considered after all design improvement efforts are exhausted. To further emphasize this priority, it may be helpful to designate each recommended action with a "D" if it is a design action, or a "P" if it is a process-related action.

When all of the recommended actions are identified, they may be numerous and complex. Therefore, it may be helpful to number them. If, however, there are no recommended actions for a failure mode, then a "No Recommended Actions at this Time" statement should be entered in this column. This confirms that the failure mode was analyzed and not incompletely addressed.

Line (20) Area/Individual Responsible and Completion Date

The "Area or Individual Responsible and Completion Date" column documents who is assigned the responsibility to review the recommended actions and, if appropriate, implements the necessary changes (e.g., C. Robinson, 2008-09-8 [year-month-day]).

Line (21) Actions Taken

The "Actions Taken" column briefly documents the actions taken to reduce a failure mode's high risk and their completion date. An EFMEA is of limited value without positive actions to eliminate potential injury, safety issues, government regulation violation, and machine downtime or prevent part defects. Therefore, a commitment to implement high failure risk solutions is essential. Because the primary design responsibility belongs to the supplier, all FMEA updating remains the supplier's responsibility even after installation at a customer's facility or after Job #1. It is important to remind the reader here that any action taken in this column must be from the pool of choices identified in the recommended action column (#19).

Line (22) Revised RPN

The Revised RPN number is calculated from the new Severity, Occurrence, and Detection values resulting from the implemented "Actions Taken" column. These new values are established by the EFMEA team. If there were no recommended actions, then the columns are left blank.

The EFMEA team engineer then reviews the revised RPNs to determine if further design actions are necessary. If they are, then steps 19 through 22 are repeated on a new EFMEA form, and a new revision number is assigned.

Creating Fault Tree Analysis (FTA) Diagrams

Fault tree analysis (FTA) is an effect-and-cause diagram. It is a method used to identify the root causes of a failure mode with symbols developed in the defense industry. Currently, it is used heavily in commercial manufacturing. It is an approach from top to bottom, whereas the EFMEA is a bottom to top approach.

The purpose of FTA is to structure a root cause analysis of known failure modes that are not yet fully understood. If it is used in conjunction with EFMEAs, these failure modes are identified in the "Failure mode column."

The benefits of an FTA are that it establishes a great prescriptive method for determining the root causes associated with failures. It also aids in establishing a troubleshooting guide for maintenance procedures.

Special note: As an alternative to FTA, some engineers prefer to use the Ishikawa (Fishbone) diagram. This diagram complements the EFMEAs by representing the relationship of each root cause to other failure-mode root causes. However, many feel that the FTA is better suited to understanding the layers and relationships of causes. After using both of these methods, if you prefer one over the other, there is no reason not to use either one of them. The goal always remains the same: use the most efficient method for the problem at hand.

To help accomplish root cause analysis, the FTAs can be simplified, as in the five basic symbols illustrated and described in the following sections.

The Failure/Causal Event Symbol

TOP-LEVEL
FAILURE/CAUSAL
EVENT

The Failure/Causal event symbol, depicted by a rectangle, represents a top-level event or a subsequent-level event below an AND or OR gate requiring further definition.

The OR Gate Symbol

The OR gate is depicted by a circular polygon with a concave base. When used, it signifies that if any of the subsequent events below it occur, then the next-highest event above it also occurs.

The AND Gate Symbol

The AND gate is similar to the OR gate: it consists of a circular polygon with a straight horizontal base. When used, it signifies that all of the subsequent events below it must occur before the next-highest event above it can occur.

The Diamond Symbol

The Diamond represents an undeveloped event. This usually results from a root cause identification that is not within the scope of the FTA team to address.

The Circle Symbol

The Circle represents an identified root cause within the scope of the FTA team to address. It can also denote basic events that are independent of each other.

While using the standard symbols, it is important that a wide variety of people understand the results. Therefore, it should be remembered that the main goal is to understand the root causes and their relationships to each other.

Example: Real-World Rear-End Collision

Figure 12.5 offers an example in which the failure mode is the "rear-end collision." Once we identify the failure, each of the FTAs represents the events that immediately precede the next upper-level event. Probabilities of occurrence as a ranking method can be added as reflected on the EFMEA form or actual factory experience.

The second level, via OR gates in this example, represents the next level of causes. For example, cause 2 is the result of either:

- The operator response was inadequate (80%).
- The brake was inadequate (20%).

Note that at this level, cause 1 "Operator Response Inadequate," is in a Diamond, representing that it is beyond the scope of this FTA team to address. The other cause, "Brake Inadequate," has a subsequent OR gate and two root causes:

- Low adhesion (85%)
- Brake malfunction (15%)

At this level, it was the FTA team's conclusion that "low adhesion" is not something the team can address. It is beyond the team's responsibility to analyze the reasons for slow reflexes; therefore, it is placed in a Diamond. However, "Brake Malfunction" is within the scope of the team's responsibility; therefore, it is placed in a circle.

By reviewing police reports on these types of accidents and calculating event occurrence probabilities, you can calculate the number of collisions per year that could be avoided. This then establishes direction, focus, and justification for spending time and money!

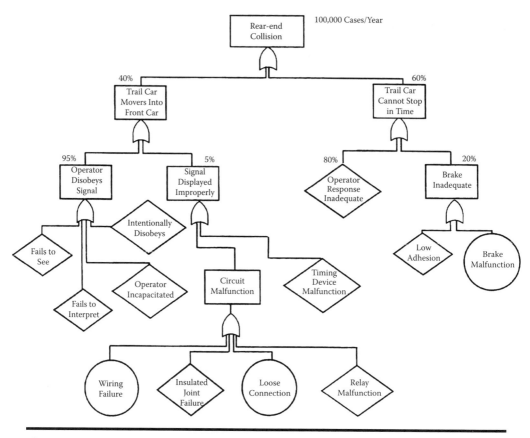

Figure 12.5 FTA example: rear-end collision.

Hints on Performing Fault Tree Analysis

1. Make sure the starting point (top-level event) is the basic fault, problem, or symptom that is seen by the operator running the machine.
2. Avoid jumping to solutions! This can result in overlooking possible causes. Always consider that there may be other root causes needing identification before the upper-level causes can be understood.
3. Know when to stop one of the branches: this can be tricky. Only go down to the level that is understood by everyone on the team. For instance, if a failure is identified down to a board level, it may be decided to stop and delegate further analysis to the board designers if it is critical enough.
4. Make sure all bottom levels end in circles or Diamonds.
5. When completed, number each circle, and identify a corresponding action item. This results in the compilation of an FTA "Action Item"

list. These action items should then be added to the EFMEA under the "Recommended Actions" column.
6. As mentioned, remember that emphasis should be on eliminating the failure by improving equipment design. This is always the most efficient way to correct a deficiency. However, after all design opportunities have been exhausted, consider process actions such as preventive maintenance or statistical process control (SPC).
7. If you use FTAs in conjunction with EFMEAs, identify all first-level causes. However, if you discover new first-level causes, update the EFMEA.
8. When conducting FTAs in conjunction with EFMEAs, remember that not every EFMEA first-level cause requires FTA. Therefore, it is helpful to consider the following:
    - Evaluate all of a machine's failure-mode first-level causes when the failure mode's severity rating is 9 or 10.
    - Evaluate a machine's 10 most expensive failure-mode first-level causes.
    - Evaluate a machine's 10 most common failure-mode first-level causes.

Summary

This chapter provided a detailed discussion describing the requirements of implementing Phase 2: specifically, the issues of development and design, such as reliability approaches, FMEA, EFMEA, and FTA. Chapter 13 discusses Phase 3.

Chapter 13

Requirements of Phase 3 of Implementing Equipment R&M: Build and Install

Chapter 12 discussed the requirements of implementing Phase 2, with some specific methodologies, such as reliability, failure mode and effect analysis (FMEA), equipment failure mode and effect analysis (EFMEA), and fault tree analysis (FTA). This chapter discusses a detailed approach to Phase 3 of life-cycle costing (LCC).

During the build and installation phase of the equipment reliability and maintainability (R&M) life cycle, it becomes very important for the customer and the supplier to ensure that the equipment has met the R&M performance goals defined during Phases 1 and 2, the concept and design phases (covered in Chapters 11 and 12). The build and installation phase is the last chance to detect failures before the equipment is placed into the production environment.

Most equipment installed is not tested to meet mean time between failure (MTBF) goals. Equipment currently is tested using a 24 h dry-cycle test, a capability test, or a part runoff. These testing methods do not effectively test whether or not the MTBF goals of the equipment have been achieved. Different testing strategies must be developed in order to measure the overall MTBF goals of the equipment.

So, here we will introduce a simple method of developing a testing strategy for equipment that will evaluate if the equipment will meet the MTBF goals as defined by the customer. The supplier is responsible for most of this work, so most of the information in this chapter pertains to the *supplier's* responsibilities; however, the customer's responsibilities are also provided at the very end of the chapter.

Selecting Machinery Parts

Parts for the equipment should be selected to optimize overall equipment reliability. Characteristics of reliability-sensitive or critical components should be defined by the manufacturer to ensure reliability at the lowest-tier supplier. The R&M plan should identify how machinery part suppliers are selected.

Completing Tolerance Studies

Tolerance studies should be completed by the supplier to ensure that electrical and mechanical tolerance stacking does not cause premature failures or wear under certain operating conditions.

Performing Stress Analysis

Stress analysis should be implemented by the supplier on areas that have been shown in the past to have failure rates above normal. It needs to be conducted on the relationship between failures and the environmental conditions that the equipment will be subjected to in the plant environment.

Conducting Reliability Qualification Testing (RQT)

Suppliers may be required to conduct tests to prove the performance of the equipment as it is related to R&M specifications. If the RQT is a method to verify performance, then it must be identified early in the life cycle of the equipment, and all parties must agree to the testing procedures necessary to obtain the R&M data.

The RQT method of reliability testing will measure if the equipment meets the overall MTBF values established in the performance specifications. During this test, when a failure is observed, the failure must be returned to the original operating condition. All failures must be addressed and recorded during this testing process.

Collecting Reliability Data at the Supplier's Facility

As an alternative to dedicated acceptance testing of the equipment on the supplier's floor, a data collection system can be developed to collect failure data relating to the equipment. This data collection system can be used to indicate reliability capabilities of the equipment over the period of measurement. Failure data and process validation tests can be used as a benchmark for the equipment. This can be used for continuous improvement on the equipment.

Collecting Reliability Data at the Customer's Plant

Reliability data should also be collected during the acceptance testing at the customer's plant to verify that reliability parameters have not been degraded during shipping and installation of the equipment.

Performing Root Cause/Failure Analysis of Equipment

Suppliers should be responsible for ensuring that a root cause analysis of equipment failures will be performed by themselves or by their associated component suppliers. The results of this analysis are then fed back to the customer, so that an action plan can be put in place to resolve the failures. How the failures are resolved should be documented as part of the R&M plan.

Eliminating Testing Roadblocks

There are many different types of roadblocks that are raised to get around implementing a reliability testing strategy—for example:

- "It costs too much to test."
- "It takes too long—testing typically is done prior to shipping the equipment."

- "There are no parts available for testing."
- "Our equipment will never pass the test."
- "We cannot test in real-world situations."

These roadblocks need to be systematically eliminated by developing a testing strategy, described in the following sections.

Do Not Test Everything

Use the design reviews to ensure that a high level of confidence has been incorporated into the design. Ensure that all failures have been addressed during the EFMEA phase of the project. Ensure that an FTA has been developed, with detailed action plans.

Reduce the Test Time

The testing strategy should include methods of accelerating the testing process, automating it, and reviewing test data from suppliers.

Establish Confidence through Design

Testing should not be the answer to increasing the reliability of equipment. The design should be error free, thus eliminating failures.

Perform Simultaneous Engineering

This will increase the probability of passing the reliability tests in a short period of time.

Conduct Reliability Testing

There are many different types of tests that can be conducted to determine equipment reliability. These tests could be

- Life testing
- Burn-in
- Marathon runs
- Accelerated testing
- Test to failure
- Test to a standard (boggy—i.e., a minimum standard of a test performance)

Common Testing Programs

The reliability testing phase of the R&M program is the last chance for the customer and supplier to ensure that the equipment meets all of the R&M specifications. The common testing programs are described in the following sections.

Conduct a 24-Hour Dry Cycle Run

The 24-Hour Dry Cycle Run is an important testing program for all equipment. The "infant mortality" phase can reveal a high amount of failures related to workmanship and, by using this test, the supplier can catch these failures at the supplier's facility. All parameters associated with the test should be identified before the test is conducted. This will ensure that all parties understand the deliverables for this test. A checklist should be provided to the supplier to ensure that all characteristics, as listed in the following text, will be addressed. The test should include the following operations:

- 24 h of continuous operation without failure or intervention
- Verification of all functional specifications
- Verification of proper operation of safety devices
- Verification of all control operation and sequencing
- Verification of pressures, temperatures, current-draw power factors, etc.
- Verification of cycle times

Positional repeatability should also be verified during the dry run, and the test should be conducted with a simulated load. The repeatability test is especially important for the certification of robots and other positional machines.

Conduct a Vibration Measurements Test

At the tryout of the equipment, vibration measurements should be conducted. This test should be mandatory because the parts quality can be affected by the machine's vibration. Vibration measurements should be developed to focus on

- Motors
- Bearings
- Gears
- Rotating shafts

When a preventive maintenance program is established for the equipment, a baseline of vibration signatures should be developed as a reference for future measurements of the equipment.

Perform a Dimensional Prove-Out Test

The dimensional prove-out test should be conducted to verify the dimensions and tolerances of the machine. These dimensions could have an impact on the performance of the machine.

This prove-out should be conducted at the supplier facility and with the customer engineer present. This test should be conducted with the machine assembled. Positional repeatability can also be conducted during this test in order to ensure that positional devices are able to return to the original programmed positions.

Conduct a Preliminary Process Capability Study

The preliminary process capability study is used to determine if the machine can contribute to a capable process. This test should be conducted at the supplier's facility and requires that tryout parts be made available.

This study should be conducted based on a minimum of 60 parts being available. This study can also be run with smaller lot sizes of 10–30 parts. These short runs will enable the supplier and the customer engineer to predict if instability or incapability of the equipment can be projected. If the equipment is determined to be unstable for the process operation, then it should be repaired before the study is conducted again.

When certified production gauges or measuring devices are available, 20–30 parts from the initial short runs may be used to verify the repeatability and reproducibility (R&R) of the machine or equipment. For a valid capability study, the R&R value should not exceed 10% for the tolerance being measured. When gauges are not available, a coordinate measuring machine (CMM) can be used. (The actual number of parts may be different for individual customers. The supplier must be aware of the customer's requirements and follow them.)

Test for Overall Equipment Effectiveness (OEE)

All new and refurbished equipment and machines that are ordered by the customer should have—at least—an 85% OEE value for acceptance or one that is established by the organization. Check to ensure that a value has been established.

Perform Reliability Qualification Testing

One of the most acceptable tests that can be performed to determine reliability claims is the RQT: this is the only true reliability test. The RQT is a cost-effective method to determine reliability claims. The RQT can be used to

- Evaluate prototypes prior to assembling the failure into the equipment.
- Establish test plans for subsystems.
- Develop test plans for integrated systems.
- Determine acceptance requirements.

RQT provides a measurable statistical confidence factor that the MTBF is equal to or greater than a specified MTBF goal. To determine the MTBF of existing equipment, data must be collected and recorded about the uptime and failures associated for that piece of equipment for a designated amount of time.

The RQT method in the following example is identified as a one-sided confidence limit calculation. The calculation will identify with an 80% confidence factor where the lower limit MTBF value is.

Establishing Existing Equipment's MTBF Value

The RQT method will give a measurable confidence as to whether or not the MTBF is at least at a certain point. The value of the MTBF can be calculated using the following equation:

$$\text{MTBF} = \frac{2(T)}{\chi^2 \alpha, 2r + 2}$$

where
  T = Total test time for observation of failures
  χ^2 = Chi-squared value
  α = Risk level for the test. This value is equal to 1 − Confidence value in the test
  $2r + 2$ = Degree of freedom as located on the chi-squared table (see partial chi squared—Table 13.1)
  r = Number of failures observed during the test period

Table 13.1 Partial Chi-Squared Table

α	Chi-Squared Table		
D. F	.10	.20	.30
2	4.61	3.22	2.4
4	7.78	6	4.88
6	10.64	8.58	7.26
8	13.36	11.04	9.52
10	15.99	13.4	11.8
...	...	...	...

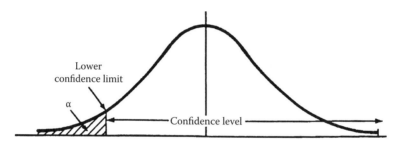

Figure 13.1 One-sided confidence value.

This equation is identified as the one-sided confidence equation. A graphical format is shown in Figure 13.1. The confidence limit is the probability that the MTBF—the mean of the curve—lies above the lower limit value. This equation will also identify the lower limit of the MTBF of the equipment or machinery under investigation. This equation can be used for new and existing equipment on the plant floor.

Example

A machine has been tested for 80 h of operation: during that period of time, 4 failures were observed. With an 80% confidence in the test, what will the lower limit value of the MTBF be?

$$T = 80 \text{ h}$$

$$\alpha = 1 - 0.8$$

$$r = 4$$

$$\text{MTBF} = \frac{2(T)}{\chi^2\alpha, 2r+2} = \text{MTBF} = \frac{2(80)}{\chi^2.2, 2(4)+2} = \frac{160}{13.4} = 11.94$$

This result indicates that the lower limit of the MTBF of this equipment is at least 11.94 h.

Validating MTBF Parameters

The following equation can be used to calculate the test time required to support the MTBF claims of the equipment. The test time equation is as follows:

$$\text{Test time} = \frac{\text{MTBF}(\chi^2\alpha 2r+2)}{2}$$

where
 Test time = Length of time needed to measure MTBF claims of equipment
 MTBF = The known value of MTBF for the equipment
 χ^2 = Chi-squared value
 α = Risk level for the test. This value is equal to 1 – Confidence value in the test
 $2r + 2$ = Degree of freedom as located on the chi-squared table (see Table 13.1: chi-squared table)
 r = Number of failures observed during the test period

Example

Our existing drill press has an MTBF value of 11.94. The new drill press has an MTBF goal of 100 h. What is the length of the test time needed to prove this MTBF value with an 80% confidence limit?

$$\text{Test time} = \frac{\text{MTBF}(\chi^2\alpha 2r+2)}{2} = \frac{100(\chi^2.2, 2(0)+2)}{2} = \frac{100(3.22)}{2} = 161 \text{h}$$

This equation illustrates that the test time for this equipment should be at least 161 h. If, during that period of time, no failures have been observed, the test has identified that the equipment has at least an 80% confidence limit of 100 MTBF h.

On the other hand, if a failure is observed at any point during the 161 h of testing, then a new test time must be established. This does not necessarily mean that the test starts over; rather, additional failure numbers are inserted into the formula to determine the additional test time required to meet the MTBF goal. Table 13.2 illustrates this point.

Table 13.2 Testing for an 80% Confidence Factor for a 100 h MTBF Goal

Number of Failures	Test Time Required (h)
0	161
1	300
2	429
3	552

RQT Assumptions

When implementing Reliability Qualification Testing, there are several assumptions that must be understood for the testing process to be carried out properly:

- Testing should be conducted when the equipment is out of the infant mortality phase.
- When failures are addressed, the equipment is returned to its original reliability condition.
- If the machine is redesigned to affect reliability, the test must be started over again.
- If one subsystem fails, it does not affect the other reliability conditions of the equipment.

These are big assumptions regarding the RQT philosophy. It must be understood by all the parties that the testing process is there to measure the MTBF of the equipment and that the readiness of the equipment must be determined before the test is conducted.

RQT Test Plan

The RQT test plan is a simple and straightforward approach to developing the MTBF rates for the equipment. The test plan consists of the following steps:

1. Identify the system or subsystem specifications associated with reliability.
2. Apportion the MTBF to all subsystems within the equipment.
3. Prioritize the MTBF of each subsystem to identify the lowest MTBF value (weak link concept).

4. Look for existing test data from other plants, suppliers, and other databases.
5. Look for increased sample size to test.
6. Strive to simulate real-world conditions during the test.
7. Determine if the equipment can be accelerated during the testing session.
8. Look for opportunities for automating the testing.
9. Calculate the acceleration factor for subsystems.
10. Calculate the test time for the system.

All of these or just some of these steps may be used. It is up to the customer engineer and supplier to determine how the RQT should be conducted.

Using RAMGT Software

The RAMGT (Robust Aerofoils for Modern Gas Turbines) software package provides a calculation method for identifying test times. This process will enable the calculation of test times with various parameters of the test identified.

To calculate the tests, the software requires you to input the following:

- Required MTBF: Enter the value of MTBF required for the system, subsystem, or components that will be tested.
- Normal cycles per hour: Enter the normal cycles per hour that the system, subsystem, or component will be subjected to on a normal basis.
- Accelerated cycles per hour: Enter the number of accelerated cycles per hour that the equipment can be adjusted to during the testing period.
- Test MTBF: The resulting calculation illustrates the total test time for the system, subsystem, or component during the testing period.

All of the preceding four items are examples of tests that can be performed on the equipment prior to being shipped. These tests should be conducted at the supplier's facility to ensure that the equipment meets the design R&M specifications.

Establishing System Requirements

The systems requirements always are defined by the customer. It is up to the supplier to meet them. However, before the requirements are finalized, it is the responsibility of the supplier to make sure that all the requirements can be met. Therefore, the system requirements related to the R&M

specifications should be reviewed to identify that all goals are established and are achievable. These goals should be based on specific requirements as much as possible. For example:

- MTBF > 100 h
- mean time between assists (MTBA) > 2 h
- mean time to repair (MTTR) < 1 h
- 1st run % capability > 99%
- Noise < 20 dbA
- Safety (e.g., governmental regulations)
- Environmental (e.g., allowable limits for oil mist)
- Quality requirements (e.g., system performance specifications)

In this stage, the apportion MTBF requirements to subsystems should be identified. The reason for apportioning MTBF goals to the subsystems is to determine the "weak link" of the various subsystems associated with the equipment. For example:

- System MTBF = 100
- Loading MTBF = 375
- Positioning MTBF = 275
- Processing MTBF = 575
- Unloading MTBF = 510

Select Area for Test Focusing

There is typically no need to test every subsystem or component in a machine. In order to develop an efficient means of testing to save money, time, and resources, the following steps should be implemented:

- Ensure performance of the worst subsystem: If the design is not proven, spend additional time during the testing period to validate MTBF values.
- Refer to EFMEA for high risk priority number (RPN) values: Use the high RPN number to focus on developing the necessary testing strategies. The test again should focus on the "weakest link" in the subsystems.
- Refer to the FTA to ensure that all existing problems have been addressed: The FTA will determine and verify that the major problems have been addressed. This means that the problems should not occur again.

- Focus on new nontested designs: This is the area where new designs have been incorporated into the machine. New designs will typically have the lowest confidence level of the user.
- Identify areas that have high confidence, old designs, experience, and other customer input: Information in this area will increase the confidence that the design is proven.
- Delegate testing to suppliers when appropriate and necessary: Suppliers should take an active part in the testing process. Suppliers may have databases on failures or tests on their products.

Look for Existing Data

Data from the plant floor is the most important information that can assist the supplier in testing the equipment. The sources for this information can be as follows:

- Customer's experience with identical subsystems or systems. This can assist in removing the reengineering of some of the subsystems.
- Suppliers will occasionally allow you to talk with other customers to inquire about equipment performance data.
- Use other database sources such as
  - Government Industry Data Exchange Program (GIDEP)
  - Rome Air Development Center (RADC)
  - Supplier's test data
- Plant floor information and data

Look for Increased Sample Size

If possible, look for an increased sample size for the testing process. When several subsystems are built by the supplier, they can be tested together to reduce the test time of the equipment. Using more than one subsystem will allow the test time to be accumulated on the subsystems. That the subsystems can be tested at the same time is a big assumption. The systems must have

- Identical components all meeting the same reliability specifications
- A constant failure rate for each subsystem
- A return to original reliability condition after each failure during the test period for the subsystems

Example

There are two identical subsystems to test. How long should the test be run if you want an 80% confidence level where the MTBF is at least 100 h?

$$T_o = [100(3.22)]/2 = 161 \text{ h}$$

$$T = 161/2 = 80.5 \text{ h}$$

This means that each subsystem will be required to operate failure-free for 80.5 h. If a failure occurs during that test period, then the test time would have to be calculated for one failure.

The following concerns are developed using this testing method:

- Are both subsystems designed and built the same way?
- Are both subsystems constructed with the same quality control?
- Are both subsystems constructed with the same reliable components?

Strive to Simulate Real-World Conditions

It is important to develop real-world conditions during the test. This will allow the equipment under test to come as close as possible to the plant environment. This will allow failures to be detected at the supplier before being shipped to the customer. Areas to simulate real-world environments are described in the following sections.

Harsh Environment

It is valuable to simulate the environment that the machine will be subjected to in the plant, such as

- Temperature variations
- Humidity
- Vibration
- Oil mist
- Dirt
- Metal chips
- Component/material contamination

Processing Parts

Tryout parts may be limited, or there may be none at all during the testing process. This also can be very expensive to build parts for testing the equipment at the supplier. Therefore, it is the responsibility of the customer engineer to come up with the number of parts that will be required to operate the test. If the parts are not available, then simulation of these parts in fixture should be developed to meet the testing requirements.

Imperfect Operators

Simulate the imperfect operator whenever possible. Imperfect operators will help to identify

- Part loading failures
- Failure recovery procedures
- Interface failures

Accelerated Testing Considerations

Determine if the testing can be accelerated during the test period. This will reduce the amount of test time needed to complete the testing activity. Acceleration can be accomplished by the following methods:

- Increased voltage levels
- Increased temperature levels
- Increased stress
- Increased cycling frequency (e.g., if the normal rate is 25 cycles/h for 100 h, an accelerated rate may be 50 cycles/h for 50 h. In this example, acceleration testing can cause the test time to be reduced from 100 h to 50 h. Therefore, this can translate into a considerable cost savings!)

Rules of accelerated testing:

- Do not introduce undue stress and strain on the accelerated components, subsystems, or equipment.
- Do not cause failure modes that would not normally be seen.
- Do not change the physical properties of the components, subsystem, or equipment under test.

Automating the Test

Automating the test means that the test can operate without the intervention of personnel to monitor the testing data. Data can be collected and stored for retrieval at a later time. The following are methods that can be used to automate the testing process:

- Automatic test programs—Programs can be written that will cycle the equipment through various operations. The data can be collected and stored in the program for later evaluation.
- Test fixtures/jigs—These can be used to simulate how the equipment will interface with other equipment or subsystems. They can also be used to simulate real production.
- Sensors/gauges—These devices can be used to collect and store the data from the testing operation. The downside of this operation is that the reliability of the sensors and gauges must be determined in order to collect proper data.
- Recycling mechanisms—These devices are developed to recycle the materials that are being used with the equipment during the test period. These mechanisms will keep someone from having to be with the equipment during the entire test period.

The benefits of test automation are as follows:

- Reduces test time by running the tests on weekends and nights
- Frees employees from performing the testing operation
- Promotes standard testing practices
- Uses larger sample sizes, which results in higher confidence levels in the test results.

Acceleration Testing for Subsystems

Subsystems need to be tested at the supplier's facility as they are manufactured. Testing the subsystems first will provide

- Free exercising of subsystems
- Early failure detection before system integration
- Quicker completion of subsystem due to acceleration

Integrated System Testing

Integrated system testing can be accomplished when all the subsystems have completed the testing process. The integrated system testing must focus on the following areas:

- Electrical and mechanical interfaces.
- System failures.
- Less acceleration on the system testing.
- Test duration time is critical.
- Time (normal rate).
- Parts required (real parts).
- Cost (manpower, parts, etc.).
- Test goals can be lowered due to the relationship between uptime and MTBF.

Testing Guidelines

Test plans should follow these guidelines:

- The RQT method will be used to structure the test.
- All tests should contain at least an 80% confidence limit.
- If any part of the system is redesigned during the test, the test should start over again.
- All failures observed during the test should be recorded.
- Environmental conditions should be duplicated during the test.

Subsystem Testing

When developing subsystem testing, the following guidelines should be addressed:

- Acceleration rates for subsystems can be accelerated to four times normal cycle rates (if possible).
- Automation of testing should be implemented, such as software developed to route parts through subsystems.
- Automation should be used to recycle parts through the test, requiring less parts and manpower during the testing period.
- Apply sensors to collect data during the testing period. Test jigs can be used to ensure that the part is positioned accurately.

Test Progress Sheets

Test progress sheets should be developed to track and record the following information during the testing period:

- Identify the failure.
- Date of test and date of failure.

- Time of test and time of failure.
- Running time since last failure.
- Test operator's name.
- Record of what failures the operator observed. Identify the root cause of the failure and what can be done to fix it.
- Duration of the repair of the failure, identification of steps leading up to the failure, quality of the part before the failure, and how the machine was set up prior to the failure.
- Alarms displayed at the time of the failure.
- Recommended followup or repair.
- Number of hours of extended testing required to pass the test.

Reliability Growth

Reliability growth can be defined as the improvement as a result of identifying and eliminating machinery or equipment failure causes during machine test and operation. It is the result of an interactive design between the customer and the supplier. The essential components for developing a growth program are

- In-depth analysis of each failure, effects of the failure, and cause of the failure that occurs after installation
- Upgrading of equipment and training people, thus eliminating root cause failures
- Managing failures by redesign of equipment, maintenance, and training
- Upgrading the equipment when necessary
- Continuous improvement of the equipment

Reliability growth will continue even after the equipment is installed in the plant. However, it will take an effort on the part of the R&M teams and the supplier in order to develop a solid reliability growth program.

Equipment Supplier

During installation and operation of the equipment, the supplier should provide the customer with recommendations on how the reliability of the equipment can be improved. This action can be accomplished by visiting, on a regular basis, the equipment on the plant floor and interacting with the customer's R&M teams to conduct root cause analysis on the equipment.

The participation of the supplier at this point would greatly enhance the reliability growth of the equipment on the floor and future equipment the supplier will be providing. The results of this growth can also affect the mean time to repair (MTTR) numbers associated with the equipment. Any new improvements should be logged and included as part of new designs for the equipment.

The Customer's Responsibility during Building and Installation of Equipment

The responsibility of the customer is to keep accurate failure history logs on the equipment. This log should identify the following:

- The classification of failure
- The time of failure
- The root cause of failure
- The corrective action for the failure

How to Implement a Reliability Growth Program

- Set the overall reliability goals for the equipment. These goals can be divided into short- and long-term goals.
- Review the overall equipment effectiveness (OEE) calculations for the equipment. Set new goals for availability, performance, and quality for the short and long term.
- Establish a universal maintenance tagging system for identification of failures and their corrective actions.

Summary

This chapter discussed Phase 3 of the LCC: specifically, the responsibilities of the supplier for the build and install phase. The customer's requirements for this phase were also described. Chapter 14 discusses Phases 4 and 5.

Chapter 14

Requirements for Implementing Equipment R&M in Phase 4 (Operations and Support) and Phase 5 (Conversion and/or Decommission)

Chapter 13 discussed the issues of build and install, which of course, is Phase 3. This chapter discusses the last two phases of life-cycle costs (LCC). Note that in Phases 1–3 of the LCC, we typically try to identify nonrecurring costs. In contrast, in Phases 4 and 5, we focus on equipment support costs.

Introduction to Life-Cycle Costs (LCC)

Chapter 10 introduced the concept of LCC; this chapter addresses the issue of LCC in a more specific manner, from a *total cost* of ownership perspective. The total cost of ownership is identified as the LCC of the equipment. These costs are divided into two cost factors:

- Nonrecurring costs
- Support costs

The purpose of LCC analysis is to explore various alternatives to identify the most cost-effective production machinery or equipment for a specific application. Applying the LCC concept simply means identifying and adding up costs associated with the system's life cycle (i.e., the total operational life). As already mentioned, the application is based on the five phases covered in Chapters 11–14:

- Phase 1 Concept
- Phase 2 Development and Design
- Phase 3 Build and Install
- Phase 4 Operation and Support
- Phase 5 Conversion and/or Decommission

For example, if a luxury taxi service has a new specification that is only met by two cars, car A or car B, a purchase decision must be made. However, it should not be made based on performance alone, or even purchase price, but on the LCC.

This decision should be based on these or similar types of questions:

- Which has the lowest acquisition price?
- Which has the lowest delivery costs?
- Which has the lowest operating costs?
- Which has the lowest spare parts costs?
- Which has the lowest repair time?
- Which requires the fewest number of support personnel?
- Which has the lowest personnel training costs?
- Which has the best fuel economy?

These costs present a very simple LCC (L) model. In a formula notation, they may be represented as

$$L = A + O + M \pm C$$

where
  A = Allocation costs
  O = Operation costs
  M = Maintenance/decommission costs

Although both cars meet the performance specification, it is clear after considering these questions and investigating the relative costs that car A, at 20% of the price, has the lowest LCC.

Therefore, it is always important to consider the LCC before purchasing a component. Furthermore, unlike this example, often a piece of equipment that appears to be a bargain, based on its acquisition price, is much more expensive after considering the LCC. It is clear that achieving the best LCC is one of our greatest competitive advantages.

Calculating Nonrecurring Costs for Acquiring Equipment

Nonrecurring costs typically consist of:

- System Concept and Design
- Design and Development
- Manufacturing, Build, and Install

Some of these costs may carry over from Phases 1–3; however, the bulk of the costs are in Phases 4–5. Therefore, as with all reliability and maintainability (R&M) phases, it is important to document a component's design and performance, so that it can be analyzed to determine the best LCC. Also remember that the nonrecurring costs are typically found in the *acquisition* costs of the equipment, such as those described in the following paragraphs. The acquisition model is simply a total of all nonrecurring costs:

$$A = P + AE + 1 + T + C + TR$$

Purchase price (P): The purchase price is associated with the production of machinery and equipment, excluding transportation costs. The purchase price should account for currency differences and cost of money associated with payment schedules.

Administration and engineering costs (A and E): These costs are associated with the runoff costs, travel costs to the site, and personnel associated with these administration and engineering costs.

Installation costs (I): The installation costs are a one-time cost associated with installation of the production equipment.

Training costs (TR): The training costs are associated with training of the work force to operate or maintain the production equipment or machinery. The largest cost here should be considered the replacement costs of workers that need to back-fill personnel when they are in training.

Conversion costs (C): The conversion costs are associated with the conversion of the production machinery and equipment to a new product during the life cycle of the equipment or machinery.

Transportation costs (T): The transportation costs are associated with the movement of the production equipment or machinery from the supplier's floor to the user's floor.

Calculating Support Costs for Continued Operation of Equipment

Support costs typically consist of

- Operation and support
- Support costs for conversion and decommission

Operating costs are determined by several factors that are addressed in the LCC analysis. These costs are identified in the following paragraphs. The operating costs are the total of all support costs:

$$\text{Operating cost } (O) = D + U + Cc + W + Lp + Sp$$

Direct labor (D): Direct labor costs are the total cost of direct labor to operate the production machinery and equipment over the life cycle of the equipment.

Utilities (U): The utilities cost is the total utilities consumption cost over the life cycle of the equipment. This includes air, electricity, steam, gas, and water. These costs can vary greatly and should be considered if the utility is scarce.

Consumables (Cc): These are the costs of all consumable items used over the life cycle of the equipment. Cost items to consider are coolant, filters media, etc.

Waste handling (W): The costs associated with waste handling are the collection and dumping of waste products produced with the equipment. These costs should be considered over the life cycle of the equipment.

Lost production (Lp): This cost is identified due to the equipment failure and appropriate downtime costs. These downtime costs should be agreed on by all parties who are working on the LCC model.

Spare parts maintenance (Sp): This cost is associated with carrying and maintaining a spare parts inventory to support the equipment under LCC investigation.

Calculating Equipment Maintenance Costs (M_C)

The costs associated with maintaining the equipment over the life cycle are divided into two categories. These costs are identified as scheduled maintenance costs and unscheduled maintenance costs:

$$M_C = S_M + U_M$$

Calculating Scheduled Equipment Maintenance (S_M) Costs

Scheduled Maintenance is divided into four areas:

$$S_M = L_E + P_M + C_R + F_{LC}$$

- *Life of the equipment (L_E)*: These are the five phases of the equipment life cycle. They include the time from concept to decommissioning of the equipment. This time is typically measured in years.
- *Preventive maintenance schedule (P_M)*: The preventive maintenance is the schedule of recurring maintenance actions that will assist in the equipment meeting its life-cycle goals. The schedule for these preventive maintenance actions is provided by the supplier of the equipment.
- *Cost of repair (C_R)*: These costs are associated with the labor and parts necessary to rectify the defective equipment.
- *Fixed labor costs (F_{LC})*: These costs are associated with the cost of keeping a skilled labor pool to service unscheduled breakdowns of the equipment.

Calculating Unscheduled Equipment Maintenance (U_M) Costs

Unscheduled maintenance is associated with the cost of materials and labor needed to return the equipment to its operational condition when a breakdown occurs. These breakdowns are obviously not scheduled during the expected life of the equipment. Unscheduled breakdowns are classified into four areas:

$$U_M = L_E + U_B + A_{CR} + P_{PY}$$

- *Life of the equipment (L_E):* These are the five phases of the equipment life cycle. They include the time from concept to decommissioning of the equipment. This time is typically measured in years.
- *Unscheduled breakdowns (U_B):* These costs are associated with catastrophic and wear failures of the equipment.
- *Average cost of repair (A_{CR}):* These costs are associated with the average cost of repair actions caused by unscheduled breakdowns. These costs are averaged over several breakdowns.
- *Parts per year (P_{PY}):* Parts per year is the cost of all repair parts consumed in one year.

The Impact of Life-Cycle Costs (LCC) on Concept and Design

It is important to note that concept and design constitute the majority of a component's LCC. Unfortunately, however, as previously mentioned, most of these costs are determined during the System Concept and Definition stage. For example, studies have shown that up to 95% of a component's LCC is determined during the Concept and Design/Development phase of the equipment's life cycle. Therefore, once the equipment has reached this stage, only about 5% improvement remains for lowering the LCC. Experience shows that R&M deployment during these early design stages results in only incremental spending increases. However, significant savings to the customer are realized in the Operation and Support stages (see Figure 14.1).

Figure 14.1 helps verify that by the time the "Critical Design Review" is complete, 95% of the LCC is determined. However, only about 12% of the design budget has been spent. Therefore, only 12% of the design costs are determining 95% of the LCC. As seen earlier, investing more effort early in the Concept and Definition stage allows better identification of the best LCC. It is at this stage that decisions are made, such as determining general configuration, materials, and suppliers. Although R&M deployment initially requires a greater engineering effort, the benefits are reaped when the equipment becomes operational (see Figure 14.2.)

Requirements for Implementing Equipment R&M in Phase 4 and Phase 5 ■ 371

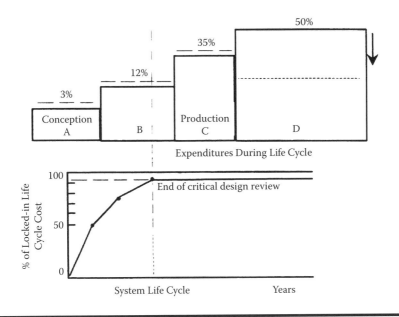

Figure 14.1 Impact of LCC on equipment.

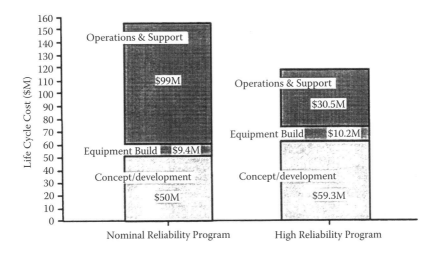

Figure 14.2 Comparable program LCC.

Although the benefit of R&M can be clearly seen with reports such as the one in Figure 14.2, it is equally important to document information important to a component's LCC. This is how you are able to evaluate all of a design's LCC aspects as well as calculate the impact proposed changes will make. It is also important to identify the areas representing the greatest opportunity for improvement.

Improving LCC

Once the LCC is identified, it is important to review for potential improvement. To document this, you need to calculate the following:

- Cost of the design changes
- Possible increased equipment costs
- Possible construction time increases
- Projected labor savings

Documenting the LCC Process

The LCC method can be used to determine the overall cost difference between equipment that is being ordered. The LCC can identify that equipment that has a greater acquisition cost may be cheaper in the overall LCC structure. Here are the steps to calculate the LCC:

1. Identify the operational requirements of the equipment and the costs associated with the following parameters:
   - Machine lifetime in years
   - Annual part quantity required
   - Operating hours per month
   - Maintenance cost per hour
   - Inventory carrying costs per year
   - Crisis downtime per hour
   - Floor space cost per square foot
   - Electrical cost per kilowatt hour
   - Shipping cost per pound
   - Inflation rate per year
   - Interest rate per year
2. List all of the R&M performance parameters per machine under evaluation. These parameters should include
   - Quality rate
   - mean time between failure (MTBF) hour
   - mean time to repair (MTTR) hour
   - Time to retool—hour
   - Cycle time—parts per hour
   - Machine weight—pounds
   - Foot print—square footage

3. List all of the acquisition costs associated with the equipment. These costs include all of the one-time costs associated with the purchase of the equipment.
4. List all of the maintenance costs associated with the equipment. These costs should include the scheduled and unscheduled maintenance costs of the equipment.
5. List all of the operating costs associated with the purchase of the equipment under investigation. These costs should include all contributing costs needed to operate the equipment.
6. Total the overall costs categories, and identify the LCC of the equipment.

Overall Items of Concern for New Machinery

Many organizations that are acquiring new machinery require that acquisition also have an attachment that defines the R&M requirements. This attachment can be a contract clause as part of the purchase agreement, and it should read something like the following provision:

> Suppliers must include with the proposal a completed R&M Matrix, supported by an R&M Program Plan, in accordance with the latest Customer's Manufacturing Engineering R&M Specifications.

The individual customer's Manufacturing Engineering R&M Specification should be developed to provide suppliers and the specific customer's asset-acquiring engineers with a tool to describe the R&M quantitative, programmatic, engineering, and continuous improvement tools of the equipment life cycle. The specification uses a matrix to tailor the R&M activities consistently with the end item being procured by that individual customer.

This should be done at the start of each new program. This means that the R&M matrix should be reviewed by the supplier and the customer asset-acquiring engineer together. The amount of R&M activity specified in the matrix should depend on equipment complexity, commodity versus new design, history of prior equipment performance, and common sense that leads to a cost-effective R&M solution. Some general items that should be covered are described in the rest of this chapter.

Develop a Life-Cycle Costing (LCC) Model

The supplier shall develop a LCC model for the conceptual design and any proposed or requested alternatives, and continually update such models until they are no longer needed to evaluate options. The models' purpose is to allow evaluation of the approach that provides the best LCC for the equipment that will be acquired by the asset-acquiring engineer from the customer.

Conduct Periodic R&M Reporting

The supplier shall report on the progress of the R&M effort through periodic management and engineering review meetings. Reporting must include

- Identification of the program and equipment
- Identification of the specific R&M activity
- Estimated R&M activity completion date and review complete status with respect to this date
- Key findings, results, and conclusions
- Thorough documentation of all recently completed R&M activities
- Inhibitors to complete the designated R&M task
- Action plans to resolve any R&M task that is behind schedule

Continuously Review Field History Data

Suppliers shall continuously review current field data (warranty claims, service reports, etc.) that demonstrates their understanding of achieved performance levels, Things Gone Right and Things Gone Wrong, equipment failure causes, and their present "Top 10" field problems that are pertinent to the current program. In addition, suppliers shall provide examples of lessons learned from previous designs that will be incorporated into the current design.

Conduct a Failure Mode and Effects Analysis (FMEA)

Cross-functional teams shall prepare a comprehensive FMEA to eliminate failure modes that may affect safety, equipment performance, and compliance with government regulation and customer specifications. Although most people use the FMEA as a one-time document and forget it, the FMEA must serve as a living document during the equipment life cycle, support the development of maintenance manuals, and guide problem diagnosis and corrective action during in-field operation.

Conduct Machinery FMEA

Original equipment manufacturers (OEMs) shall follow the defined format (as described in Chapters 5 and 12) for assembly-level equipment (e.g., for full machine stations).

Conduct Design FMEA

OEMs and SCS (Standard Component Suppliers) shall follow the defined format (as described in Chapters 5 and 12) for all equipment (e.g., transfer mechanisms, spindles, switches) exclusive of assembly-level equipment.

Perform Electrical Stress Analysis and Derating

Electrical stress analysis shall be performed on all circuitry within the equipment in order to ensure that electrical derating criteria are not violated. Voltage derating shall be based on the peak voltages that may be present in a circuit. Current or power derating shall consider the maximum current or power level averaged over a time period appropriate for the thermal mass involved. All derating criteria are to be based on the component rating at the maximum allowable equipment temperature and tolerance stackups.

Analysis shall be performed during initial design (we are discussing it here because changes may occur and design changes may be proposed). All electrical/electromechanical/electronic components shall not exceed their rated voltage, current, or power ratings at their rated operating temperature. This includes not only plant ambient temperature and internal electrical panel temperature, but internal component temperature rises as well.

Perform Mechanical Stress Analysis to Check Design Margins

Mechanical stress analysis shall be performed on critical components of the equipment in order to ensure that design margins are not violated. The design shall be evaluated at the maximum stress levels encountered, whether steady state or transient. Adequate design margins shall be applied to all mechanical components. These design margins shall be applied with respect to the maximum expected mechanical loads on the equipment and tolerance stackups. The supplier shall determine and inform the customer of the design margins used for each class of equipment component.

Perform Thermal Analysis

All electrical enclosures housing electrical and electronic components supporting manufacturing operations will require thermal analysis to ensure that thermal requirements are met. Electrical enclosures must be designed to ensure that under worst-case full-load conditions, the maximum temperature within the enclosure is no more than 44°C (which is a standard for thermal analysis).

Applications Engineering

The customer shall assume the responsibility for the proper application of all standard components. Proper application includes consideration of the functional requirements of the component, the environment where the component is used, derating criteria, and specific application guidelines as defined by the standard component supplier (SCS). The SCS is required to publish equipment-specific application guidelines (including problem diagnosis guidelines) and clear product mechanical/electrical derating criteria for industrial applications. The supplier will provide written component application agreement (CAA) letters from the SCS, detailing the means by which this will be accomplished. Each CAA letter must clearly demonstrate a mutually respectful and productive partnership between both parties, as well as a thorough engineering analysis proving the proper application of components in the equipment designs.

Compliance shall require a complete, documented review of all component applications, per the agreement letters, by the SCS on all completed build equipment on the supplier floor. Component application deficiencies shall be documented by the SCS and submitted formally to the customer. Corrective action plans, including timing, for any identified deficiencies shall be prepared and communicated to all parties concerned as soon as possible.

R&M Design Review Guidelines

Each supplier is required to utilize the R&M Design Review Guidelines (see Appendices B and G). These guidelines provide a thorough checklist approach toward assessing inherent equipment R&M for each piece of provided equipment. Identified deficiencies must be formally documented, including resolution and timing plans; deficiencies may be waived only upon formal request by the supplier and subsequent approval by the customer.

Conducting R&M Testing and Assessment of Results

The supplier shall conduct a thorough reliability testing program to adequately demonstrate that the equipment meets all requirements pertaining to all of the following:

- Performance
- R&M
- Vibration or machine condition signature analysis, if specified by the customer
- Electrical panel thermal characterization
- Fault diagnosis
- Software reliability
- Parts quality

All unscheduled equipment stoppages will also be documented, and an analysis will be conducted to determine the root cause of the problem. Identified major design defects must be corrected prior to test continuation and customer's acceptance.

SCS (standard component suppliers) shall also provide life test results and R&M test findings based on nominal and extreme operating conditions (such as high or low temperature extremes, humidity, vibration, and shock). The test findings must definitively support the equipment's industrial application within a plant.

Provide an Equipment Maintainability Matrix

For all routinely generated faults and alarms captured by the controller, a prioritized matrix of the most probable causes of the malfunction must be provided by the customer. These causal factors or failure mechanisms will be derived from the FMEA and may be displayed on the equipment control monitor to enhance troubleshooting.

Performing Total Productive Maintenance (TPM)

The supplier must provide well-documented and visually graphic preventive maintenance procedures for all components other than those specified as a part of the standard components program. Preferably, they will develop and submit single-point lessons for unique operator or skilled-trade preventive maintenance (PM) activities.

Documenting Things Gone Right/Things Gone Wrong (TGR/TGW)

Suppliers must record and document TGR/TGW throughout their program participation. The TGR/TGW listing must be specifically oriented toward equipment R&M and must be prepared in the required reporting format. Particular emphasis should be placed on TGR to ensure that successful R&M features are repeated on subsequent programs and serve as the benchmark for future programs.

Developing an R&M Matrix

As described in Chapter 11, the overall goal of the R&M program is focused on three main areas:

1. Increase the machines' uptime.
2. Reduce the repair time when a failure occurs.
3. Minimize the overall LCC of the equipment.

Therefore, the completion of the matrix should be based on

- Complexity of the new equipment
- Amount of new technology of the equipment

Table 14.1 indicates typical specifications that should be completed within the overall R&M program. Specifically, it illustrates at what phase each specification should be completed within the machinery life cycle. This matrix is an effective tool that should be used to manage the overall R&M program.

Table 14.1 Typical Specification Matrix

	R&M Specification Matrix			
	Phase 1	Phase 2	Phase 3	Phase 4
Specification	Concept	Design/ Development	Build/ Install	Operation/ Support
R&M Program				
Supplier R&M Program Plan	X			
Quality Control	X			
Life-Cycle Costing (LCC) Model Development and R&M Optimization	X			
Periodic R&M Reporting Requirements	X			
R&M Design Reviews	X			
Formal R&M Evidence Book	X			
R&M Program Review, 6 months after Job #1				X
R&M Engineering				
Field History Review		X		
Failure Mode and Effects Analysis (FMEA)		X		
Electrical Stress Analysis and Derating		X		
Mechanical Stress Analysis and Derating		X		
Thermal Analysis		X		
Applications Engineering		X		
R&M Design Review Guidelines		X		
R&M Testing and Assessment		X		
Maintainability Matrix		X		

(Continued)

Table 14.1 Typical Specification Matrix (*Continued*)

Specification	Phase 1 Concept	Phase 2 Design/ Development	Phase 3 Build/ Install	Phase 4 Operation/ Support
R&M Continuous Improvement Activities				
Failure Reporting, Analysis, and Corrective Action System (FRACAS)			X	
Machinery and Equipment Performance Data Feedback Plan				X
R&M Continuous Improvement Activities				X
Ford Total Productive Maintenance Activities (FTPM)		X		
Things Gone Right/Things Gone Wrong (TGR/TGW)		X		

Developing an R&M Checklist

An R&M checklist is developed as part of the R&M program to evaluate at the time of buy-off that all R&M specifications have been met. The checklist can be divided in categories related to the R&M project. Several R&M categories are listed as follows:

- Machine access
- Machinery conditions
- Switches
- Coolant protection
- Assembly/disassembly/adjustment
- Gauges
- Pipes and hoses
- Guarding
- Machine general issues
- Miscellaneous

Table 14.2 A Typical R&M Activities List

	Component	Supplier	User
S	Safety	X	X
RM	Continuous Improvement Monitor		Me
RM	Equipment Failure Mode and Effect Analysis (FMEA)	X	X, Me, Ma
RM	Process FMEA		
RM	Design Reviews	X	X, Op, Me, Ma
RM	Machine Data Feedback Plan	X	X, Me, Ma, Op
RM	Mean time between failures (MTBF)	Xe	Xe, Me, Ma
R	Fault Tolerance	X	X
R	Fault Tree Analysis		
R	Life In Terms of Throughput	X	X, Me
RM	Overall Equipment Effectiveness	Xe	Xe, Me, Ma, Op, Pu
R	Reliability Block Diagram	X	X, Me
R	Failure Mode Analysis		
RM	Quality Function Deployment	X	X
RM	Environment	X	X
RM	Life-Cycle Costing	Xe	Xe, Me, Fi
RM	Mean Time to Repair	Xe	Xe, Me, Ma
RM	Validation Process		
M	Accessibility		
M	Built-in Diagnostics	X	X
M	Captive Hardware		
M	Color Coding		
M	User Training		
M	Maintenance Procedures		
M	Modularity		
M	Spare Parts Inventory		

(Continued)

Table 14.2 A Typical R&M Activities List (*Continued*)

	Component	Supplier	User
M	Standardization		
SM	Ergonomics	X	X
RM	Computer/Process Simulation	X	X

Legend for table

Function in User's Organization R—Indicates the checklist item affects reliability
Manufacturing Engineer = Me M—Indicates the checklist item affects maintenance
Maintenance = Ma S—Indicates the checklist item affects safety
Purchasing = Pu X—Indicates the checklist item is recommended for use during the program phase
Machine Operator = Op Xe—Indicates the checklist item can be estimated during the program phase
Plant Manager = Pm Xm—Indicates the checklist item can be measured during the program phase

Below each of these areas, specific questions are addressed to identify if the machinery meets the original R&M specifications. A typical listing of R&M activities is shown in Table 14.2.

Performing Design Reviews

The purpose of the R&M design reviews is to provide an in-depth review of the evolving design. This review should be supported by drawings, process flow descriptions, engineering analyses, reliability design features, and maintainability design considerations.

There are two formal design reviews required of the supplier during design, development, and fabrication phases of the project. The supplier is required to follow a specific R&M design review outline. An example is provided in Appendices B and D.

Preliminary Design Review (PDR): The first R&M design review, or PDR, shall be conducted at the conclusion of simultaneous engineering and will review the complete conceptual design and the R&M trade-off analyses generated for alternative designs. This review is conducted when no simultaneous engineering teams are required.

Critical Design Review (CDR): The second R&M design review, or CDR, shall be conducted at the conclusion of detail engineering activities. The CDR will focus on the final engineered design and all R&M design assurance activities conducted to support this design.

Developing Runoff Assessment Techniques

The equipment qualification requirements as detailed by the E supplement to QS9000 and the ISO/TS 16949 have outlined several steps that are necessary to ensure that equipment will start up with the least amount of failures. These runoff assessments are developed to meet the needs of the equipment, tooling, robot, and machinery. Specifically, all identifiable problems shall be identified and eliminated prior to the subsystems being integrated into larger systems. The purpose of this procedure is to

- Reduce or eliminate failures at start-up.
- Improve the quality of equipment to meet the customer's requirements.
- Resolve hardware and software problems prior to equipment launch.
- Confirm that cycle times are correct to meet customer's productivity requirements.
- Verify the reliability of tooling and equipment.

Table 14.3 illustrates the equipment qualification methods for the customer and the supplier.

Conducting a 50/20 Dry-Cycle Run

The 50 h quality test applies specifically to robots: this test requires that robots cycle for 50 h under the following conditions:

- Maximum rated load
- Full rotation through all axes
- All motions at maximum rated velocity
- Motion to encompass the specific envelope

The 20 h continuous uptime, or two 10 h periods of continuous uptime, applies to all automated machinery, tools, and equipment. It also includes all robots within equipment systems.

Table 14.3 A Typical Equipment Qualification Method

Location	Equipment Qualification Methods	Study Time of Agreed-Upon Production Rates	Quantity of Pieces
At supplier	50/20 Dry Run	50 h robots 20 h all others	None
	Phase 1 of Initial Process Performance Preliminary Evaluation	As required by the customer	As required by the customer
	Phase 2 Initial Process Performance Evaluation—Pp	As required by the customer	As required by the customer
	Phase 3 Initial Process Performance Evaluation—Ppk	As required by the customer	As required by the customer
	Reliability Verification	As required by the customer	
At customer plant	20 Dry-Cycle Run	20 h	None
	Short-term process study	As required by the customer	As required by the customer
	Long-term process study	25 working days (200 hours)	As required by the customer

Both 50 and 20 h runs are to be performed by the supplier prior to shipment of completed tooling or equipment. The purpose of both runs is to reduce failures during the launch period. It also offers an opportunity to monitor the repeatability of the machinery.

It must be emphasized here that the objective of these procedures is to demonstrate that the machinery is free of cycle variations and that the process does not drift over time.

A typical 50 h quality test established for robots is

- At the customer's discretion, the OEM's test data may be used in lieu of the on-site 50 h quality test.

The 20 h continuous dry-cycle run applies to all machinery/equipment that has robots included in these systems. The following steps of this procedure govern both the 50 and 20 h applications:

- All tests will be conducted at the supplier's facility.
- Customer personnel will be on site at the beginning of the test and may provide assistance.
- The supplier will supply personnel to run and service the equipment during the test.
- The system supplier shall be responsible for all components during all test runs. This shall include preventive maintenance actions required by the Scheduled Maintenance Procedures.
- Time required for testing should be part of the overall delivery schedule of the equipment.
- Failures during the 20 h test will require a restart of the test and documentation of the failure and analysis of the root cause.

Summary

This chapter discussed Phases 4 and 5 of the LCC model, specifically covering the issues of nonrecurring costs and support costs, as well as some of the continuing items of concern during the conversion and decommission. Chapter 15 introduces and discusses the Weibull distribution.

Chapter 15
Weibull Distribution

Chapters 10–14 have addressed the issues and concerns of maintenance. This chapter introduces the Weibull analysis as one of many methodologies—although it is the most common—for performing life data analysis in machinery, equipment, as well as products that are produced by those machines or equipment. Life data analysis is sometimes called "Weibull analysis" because the Weibull distribution, formulated by Professor Wallodi Weibull, is a popular distribution for analyzing life data.

Performing Weibull analysis enables you to make predictions about the life of all machines, equipment, and their products in the population by "fitting" a statistical distribution to life data from a representative sample of units. You can then use the parameterized distribution for the dataset to estimate important life characteristics of the product, such as reliability or probability of failure at a specific time, the mean life for the product, and the failure rate.

Here are the steps involved in life data analysis:

- Gather life data for the machine, equipment, or product.
- Select a lifetime distribution that will fit the data, and model the life of the machine, equipment, or product.
- Estimate the parameters that will fit the distribution to the data.
- Generate plots and results that estimate the life characteristics, such as the reliability or mean life, of the machine, equipment, or product.

This life data analysis process and some suggestions for additional research on the subject are presented in this chapter. (There are many software packages available to do the analysis, such as ReliaSoft's Weibull++, Nutek, Inc., and others.)

Analyzing Life Data

The term *life data* refers to measurements of the life of machines, equipment, or products. Product lifetimes can be measured in hours, miles, cycles, or any other metric that applies to the period of successful operation of a particular product. Because time is a common measure of product life, life data points are often called "times to failure," and product life will be described in terms of time throughout the rest of this chapter.

There are different types of life data and because each type provides different information about the life of the product, the analysis method will vary depending on the data type. For example:

- With *complete* data, the exact time to failure for the unit is known (e.g., the unit failed at 100 h of operation).
- With *suspended or right-censored data*, the unit operated successfully for a known period of time and then continued (or could have continued) to operate for an additional unknown period of time (e.g., the unit was still operating at 100 h of operation).
- With *interval and left-censored data,* the exact time to failure is unknown, but it falls within a known time range. For example, the unit failed between 100 h and 150 h (interval censored) or between 0 h and 100 h (left censored).

Calculating Statistical Distributions for Equipment Lifetimes

Statistical distributions have been formulated by statisticians, mathematicians, and engineers to mathematically model or represent certain behavior. The probability density function (pdf) is a mathematical function that describes the distribution. The pdf can be represented mathematically or on a plot where the x-axis represents time, as shown in Figure 15.1.

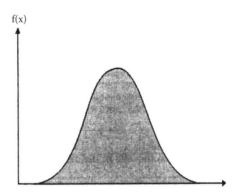

Figure 15.1 A representation of the pdf.

The equation

$$f(t) = \frac{\beta}{\eta}\left(\frac{t-\gamma}{\eta}\right)^{\beta-1} e^{-\left(\frac{t-\gamma}{\eta}\right)^{\beta}}$$

gives the pdf for the three-parameter Weibull distribution. Some distributions (such as the Weibull and lognormal) tend to better represent life data and are commonly called *lifetime distributions* or *life distributions*. The Weibull distribution can be applied in a variety of forms (including one-parameter, two-parameter, three-parameter, or mixed Weibull) and other common life distributions, including the exponential, lognormal, and normal distributions. You, as the analyst, should choose the life distribution that is most appropriate to each particular dataset, based on past experience and goodness-of-fit tests.

The Weibull distribution is very versatile in practice because it is suitable for analysis of fatigue test results, but in addition, it

- Is based on nonparametric statistics (in other words, it does not depend on the mean or standard deviation), and it can help analysts determine a likely distribution of failure data.
- Provides quick analysis of reliability data (Reliability = (1 − Cumulative Failure Distribution).
- Permits predictions, even with small sample sizes (≥6).
- Provides associated predictive risks in terms of confidence bands.
- Allows use of suspended (i.e., censored) tests to improve reliability estimates. (Note: Remember that some failures are required. Testing to bogey alone is not sufficient to do the analysis).

Estimating the Parameters of Equipment Lifetime Distribution to Analyze Failures of Machinery or Equipment

In order to "fit" a statistical model to a life dataset, you need to estimate the parameters of the life distribution that will make the function most closely fit the data. The parameters control the scale, shape, and location of the pdf function. For example, in the three-parameter Weibull distribution (shown in the previous equation):

- The scale parameter, η (eta), defines where the bulk of the distribution lies.
- The shape parameter, β (beta), defines the shape of the distribution.
- The location parameter, γ (gamma), defines the location of the distribution in time.

Several methods have been devised to estimate the parameters that will fit a lifetime distribution to a particular dataset. Some available parameter estimation methods include

- Probability plotting
- Rank regression on x (RRX)
- Rank regression on y (RRY)
- maximum likelihood estimation (MLE)

The appropriate analysis method will vary depending on the dataset and, in some cases, on the life distribution selected.

Other life distributions have one or more parameters that affect the shape, scale, and location of the distribution in a similar way. For example, the two-parameter exponential distribution is affected by:

- The scale parameter, λ (lambda)
- The location parameter, γ (gamma)

The shape of the exponential distribution is always the same.

The three plots described in the following sections (and illustrated in Figures 15.2–15.4) demonstrate the effect of the shape, scale, and location parameters on the Weibull distribution pdf.

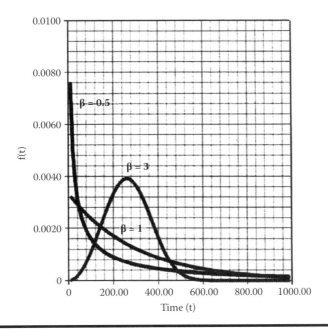

Figure 15.2 Shape parameter, β (beta), of the Weibull distribution.

Estimating the Shape Parameter

The plot in Figure 15.2 demonstrates the effect of the shape parameter, β (beta), on the Weibull distribution. When beta = 1, the Weibull distribution is identical to the exponential distribution (constant hazard rate). When beta = 3.5, the Weibull distribution approximates the properties of the normal distribution.

Estimating the Scale Parameter

The plot in Figure 15.3 demonstrates the effect of the scale parameter, η (eta), on the Weibull distribution. Quite often, the scale parameter or the characteristic life is designated as θ (theta). θ or the η value is the point in time that corresponds to 63.2% cumulative failure (1-e^{-1}). The cumulative distribution shows the cumulative frequency of failures with respect to distance, time cycles, and so on. It indicates that by the time all items have reached life, a certain percentage will have failed. The cumulative distribution function is defined as

$$F(t) = 1 - e^{-\left(\frac{t}{\theta}\right)^{\beta}}$$

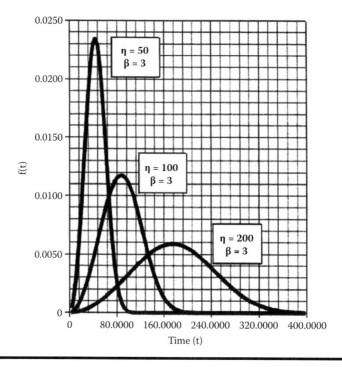

Figure 15.3 Scale parameter, η (eta), of the Weibull distribution.

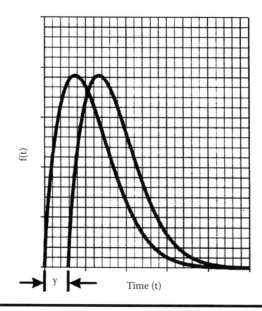

Figure 15.4 Location parameter, γ (gamma), of the Weibull distribution.

The Weibull cumulative distribution can be represented graphically as a straight line when plotted on Weibull graph paper. Weibull graph paper has been designed to simplify interpretation of results by using specially arranged scales for "Life" and "Percentage Failed" to achieve a straight line characteristic.

Derivation of Weibull plots:

$$F(t) = 1 - e^{-\left(\frac{t}{\theta}\right)^\beta} \;;\; 1 - F(t) = e^{-(t/\theta)^\beta} \;;\; \frac{1}{1 - F(t)} = e^{(t/\theta)^\beta}$$

Now take natural logs of both sides and then logs of both sides again, and you end up with

$$\ln \frac{1}{1 - F(t)} = (t/\theta)^\beta;\; \ln \ln \frac{1}{1 - F(t)} = \beta \ln t - \beta \ln \theta$$

This is now of the form Y = MX + C, which, of course, is the description of the straight line:

$$Y = \ln \ln \frac{1}{1 - F(t)};\; M = \beta;\; X = \ln t;\; \text{and}\; C = \beta \ln \theta$$

This transformation allows the engineer to plot cumulative percentage failure versus time on Weibull paper. The slope of the plotted line estimates the parameter β. A horizontal line can be drawn from the y-axis at 63.2% to intersect a point on the plotted line. The x-coordinate of that point corresponds to θ.

Estimating the Location Parameter

The plot in Figure 15.4 demonstrates the effect of the shape parameter, γ (gamma), on the Weibull distribution.

Calculating Results and Plots

Once you have calculated the parameters to fit a life distribution to a particular dataset, you can obtain a variety of plots and calculated results from the analysis, including:

- *Reliability given time:* The probability that a product will operate successfully at a particular point in time. For example, there is an

88% chance that the product will operate successfully after 3 years of operation.
- *Probability of failure given time:* The probability that a product will fail at a particular point in time. Probability of failure is also known as *unreliability*, and it is the reciprocal of the reliability. For example, there is a 12% chance that the product will fail after 3 years of operation (and an 88% chance that it will operate successfully).
- *Mean life:* The average time that the products in the population are expected to operate before failure. This metric is often referred to as *mean time to failure* (MTTF) or *mean time before failure* (MTBF).
- *Failure rate:* The number of failures per unit time that can be expected to occur for the product.
- *Warranty time:* The estimated time when the reliability will be equal to a specified goal. For example, the estimated time of operation is 4 years for a reliability of 90%.
- B_X *life:* The estimated time when the probability of failure will reach a specified point (X%). For example, if 10% of the products are expected to fail by 4 years of operation, then the B_{10} life is 4 years. (Note that this is equivalent to a warranty time of 4 years for 90% reliability.)
- *Probability plot:* A plot of the probability of failure over time. (Note that probability plots are based on the linearization of a specific distribution. Consequently, the form of a probability plot for one distribution will be different from the form for another. For example, an exponential distribution probability plot has different axes from those of a normal distribution probability plot.)
- *Reliability versus time plot:* A plot of the reliability over time.
- *Pdf plot:* A plot of the pdf.
- *Failure rate versus time plot:* A plot of the failure rate over time.
- *Contour plot:* A graphical representation of the possible solutions to the likelihood ratio equation. This is employed to make comparisons between two different datasets.

Use Confidence Bounds to Quantify Uncertainty in Analyzing Data

Because life data analysis results are estimates based on the observed lifetimes of a product's sample, there is uncertainty in the results due to the limited sample sizes. Confidence bounds (also called *confidence intervals*) are used to quantify this uncertainty due to sampling error by expressing the confidence that a specific interval contains the quantity of interest. Whether or not a specific interval contains the quantity of interest is unknown.

Confidence bounds can be expressed as two sided or one sided:

- Two-sided bounds are used to indicate that the quantity of interest is contained within the bounds with a specific confidence.
- One-sided bounds are used to indicate that the quantity of interest is above the lower bound or below the upper bound with a specific confidence.

Depending on the application, one-sided or two-sided bounds are used. For example, the analyst would use a one-sided lower bound on reliability, a one-sided upper bound for percentage failing under warranty, and two-sided bounds on the parameters of the distribution. (Note that one-sided and two-sided bounds are related. For example, the 90% lower two-sided bound is the 95% lower one-sided bound, and the 90% upper two-sided bound is the 95% upper one-sided bound.)

Introduction to Failure Data Modeling

The time to failure for a population of devices follows some distinctive pattern that is unique to that particular device. We model the time to failure with a pdf. The Weibull distribution is a very versatile and popular model for time to failure for mechanical components. In order to illustrate this modeling process, consider the histogram of Figure 15.5, representing a plot of 50 devices tested to failure. This histogram is an empirical pdf obtained from test data. The best-fit Weibull pdf model for this data is shown in Figure 15.6. This pdf attempts to fit the pattern of failure.

Use of a Probability Density Function Model

A pdf provides a means for evaluating the probability of a failure occurring in a particular time interval. This probability is found by calculating the area over the time interval and under the curve. That is,

396 ■ *The OEE Primer*

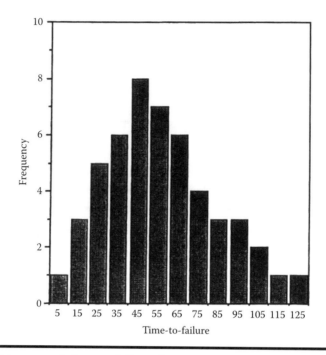

Figure 15.5 Histogram of time-to-failure data.

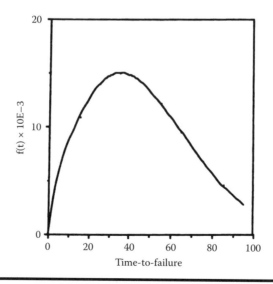

Figure 15.6 Best-fit probability density function.

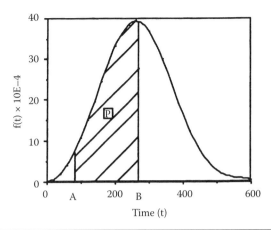

Figure 15.7 Probability calculation.

$$P = prob(A \leq t \leq B) = \int_A^B f(t)\,dt$$

where
F(t) = the pdf for the time-to-failure random variable t (see Figure 15.7).
One can say that there is a 100P% chance of failure in the time interval ($A \leq t \leq B$).

The cumulative distribution can be found from any pdf, and this gives us the total fraction (or percentage) failing prior to a time t. The cumulative distribution, F(t), is found from the pdf by

$$F(t) = \int_{-\infty}^{t} f(t)\,dt$$

The reliability function R(t) is related to F(t) by

$$R(t) = 1 - F(t)$$

While F(t) gives the fraction failing, R(t) gives the fraction surviving at time t. These relationships are shown in Figures 15.8 and 15.9.

Calculating Rate of Failure and Hazard Functions

The failure rate is the rate at which failures occur in a certain interval of life for those devices that are surviving at the start of the interval. The failure rate is calculated from life data as

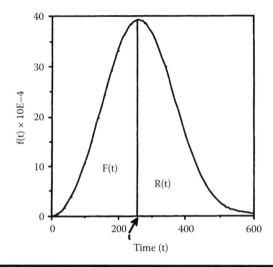

Figure 15.8 Relationships between the pdf, *F(t)*, and *R(t)*.

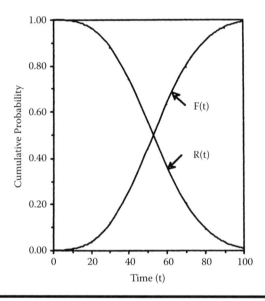

Figure 15.9 Relationship between *F(t)* and *R(t)*.

$$\text{Failure rate} = M/N \times T$$

where
- M = number of failures in the lifetime interval
- N = number of survivors at start of the time interval
- T = length of lifetime interval

While theoretically, $h(t) = f(t)/R(t)$

This can be easily rationalized by considering that, for an original number of devices, N, the number surviving at some arbitrarily selected time, t, would be $N \times R(t)$.

While the number of the original devices failing in an interval of time, Δt, would be

$$N f(t) \Delta t$$

So, we have

$$h(t) = \frac{Nf(t)\Delta t}{NR(t)\Delta t} \text{ or, } h(t) = f(t)/R(t)$$

This means that every pdf model has an implied hazard function (theoretical failure rate). The failure of a population of fielded products is due to many factors (or forcing functions) such as

- The product design
- The manufacturing variability
- The variability in customer usage
- The maintainability policies as practiced by the customer
- The environments encountered

The failure rate can change over the life of the product due to different dominating forcing functions. For example, if there is considerable variability in manufacturing and bad product slips through the system, then a high early failure rate can be expected (called *infant mortality*). As the inferior products drop out, the failure rate will stabilize at its lowest level until the onset of wearout. Later in the product life, the failure rate will increase due to wearout. This change in failure rate over the life of a product gives rise to the bathtub failure rate curve, as shown in Figure 15.10.

The following example of failure data demonstrates these phenomena.

Example

The failure characteristics of a particular brand of fan were of interest. The fan was for a computer cooling application. In a large mainframe installation, all fans were replaced with this new brand. In order to keep accurate records, as a "new"

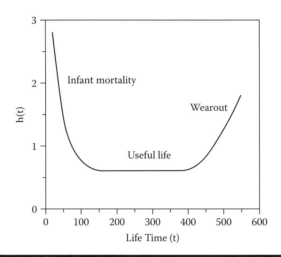

Figure 15.10 Bathtub hazard function.

Table 15.1 Fan Data

Week	No. Failed	Week	No. Failed	Week	No. Failed
1	7	13	0	25	4
2	3	14	0	26	7
3	3	15	0	27	9
4	2	16	0	28	12
5	1	17	1	29	9
6	0	18	0	30	10
7	0	19	0	31	18
8	0	20	2	32	10
9	0	21	0	33	19
10	0	22	0	34	21
11	0	23	0		
12	0	24	3		

fan failed, it was replaced with the "old" brand of fan. Thus, all originally (new) installed fans were monitored. A total of 214 fans were installed at the start of this project. The fans operated 24 h/day. The data is shown in Table 15.1.

Looking at the failure rate gives us an idea of what is causing the failure. If the failure rate is initially high and decreasing, this is infant mortality. The cause for this kind of failure must be found in the production and delivery system. For example, metal cuttings are occasionally left in internal combustion engines. This causes premature bearing wear, oil pump failure, and other nasty problems.

Weibull Probability Density Function

The Weibull distribution can take on various shapes to model different failure phenomena. Some general shapes of the Weibull pdf are shown in Figure 15.11.

Three-Parameter Weibull

The pdf for the three-parameter Weibull is given by

$$f(t) = \frac{\beta}{(\theta - \delta)^\beta}(t - \delta)^{\beta-1}e^{-(t-\delta/\theta-\delta)^\beta} , t \geq \delta$$

where
  t = time to failure ($t \geq \delta$)
  δ = minimum life parameter ($\delta \geq 0$)
  β = Weibull slope parameter ($\beta > 0$)
  θ = characteristic life ($\theta > \delta$)

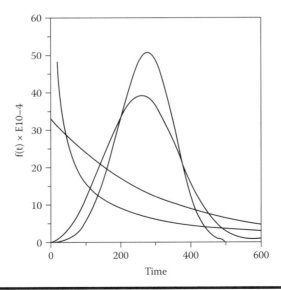

Figure 15.11 Weibull distribution.

The peakedness of the Weibull pdf is determined by the value of the parameter β. Higher values of β give more peaked curves. This parameter is sometimes called the *shape parameter*.

The reliability function is

$$R(t) = e^{-(t-\delta/\theta-\delta)^\beta}, t \geq \delta$$

Also, for a given reliability

$$t = \delta + (\theta - \delta)[\ln(1/R)]^{1/\beta}$$

and the B_{10} life is

$$B10 = \delta + (\theta - \delta)[\ln(1/0.90)]^{1/\beta}$$

This is the time at which 10% of the population will fail. The B_{10} life was a terminology initially used to describe bearing reliability.

Two-Parameter Weibull

For the two-parameter Weibull, the pdf is given by

$$F(t) = \frac{\beta}{\theta^\beta} t^{\beta-1} e^{-(t/\theta)^\beta}, t \geq 0$$

and the reliability function is

$$R(t) = e^{-(t/\theta)^\beta}, t \geq 0$$

Some plots of this reliability function are shown in Figure 15.12. For a given reliability R,

$$t = \theta[\ln(1/R)]^{1/\beta}$$

and the B_{10} life is

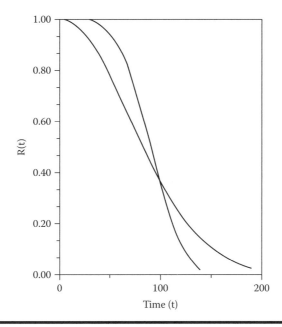

Figure 15.12 Two-parameter Weibull reliability functions.

$$B_{10} = \theta[\ln(1/0.90)]^{1/\beta}$$

The hazard function (theoretical failure rate) is

$$h(t) = \frac{\beta}{\theta^\beta} t^{\beta-1}$$

Various plots of this hazard function are shown in Figure 15.13. From the Weibull hazard function, it can be seen that the failure rate is constant when $\beta = 1$, and if we substitute this value of β into the Weibull reliability function, we have $R(t) = e^{-t/\theta}$, which is the exponential distribution.

The value of Beta (the Weibull slope) can give us an idea of the mechanism of failure. This is described in Table 15.2 for various important ranges of values of β. Here, it can be seen that a $\beta < 1$ implies infant mortality, whereas $\beta > 1$ implies wearout.

Characteristic Life

Given that $R(t) = e^{-(t/\theta)^\beta}$, then for $t = \theta$, we have

$$R(\theta) = e^{-(\theta/\theta)^\beta} = e^{-1} = 0.368$$

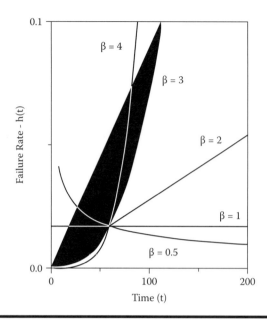

Figure 15.13 Weibull hazard functions.

Table 15.2 Weibull Slope Versus Failure Mechanism

Weibull Slope (β)	Failure Mechanism
$\beta < 1$	Decreasing failure rate Infant mortality
$\beta = 1$	Constant failure rate Exponential
$\beta > 1$	Increasing failure rate Wearout, aging

which provides some motivation for the term *characteristic life*. As can be seen in Figure 15.14, at the characteristic life, the reliability is *always* 0.368.

Graphical Estimation of Weibull Parameters

The Weibull distribution is very amenable to graphical analysis. Graphical analysis is useful for visualizing failure phenomena, particularly in small-sample-size situations. Weibull paper is constructed using the cumulative distribution, $F(t)$. Recall that $R(t) = 1 - F(t)$, so we would proceed as follows:

$$R(t) = e^{-(t/\theta)^\beta} = 1 - F(t)$$

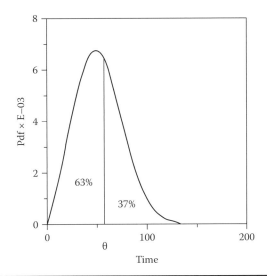

Figure 15.14 Characteristic life value and population failure.

$$\frac{1}{1-F(t)} = e^{(t/\theta)^\beta}$$

$$\ln\left[\frac{1}{1-F(t)}\right] = \left(t/\theta\right)^\beta$$

$$\ln\ln\left[\frac{1}{1-F(t)}\right] = \beta\ln(t) - \beta\ln(\theta)$$

Weibull Slope

This is now of the form

$$Y = \beta X - A$$

and we could plot on rectangular graph paper by using $\ln t$ and $\ln \ln (1/1 - F(t))$, but it is easier to transform the scales. The scale for time, t, would be a simple natural log scale, while the scale for $F(t)$ would take some special work. The value of β would be the slope of the plotted line. This slope is not affected by the logarithms and is untransformed. On Weibull paper, we could measure the slope with a ruler. If a software package is used, the slope is calculated based on the coefficient of the regression line.

Analysis of data with Weibull paper. Let us apply graphical analysis to the following data:

Order Number j	Time to Failure (h) t	Plotting Position pj
1	91	
2	122	
3	195	
4	220	
5	261	
6	315	
7	397	

The order number (j), is the ordering of a particular failure time with respect to all of the failure times. In this instance, the total number of devices placed on test was $n = 7$, and so $j = 1, 2, \ldots, 7$.

The failure times, (t), are plotted on the abscissa (x-axis), while the plotting position, (p_j), is the cumulative percentage failing in the population and is plotted on the ordinate (y-axis). We must select appropriate plotting positions.

Plotting Positions

We know that at 91 h, the fraction of the sample failing was 1/7, but this is not the fraction of the population. We must estimate the fraction of the population that might have failed by each known failure time. There are two possibilities for assigning plotting positions to the failure times: these are termed *mean rank* and *median rank* plotting positions. Almost everyone uses median rank plotting.

Median Rank — The median rank can be calculated by

$$p_j = \frac{j - 0.3}{n + 0.4}$$

where
n = sample size
j = the order number of the observation

We also have rank tables.

Mean Rank

The mean rank plotting position is also easy to calculate and is given by

$$p_j = \frac{j}{n+1}$$

where
n = sample size
j = the order number of the observation

The mean rank plotting position is useful for a quick-and-dirty estimate of the B_{10} life, for you can see that if we place $n = 9$ devices on test and wait until the first failure, then

$$p_1 = \frac{1}{9+1} = 0.10$$

So t_1 is an estimate of the B_{10} life. We will use median rank plotting, because as mentioned, it is the most popular. (Rarely will you find the mean rank method used in practice.)

Median Rank Tables

The median rank tables give the exact values for p_j for each ordered failure time. In our example, we find that:

j\n	7
1	9.428
2	22.849
3	36.412
4	50.000
5	63.588
6	77.151
7	90.572

The values are percentages and should be rounded off for plotting purposes. In rounding the values, the Weibull paper has an expanded scale below 10% and above 90%; so, values in these ranges should be rounded to the nearest 0.1%, while values between 10% and 90% can be rounded to the nearest 1%.

Procedure for plotting the Weibull:

1. Arrange the data points from the lowest to the highest value, and number the data points sequentially. The sequential numbers are called *order numbers*.
2. From the table of median ranks, assign the appropriate median rank value to each ordered data value. The median rank value is read under the sample size.
3. On a sheet of Weibull probability paper, establish a scale on the axis that will include all the failure time values. The scale is logarithmic and is laid out in major sections called *decades*. As the scale is read from left to right, the decimal point moves one place to the right at the beginning of each new decade.
4. Plot each failure time using the recorded failure value and associated median rank table as obtained from step 2.
5. By eye, draw the best straight line through the plotted points. Place more emphasis on points near the center of the line than on points toward the ends.

You can estimate the Weibull slope directly from the slope of the straight line. The characteristic life can be estimated at the 63% point.

Confidence Limits for Graphical Analysis

Let us imagine that we had two Weibull plots for the same test situation. One Weibull plot was based on a sample size of 10, while the second plot was based on a sample of 100. Which plot would we have more confidence in (for estimation purposes)?

Obviously, we have more confidence in predictions made with larger sample sizes. This leads us to the subject of confidence limits. Confidence can be shown by placing confidence limits about the Weibull plot.

j	5%	95%
1	0.730	34.816
2	5.337	52.070
3	12.876	65.874
4	22.532	77.468
5	34.126	87.124
6	47.820	94.662
7	65.184	99.270

Let us look up the confidence values for our example problem. These values must be plotted about the population line. Any plotted points not on the population line must be moved horizontally to the population line.

Constructing Confidence Limits

1. Copy the 5% and 95% rank values from the rank tables (or other suitable values).
2. Horizontally intersect the Weibull population line (plotted line) at each data point.
3. Using the 5% rank values, plot these vertically below each intersect on the population line.
4. Connect these points with a smooth curve.
5. Using the 95% rank values, plot these vertically above each intersect on the population line.
6. Connect these points with a smooth curve.

You now have constructed a 90% confidence interval.

Confidence interval for the Weibull slope. The Weibull slope indicates the phase of the bathtub curve that is associated with product life. We would sometimes like to have confidence in the value of the slope. The following method assumes that the Weibull slope is a maximum likelihood estimate, which is a statistical procedure for estimating the slope. However, we can use our graphical estimate and get some idea of the possible range of the slope. The approximate confidence limits for the Weibull slope are

$$\frac{g\hat{\beta}}{\chi^2_{\alpha/2,b}} \leq \beta \leq \frac{g\hat{\beta}}{\chi^2_{1-\alpha/2,b}}$$

where
  b is calculated from $b = 3.30n - 3.65$ and $g = b + 2$
  $\chi^2_{\alpha,b}$ values are found in chi-square tables.

Suspended Item Analysis

Suspended items are samples placed on test and that were removed from test for reasons other than failure. Suspended times can result from

- Test stand failure
- Secondary failure
- Accidents
- Field data

So, a suspended item is an item that was taken off test for reasons other than failure. Consider the situation:

No.	Time (h)	Occurrence
1	84	Failure (F_1)
2	91	Suspension (S)
3	122	F_2
4	274	F_3

We will now use a plotting method that adjusts the order number of the failed items by recognizing the pattern of suspensions and failures. The basis of this method is to consider all possible orderings of suspensions and failures and assign an average order number to the failure times.

Considering the preceding sequence, the possible sequences of S, F that could have occurred are

I	II	III
F_1	F_1	F_1
S→F	F_2	F_2
F_2	S→F	F_3
F_3	F_3	S→F

The approach is to assign an average rank to each failure time. We calculate the average ranks and then proceed to plot the points.

New Increment Method

This method provides the average rank values in an easy and systematic fashion. The new increment is

$$I = \frac{(n+1)-(\text{Previous order number})}{1+(\text{Number of items following suspended set})}$$

The median rank plotting position can be found from the formula:

$$p_j = \frac{j-0.3}{n+0.4}$$

Sudden Death Testing

The sudden death approach to life testing was proposed by Leonard Johnson in the 1950s when he was working as a research scientist at the General Motors Technical Center. This procedure helps to shorten the test duration.

Testing occurs in batches. The batch size might be determined by your test facility capacity. A batch is placed on test, and the testing continues until the first failure occurs, upon which the batch is suspended and a second batch is placed on test. This procedure is followed successively.

Let us say you have 20 test pieces. You elect to randomly divide the test pieces up into four batches of five pieces each. You then test each batch until the first failure occurs for the batch. Your results are shown in Table 15.3.

Let us plot these four failure times on Weibull paper using the rank values for $n = 4$. These rank values are found in the median rank tables and are shown in Table 15.3.

We then plot the times and ranks as summarized in Table 15.3. This is the first in five plot. The plot is shown in Figure 15.15.

The plot of Figure 15.15 represents the median rank estimate of the first in five failure times. The rank for the first in five failures is 12.9%, as found from the median rank values for $n = 5$ (the batch size). The first in five plot is the median rank (50%-tile) estimate of the distribution of the 12.9%-tile. If we go across at 50% on the Weibull paper to the plotted line and drop down to 12.9%, we have located the population line. This population line is parallel to our plotted line. So, we can easily draw in the population line. The procedure can be outlined as follows:

- Divide the test sample into equal batches.
- Test each batch until the first failure occurs.

Table 15.3 Sudden Death Data

Batch Number	Time to Failure (in Hours)	j	p_j(%)	t (in Hours)
1	56	1	16	17
2	124	2	39	56
3	294	3	61	124
4	17	4	84	294

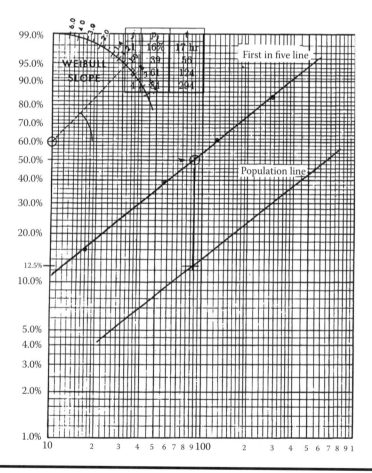

Figure 15.15 Plot of first in five and population line.

Table 15.4 Data for First Failure in Ten

j	p_i	t
1	13%	72 cycles
2	31	241
3	50	505
4	69	910
5	87	1,920

- Plot the first failure times on Weibull paper.
- Using the batch size, identify the correct percentage from the median rank tables.
- At the 50% point, go across to the plotted line and down to the correct percentage.
- Draw in the parallel population line.

Larger batches should produce the first failure quicker.

Let us try this with a second situation. In this situation, we tested 50 parts in groups of 10. The values in Table 15.4 represent the first failure in a batch of 10 parts placed on test. Here, we had five such batches. We looked up the median rank values (p_i) in the table under $n = 5$. For a sample of size $n = 10$, which is the batch size, we find that the first failure represents 6.7%. So we go across at 50%, and drop down to the 6.7% value to locate the population line. The plot is given in Figure 15.16.

Example 1: Precision Grinder

In this example, remember that
- The life distribution of a product can often be determined through life testing of some samples.
- Data from the test is used to construct a cumulative distribution function (CDF) of failures.
- Clear definition of failure is required, for example:
  - Level of function below target, for example, excessive vibration.
- Cessation of function, such as a break, crack, or malfunction.

In our situation, a random sample of 10 precision grinder wheel failures was obtained over several months of production. When a wheel fails, the number of pieces cut is recorded (in thousands). The data are

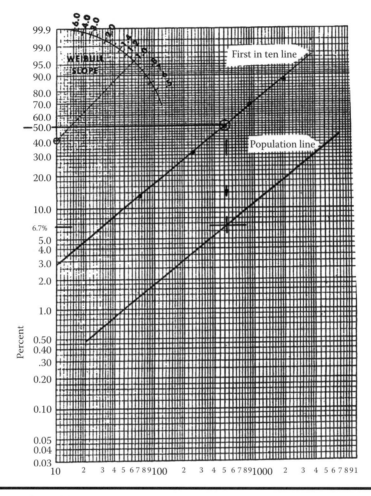

Figure 15.16 Plot of first in 10 and population line.

Pieces Cut at Failure	
Wheel A: 1.2×10^5	Wheel B: 7.2×10^4
Wheel C: 1.6×10^5	Wheel D: 2.3×10^5
Wheel E: 5.4×10^4	Wheel F: 1.6×10^4
Wheel G: 9.2×10^4	Wheel H: 5.8×10^3
Wheel I: 2.1×10^4	Wheel J: 3.8×10^4

Assumptions: The random sample is representative of production. All failures occur due to the same failure mode.

Objective: Use Weibull analysis to predict the failure characteristics of a population of items based on the failure behavior of a small sample of that population. We will create a CDF for the data in this example.

ANALYSIS

STEP 1: RANK THE DATA IN INCREASING ORDER (SHORTEST TO LONGEST LIFE)

j	Failure Times	Median Rank
1	5.8×10^3	6.7
2	1.6×10^4	16.3
3	2.1×10^4	25.9
4	3.8×10^4	35.6
5	5.4×10^4	45.2
6	7.2×10^4	54.8
7	9.2×10^4	64.4
8	1.2×10^5	74.1
9	1.6×10^5	83.7
10	2.3×10^5	93.3
n = 10		$100 \dfrac{(j-0.3)}{(n+0.4)}$

STEP 2: AGAINST EACH FAILURE TIME, ASSIGN MEDIAN RANKS

For each observation, the median rank gives a typical population percentage represented by that observation. Median ranks can be estimated by the formula $[(j - 0.3)/(n + 0.4)] \times 100$, where j is the failure order (rank) and n is the total number of samples. Actual values for these ranks can be found in statistical tables (e.g., see Appendix K).

STEP 3: PLOT THE DATA ON WEIBULL PAPER: FAILURE TIME (X) VERSUS RANK (Y)

Determine the slope b of the best-fit line. In this case, the plotted data indicates a slope of $b = 1.1$. The data plotted represents a CDF made linear by the Weibull transformation. Examine the plotted data. If the data does not appear close to a straight line, the data may need to be modified or a different statistical model may need to be applied (refer to the discussion of nonlinearity of Weibull plots). See Figure 15.17.

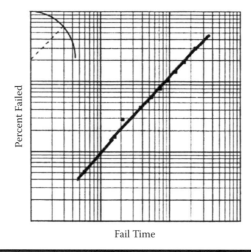

Figure 15.17 Failure time (x) vs. Rank (y).

STEP 4: Determine the Characteristic Life, θ

Characteristic life: Draw a horizontal line between the 63.2% point on the y-axis and the best-fit line. From the intersection point with the best-fit line, draw a vertical line straight down to the x-axis. This value on the x-axis is θ. In this case, θ = 8×10^4. Substitute the estimates of b and θ into the equation for the cumulative distribution to get the estimated distribution.

$$F(t) = 1 - e^{-(t/\theta)b}; \; F(t) = 1 - e^{-\left(t/8 \times 10^4\right)^{1.1}}$$

STEP 5: If Desired, Determine the Mean (Mean Failure Time) of the Estimated Weibull Distribution

This is useful for comparing two sets of Weibull data. The location of the mean of a Weibull distribution varies with the parameter b according to the functional relationship shown in the preceding graph. For the data from our example, b = 1.1. The corresponding point on the y-axis of the graph is 62% cumulative failure. On the Weibull paper, trace a line horizontally from 62% on the y-axis to the plotted line. Trace a line down vertically from the intersection point to the x-axis. This point on the x-axis is the mean failure time. In our example plot, 62% cumulative failure will correspond to 7.7×10^4 cycles. This is the mean failure time [$4 \times 10^6 < \mu < 15 \times 10^6$ at 90% C.L.]. See Figure 15.18.

How certain can we be that our estimates of reliability represent the true reliability of the population based on the test?

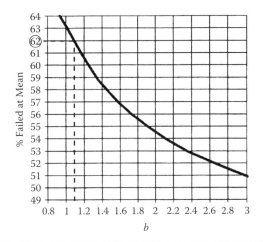

Figure 15.18 Characteristic life.

STEP 6: PLOT CONFIDENCE BANDS

Against each failure time, assign median, 95%, and 5% ranks.

Failure j Times	Median (50%)	5% Ranks	95% Ranks
1. 5.8×10^3	6.7	.5	25.89
2. 1.6×10^4	16.3	3.7	39.4
3. 2.1×10^4	25.9	8.7	50.7
4. 3.8×10^4	35.6	15.0	60.8
5. 5.4×10^4	45.2	22.2	69.6
6. 7.2×10^4	54.8	30.4	77.8
7. 9.2×10^4	64.4	39.3	87.0
8. 1.2×10^5	74.1	49.3	91.3
9. 1.6×10^5	83.7	60.6	96.3
10. 2.3×10^5	93.3	74.1	99.5

Actual values for these ranks can be found in statistical tables (see Appendix K).

Plot the 95% and 5% ranks listed in the data table. Draw a smooth curve connecting points in each of the sets of ranks. (If one is not interested in finding the percentage of not failed, then the calculation becomes [(1 percentage failed].)

In this figure, we do not use the Weibull paper; however, the shape and slope are accurate for the example.

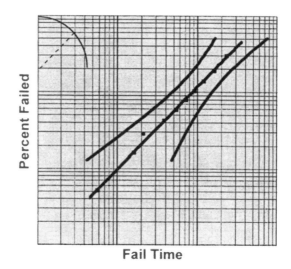

The 95% and 5% confidence curves create a 90% confidence interval about the cumulative distribution line. For example, when $t = 2 \times 10^4$ cycles, the corresponding percentages on the 95 and 5 percentile curves are roughly 40% and 5.5%. This means that there is 90% confidence that the reliability at 2×10^4 cycles is between 60% and 94.5%.

Question: What is the 90% confidence level around the B_{50} life?
Answer: $2.6 \times 10^4 < B_{50} = 5.5 \times 10^4 < 1.2 \times 10^5$ at 90% C.L.

STEP 7: ACTION PLAN

Based on our analysis, what should be the manufacturing engineers' action plan?

- Check actual versus planned lifetime operation.
- Use life data to set optimal cost-effective preventative maintenance plan.
- Make comparisons between different wheel designs over time, between batches and between materials.
- Select another wheel with least lifetime variation.

See Figure 15.19.

Example 2: Turbine Blade

Often, one is interested in estimating improvement in life of a proposed new design over a current base design. Several questions may arise:

- Is the percentage improvement indicated by the test data in fact real, or is it only the manifestation of test variability?
- What confidence does one have in concluding that the improvement is real?
- How many test samples are needed to make a statistically valid conclusion?

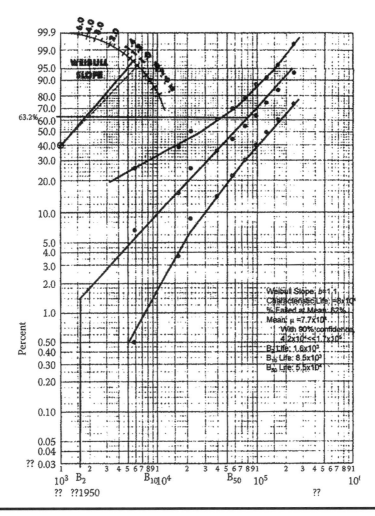

Figure 15.19 Weibull Example #1.

Let us address these questions by employing the Weibull method to turbine blade data, which are based on the example given by C. Lipson and N. J. Sheth (1973), in *Statistical Design and Analysis of Engineering Experiments,* McGraw-Hill, New York:

Design A	Design B
210	1,010
360	400
320	600
480	225
140	830
520	480
	780
$n = 6$	$n = 7$

420 ■ *The OEE Primer*

The data represent the lives of turbine rotor blades tested in the laboratory under accelerated conditions simulating an actual operation:

- Design A represents the type presently in use, which experiences some field failures.
- Design B represents a different material thought to be better, but involving a cost increase.

On the basis of the data, determine with 90%, 95%, and 99% confidences whether there exists a significant difference between the mean lives of these two designs.
Analysis:

STEP 1: Arrange the data in increasing order.
STEP 2: Assign median rank values (from table in Appendix K).

Design A		Design B	
Life hours	Median rank (%)	Life hours	Median rank (%)
		225	9.4
140	10.9	400	23.0
210	26.6	480	36.5
320	42.2	600	50.0
360	57.8	780	63.5
480	73.5	830	77.1
520	89.1	1010	90.6

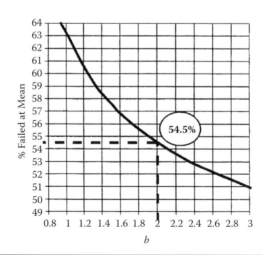

Figure 15.20 Determine Weibull slope.

STEP 3: Plot both datasets on Weibull probability paper.
STEP 4: Determine the Weibull slope B for each design.
STEP 5: Estimate the mean life for both designs.
Determining the mean
STEP 6: Look up the % mean location for a given b.
Use the table in Appendix K (Position of the Weibull mean). See Figure 15.20.
STEP 7: Read the mean life values for both designs from the Weibull plot corresponding to the percentage mean location from the table.
STEP 8: Compute the ratio of the two means (mean life ratio, MLR_{Exp})—a value greater than 1. $\mu_B/\mu_A = 650/360 = 1.8$.
STEP 9: Look up the theoretical value of MLR_{Theo} from tables in the appendix (see Appendix K). This requires calculating degrees of freedom.

$$d.f. = (n_1 - 1)(n_2 - 1) = (7 - 1)(6 - 1) = 30.$$

When using data to estimate statistical parameters, the number of data points used is the number of degrees of freedom (DoFs). Each time a parameter is estimated, a DoF is used up.

Table Confidence level MLR_{Theo}

1	90%	1.41
2	95%	1.58
3	99%	1.99

$MLR_{Exp} = 1.8$

STEP 10: If MLR_{Exp} (signal) > MLR_{Theo} (noise), then a significant difference in mean lives exists at the given confidence level (see Figure 15.21).

Example 3: Suspended Tests

An item is suspended when it is removed from test before failure (this is also known as *censoring*). Suspended item analysis is used when

- There are items in the sample that have not yet failed.
- More items are placed on test than are expected to fail during the allotted test time.
- We need to make an analysis of results before test completion.

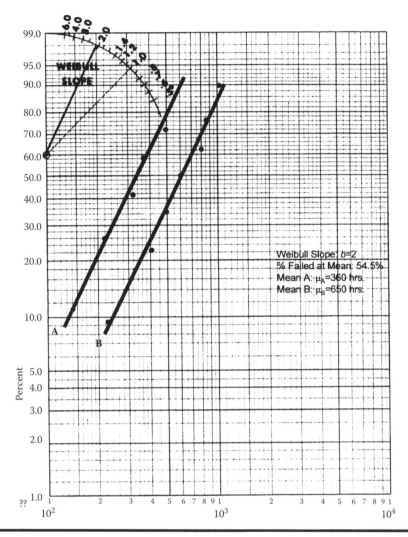

Figure 15.21 Weibull example #2.

- Some units may be malfunctioning, but it is unclear whether they have failed. These units are withdrawn and checked. If no failures are found, the items are suspended.
- All data should be used, but only the failures are plotted.
- Information from the suspended units is used to modify the ranks of the failures, which modifies the cumulative probability plotting positions. As a result, the analysis is more precise because median ranks for all items tested are used, and yields narrower confidence bands.

For example:

	Plot six failure times with modified ranks			
Life data for a shaved die	J	Failure order	Failure times	Median rank
Operating hours (in hundreds) for six dies that wore out: 13.2, 67.8, 79.0, 59, 30, 26.7	1	1	13.0	6.1%
	2	2	13.2	14.9%
	3	3	26.7	23.7%
	4		30.0	32.4%
	5		49.5	41.2%
	6		58.0	50.0%
	7	4	59.0	58.8%
	8		62.8	67.6%
	9	5	67.8	76.3%
	10		75.3	85.1%
	11	6	79.0	93.9%
Operating hours (in hundreds) for five dies replaced when production was stopped for other reasons: 58, 13, 75.3, 62.8, 49.5				

The rest of the analysis is the same as in Example 1.

Example 4: Nonlinearity in Weibull Plots

Quite often, the Weibull plot is not linear. The lack of linear fit may result from several reasons:

- The Weibull distribution is not appropriate for the data (i.e., use a different distribution).
- Mixed failure modes are present.
- Mixed populations are present (different ages, designs, etc.).
- There is some "minimum life" to which all units will live before any will fail. This can occur for two reasons:
    - Unit not stressed (e.g., storage, transportation times)

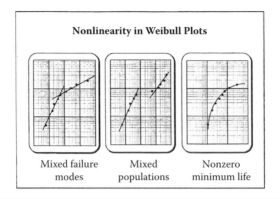

Figure 15.22 Nonlinearity in Weibull plots.

- A minimum number of time, miles, or cycles needs to occur before a failure can develop (e.g., fatigue spalling of a gear or bearing). A three-parameter Weibull can be fitted using software, or an estimate of the minimum life can be subtracted from the data, and a two-parameter Weibull fit. Be sure to add the minimum life back to any Bq values estimated from the two-parameter fitted model.
- A combination of the preceding list.

Proper analysis may require a thorough examination of the failed hardware, such as:

- Use failure mode analysis to detect mixture of failure modes.
- Trace or know background data on test hardware to detect mixture of populations.
- Perform analysis for each failure mode by considering the items failing due to other failure modes as suspended items.
- Plot mixed populations separately.

Typical graphical representation of nonlinear plots is shown in Figure 15.22.

Summary

This chapter introduced the Weibull distribution, which allows us to understand the failures of machines or equipment. This chapter provided not only the background of Weibull analysis but also showed some specific applications of dealing with life data analysis (i.e., failures). For a computer application, see Appendix L, which shows an Excel application.

Glossary

Accelerated life testing: Testing to verify design reliability of machinery/equipment much sooner than is operating typically. This is intended especially for new technology, design changes, and ongoing development.

Accelerated test methods: A test strategy that shortens the time to failure of the product under test by inducing stresses that are more severe than operationally encountered stresses (e.g., more severe than the 90th percentile loads) in a way that does not change the underlying failure mechanisms/modes that customers may potentially encounter. The technique of accelerated testing involves selection of a mathematical model, selection of accelerated stress parameters and acceleration levels, generation of test procedures, and analysis of the test data.

Acceptance test: A test to determine machinery/equipment conformance to the (qualification test): qualification requirements in its equipment specifications.

Accessibility: The amount of working space available around a component sufficient to diagnose, troubleshoot, and complete maintenance activities safely and effectively. Provision must be made for movement of necessary tools and equipment with consideration for human ergonomic limitations.

Actual machine cycle: Actual time to process a part or complete an operation. It includes all value added timing. It excludes all waste.

Allocation: The process by which a top-level quantitative requirement is assigned to lower hardware items/subsystems in relation to system-level reliability and maintainability goals.

AMADEOS: Advanced Multi-Attribute Design Evaluation and Optimization Software.

Analytical thinking: Breaking a whole into parts.
APQP: Advanced Product Quality Planning.
Associative thinking: Joining parts into a whole.
Availability, achieved: Includes preventative maintenance as well as corrective maintenance. It is a function of the mean time between maintenance actions (MTBMA), and the mean maintenance time (MMT), that is, MTBMA/(MTBMA + MMT).
Availability, inherent: The ideal state for analyzing availability. It is a function only of the mean time between failures (MTBF) and the mean time to repair (MTTR); preventative maintenance is not considered, that is, MTBF/(MTBF + MTTR).
Availability, operational: Includes preventative maintenance, corrective maintenance, and delay time before maintenance begins, such as waiting for parts or personnel. It is a function of the mean time between maintenance actions (MTBMA), and the mean downtime (MDT), that is, MTBMA/(MTBMA + MDT).
Availability: A measure of the degree to which machinery/equipment is in an operable and committable state at any point in time. Specifically, the percent of time that machinery/equipment will be operable when needed. Another way of saying it is the relationship between the time taken for repairs versus the total amount of time that the product should be available to the consumer. For total customer satisfaction, the following calculation should tend to be 1.
Availability = (total time − repair time) ÷ total time
Axiomatic design: An approach to the design of a product or process based on the independence and information design axioms that provide principles for the development of a robust design.
Background variation: Sources of variation that are always present and are part of the natural (random) variation inherent in the process itself (e.g., lighting, raw-material variation). Their origin can usually be traced to elements of the system that can be eliminated only by changing the process.
Bathtub curve: The plot of instantaneous failure rate versus time is known as a hazard curve. It is more often called a bathtub curve because of its shape. It consists of three distinct patterns of failures: failures with decreasing rates (e.g., infant mortality), failures with constant rates (e.g., accidental, sudden overload, impact), and failures with increasing rates (e.g., wearout, fatigue). The bathtub curve is actually the sum of three distributions:

Infant mortality period (beta < 1) – negative exponential
Normal period (beta = 1) (i.e., purely exponential) – constant
Wearout period (beta > 1) – positive exponential

Bayesian statistics: A method that provides a formal means of including prior information within a statistical analysis. It is a way of expressing and updating a decision-oriented learning process. A software package called ReDCAS is available to perform Bayesian analysis.

Bogey testing: Conducting a test only to a specified time, mileage, or number of cycles.

B_q life: The life at which "q" percent of the items in a population are expected to fail. For example, "B_{10} = 70,000 mi," means that 10% of the items are expected to fail by 70,000 mi (and that 90% had a life over 70,000 mi).

Built-in-Test (BIT): The self-test hardware and software that is internal to a unit to test the unit.

Built-in-Test Equipment (BITE): A unit that is part of a system and is used for the express purpose of testing the system. MTh is an identifiable unit of a system.

Burn-in: The operation of an item under stress to stabilize its characteristics.

CAE: Computer-Aided Engineering.

Capability: A measure of the ability of an item to achieve mission objectives given the conditions during the mission.

CCC: Customer Concern Code.

CDF: Cumulative Distribution Function.

CDS: Component Design Specification.

Censored data: Life data that are incomplete due to partial information. Exact failure times may not be known because some units are unfailed, some have failed at an unknown earlier point in time, some are failed due to other extraneous (e.g., test equipment) causes.

Concept design: The selection and execution of a conceptual approach in the preliminary design phase to achieve a specified function for a product or process.

Confidence interval: An interval estimate of a parameter that contains the true value of the parameter with given probability. The width of the interval is affected by the degree of confidence, sample size, and variability.

Confidence level: The probability that some specified interval will contain a quoted reliability (or other criterion), for example, the probability that engine reliability will be at least as high as a stated value.

Control factors: A term frequently used in experimental/robust design to designate those design or process variables that are controllable and may be examined for their level of impact on the performance variable of interest.

Convergent thinking: The process of narrowing down to one answer.

Core engineering: Consists of engineers that develop generic technology requirements and designs for forward models, develop common-cause resolutions, and maintain a knowledge and lessons-learned database.

Corrective (unscheduled, unplanned, repair) maintenance: All actions performed, as a result of failure, to restore a machine to a specified condition. It can include any or all of the following steps: localization, isolation, disassembly, interchange, reassembly, alignment, and checkouts.

Corrective Action System: It is a system that ensures established disciplined problem solving and approval processes are followed, resulting in an efficient and consistent internal problem solving system. It also enables the user to define and implement many different problem solving methodologies, ensuring all different problem types (e.g., supplier, customer, warranty, safety, product, audit, etc.) Can be managed as the user requires.

Corrosion: Deterioration of a metallic substance because of a chemical reaction with its environment by oxidation or chemical combination.

CRDS: Corporate Reliability Data System.

Debugging: A process to detect and remedy inadequacies.

Degradation: Degradation is the undesirable change, over time or usage, in a function of the system or component.

Delay study: A continuous study over an extended period of time (say, 2 weeks) where every incidence of downtime is recorded along with the apparent cause.

Dependability failure: Failures that cause customers to be stranded or lose vehicle function.

Derating: Using an item in such a way that applied stresses are below rated values, or (b) the lowering of the rating of an item in one stress field to allow an increase in rating in another stress field.

Design machine cycle time (process cycle time): Specifically, the shortest period of time at the end of which a series of events in an operation is repeated.

Design of Experiments (DOE): A systematic approach to estimate the effects of various factors on the performance of products or services or the outputs of processes.

Design review: A meeting of a cross-functional group of technical people assembled to probe and demonstrate the thoroughness and competency of a proposed design's ability to maximize the ultimate customer's perceived value. The intent is to recommend design changes to improve the overall design.

Design Verification (DV): A testing/evaluation discipline that is used to verify that prototype components, subsystems, and/or systems made to print and assembled with simulated or actual production processes are capable of meeting functional, quality, reliability, and durability requirements.

Divergent thinking: The process of generating many answers.

DOE: Design of Experiments.

Durability life (expected life): A measure of useful life, defining the number of operating hours (or cycles) until overhaul is expected or required.

Durability: The ability of a component, subsystem, or system to perform its intended function satisfactorily without requiring overhaul or rebuild due to wearout during its projected design life.

DV: Design Verification.

DVP and R: Design Verification Plan & Report.

DVP: Design Verification Plan.

Eight D (8D): An orderly team-oriented approach to problem solving.

Engineered system: See P-diagram.

Engineering confidence: Confidence that our DV tests actually expose products to the critical customer usage conditions that they will experience in the field at the right frequency, and that by meeting functional performance targets in these tests, our products demonstrate they will operate reliably in the real world.

Equipment: Any materials, components, subassemblies, and assemblies incorporated into machinery provided by any supplier.

Event: All incidents regardless of reasons when equipment is not available to produce parts at specified conditions when scheduled, or is not capable of producing parts or perform scheduled operations to requirements.

Exploration thinking: Exploration thinking seeks to generate new information and facilitates continuous improvement in reliability and maintainability (R&M) efforts.

Failure analysis (FA): The logical systematic examination of a failed item, its construction, application, and documentation to verify the reported failure, identify the failure mode, and determine the failure mechanism and its basic failure cause. To be adequate, the procedure must determine whether corrective action is warranted and, if so, provide information to initiate corrective action.

Failure effect: The consequence of the failure.

Failure-free life: The period of time after manufacture and before the customer gets the product when you expect no failure to occur (e.g., parts awaiting assembly, parts at the dealer, or in the service distribution system).

Failure mechanism: The physical, chemical, electrical, thermal, or other processes that result in failure.

Failure Mode Analysis (FMA): For each critical parameter of a system determining which malfunction symptoms appear just before, or immediately after, failure.

Failure mode: The manner by which a failure is observed. Generally, a failure mode describes the way the failure occurs and its impact on equipment operation.

Failure rate: The ratio of the number of units failed to the total number of units that can potentially fail during a specified interval of life units; in other words, number of failures per unit of gross operating period in terms of time, events, cycles, or number of parts. When the life unit interval is very small, this is referred to as Instantaneous Failure Rate, and is a very useful indicator for identifying various patterns of failure.

Failure, random: Failure whose occurrence is predictable only in a probabilistic or statistical sense. This applies to all distributions.

Failure: A failure occurs when a product ceases to function in the way in which the customer expects. An event when machinery/equipment is not available to produce parts at specified conditions when scheduled or is not capable of producing parts or performing scheduled operations to specification. For every failure, an action is required.

Fatigue: Damage caused by repeated mechanical stress applied to a system, subsystem, or part.

Fault tree analysis (FTA): A deductive (top-down) analysis depicting the functional relationship of a design or a process to study potential failures. Focuses on a single combination of events to describe how the top (undesirable event) may occur. Useful for identifying appropriate corrective actions.

First Time Through Capability (FTT): The percentage of units that complete a process and meet quality guidelines without being scrapped, rerun, retested, returned, or diverted into an offline repair area.

FMA: An analysis of the ways a component or subsystem had failed to meet customer expectations in the past. FMA provides an overall data-supported assessment of a commodity from current or past production. FMA quantifies top concerns reported by the customer and associated planned actions.

FMEA (Failure mode and effects analysis): An inductive (bottom-up) analysis of a product or process that (1) considers the effects of all potential ways that the design could be impacted by external and internal factors, and (2) develops countermeasures to control or reduce the effects of those factors. FMEAs are performed early in the design concept phase for a new product or process cycle in order to receive maximum benefits. FMEA should be a cross-functional team process.

FRACAS (Failure Reporting Analysis and Corrective Action System): It is a data repository of product and systems failure history that acquires data and analyzes and records corrective actions for all failures occurring during reliability test efforts.

FRACAS: It is a closed-looped Failure Reporting Analysis and Corrective Action system.

FTA: Fault tree analysis

FTT: First-Time Through Capability = [Units entering the process − (scrap + rerun + retests + repaired offline + returns)]/Units entering the process

Functional block diagrams: These are used to break a system into its smaller elements and to show the functional and physical relationships between the elements.

Gross operating time: Total time that the machine is powered and producing parts. Gross operating time = Net operating time + Scrap time.

Hazard rate: The instantaneous failure rate. The probability of failure during the current interval.

Hierarchy block diagrams: These break the product into natural and logical elements and become more detailed at each level down.

Ideal function: The ideal relationship between signal and response (e.g., Y = Beta*M). It is based on the energy transformation that achieves what the product or process is primarily or fundamentally supposed to do.

Infant mortality: Early failures attributable to defects in design, manufacturing, or construction. They typically exist until debugging eliminates faulty components. These failures often happen during burn-in or during early usage. In most cases, infant mortality is associated directly with manufacturing learning problems.

Inherent R&M value: A measure of reliability or maintainability that includes only the effects of an item design and its application, and assumes an ideal operation and support environment.

Intended function: Typically a noun–verb description of what the product or process is supposed to do (i.e., create dimension, turn vehicle). It generally only includes the response.

Key life testing: A test that duplicates the stresses under customer operation. It includes the specified extremes of loads, usage, manufacturing/assembly variability, environmental stresses, and their interactions to address a single critical failure mode or a group of failure modes that result from the same stress patterns. It may be accelerated to reduce test time.

Life cycle: The sequence of phases through which machinery/equipment passes from conception through decommission.

Life units: A measure of duration applicable to the item, for example, operating hours, cycles, distance, rounds fired, attempts to operate.

Limited life component: A component that is designed to have a specific lifetime at which point it is discarded/replaced.

Machinery failure mode and effects analysis (MFMEA): A technique to identify each potential failure mode and its effect on machinery performance.

Maintainability: A characteristic of design, installation, and operation, usually expressed as the probability that a machine can be retained in, or restored to, specified operable condition within a specified interval of time when maintenance is performed in accordance with prescribed procedures. Another way of saying this is: it is the probability that a failed component or system will be restored or repaired to a specified condition within a given period of time when maintenance is performed in accordance with prescribed procedures. The basic measure is mean time to repair (MTTR).

Maintenance action rate: The reciprocal of the mean time between maintenance actions = 1/MTBMA.

Maintenance, corrective: All actions performed, as a result of failure, to restore an item to a specified condition. Corrective maintenance can include any or all of the following steps: localization, isolation, disassembly, interchange, reassembly, alignment, and checkout.

Maintenance, preventive: All actions performed in an attempt to retrain an item in a specified condition by providing systematic inspection, detection, and prevention of incipient failures.

Maintenance: Work performed to maintain machinery and equipment in its original operating condition to the extent possible; includes scheduled and unscheduled maintenance but does not include minor construction or change work.

Mean cycle between failures (MCBF): The average cycles between failure occurrences. It is the sum of the operating cycles of a machine divided by the total number of failures.

Mean cycle to repair (MCTR): The average cycles to restore machinery or equipment to specified conditions.

Mean downtime (MDT): The average calendar time that a system is not available for use in specified periods.

Mean maintenance time (MMT): A measure of item maintainability when serviced according to its maintenance policy.

Mean time between failure (MTBF): A reliability measure that estimates the average time of operation between failures. MTBF is calculated as the total operating time on all units divided by the total number of failures.

Mean time between maintenance actions (MTBMA): A measure of the system reliability parameter related to demand for maintenance manpower. The total number of system life units, divided by the total number of maintenance actions (preventative and corrective) during a stated period of time.

Mean time to failure (MTTF): A basic measure of the average time of operation to failure of a nonrepairable system. MTTF is calculated as the total operating time on all units divided by the total number of failures. It is also known as expected life or mean life.

Mean time to repair (MTTR): The average time to restore machinery or equipment to specified conditions.

Mean: Average.

Median: The midpoint or 50th percentile.

MIS: Months in Service.

Mission profile: A time-phased description of the events and environments an item experiences from initiation to completion of a specified mission, to include the criteria of mission success of crucial failures.

Mode: Value that occurs most often, or point of highest probability.

MTTR: The average time to complete a repair. MTTR is the basic measure of maintainability.

MY: Model Year.

Net operating time: Total time that the machine is producing parts, as a first pass, to specifications.

Noise factors: Factors that cannot be controlled or that the engineer decides not to control in a robust design process. Factors whose settings are difficult or expensive to control are also called noise factors.

Nonoperating time: Total time that the machinery/equipment is up but not running due to blocking, starvation, and/or administrative time.

Normal distribution: Best-known and most widely used statistical distribution ("bell-shaped" curve). It is symmetric about its mean.

Nova C: Standardized plant quality auditing process.

Operable: The state of being able to perform the intended function.

Overall equipment effectiveness (OEE): The product of three measurements: Percentage of time the machinery is available (Availability) × how fast the machinery or equipment is running relative to its design cycle (Performance efficiency) × percentage of the resulting product within quality specifications (Yield). The overall machine effectiveness of the machinery or equipment is calculated as: OEE = Availability × Performance efficiency × Yield.

Overhaul: A comprehensive inspection and restoration of machinery/equipment, or one of its major parts, to an acceptable condition at a durability or usage limit.

P Diagram: A schematic representation of the relationship among signal factors, control factors, noise factors, and the measured response. The P stands for parameter.

Parameter design: It achieves robustness with no increase in the product/manufacturing cost by running a statistical experiment with lowest-cost components, selecting optimum values for control factors that will reduce the variability of the response and shift the mean of the response toward target, and by selecting the lowest-cost setting for factors that have minimal effect on the response.

PDF: Probability density function.

PDS: Product Development System.

Percentile customer: A level of customer usage that exceeds that generated by n percent of customers. In other words, n percent of customers will use or stress vehicles less than the n-th percentile level. Typical values of n are 90 and 95. Sometimes this term is used generically to describe severe customer usage conditions under which designs should be tested for survival. *Comment*: Designing for the n-th percentile customer is generally appropriate only for wearout-type failures. Care should be taken to identify the n-th percentile customer within the market where a particular failure mode is likely to occur (e.g., hot or cold climates, high-or low-humidity regions, etc.).

Performance efficiency: Ideal Cycle Time × Total Part Run/Operating Time.

Predicted: That which is expected at some future date, postulated on analysis of past experience and tests.

Predictive and preventive (scheduled, planned) maintenance: All actions performed in an attempt to retain a machine-specified condition by providing systematic inspection, detection, and prevention of incipient failures.

Probability: The chance of an event happening (the number of ways a particular event can happen relative to the total number of possible outcomes).

Process control methods: This is a process of training in quality tools that includes statistical process control methods.

Process: A recurring sequence of related activities that have measurable input and output.

Process time: The time for the task to be completed. It includes both value and non value added activities.

Production Validation (PV): Demonstrates that the manufacturing and assembly processes will produce the design that satisfies the product requirements.

PS: Production System.

Pugh structure selection: A structured concept selection process used by cross-functional teams to converge on superior design concepts. The process uses a matrix consisting of criteria based on the voice of the customer and its relationship to specific design concepts. The evaluations are made by comparing the new concepts to a benchmark called the datum. The process uses the classification metric of "same as the datum." Several iterations are employed, in which ever-increasing superiority is developed by combining the best features of highly ranked concepts until the best concept emerges and becomes the benchmark.

QFT: Quality Function Deployment. This is a technique to assist in translating the "voice of the customer" into operational definitions that can be used to produce and deliver products desired by customers.

QLF: Quality Loss Function.

Quality Function Deployment (QFD): A discipline for product planning and development, or redesigning an existing product in which key user wants and needs are deployed throughout an organization. QFD provides a structure for ensuring that users' wants and needs are carefully heard, then directly translated into a company's internal technical requirements from component design through final assembly.

Quality lost function: The relationship between the dollars lost by a customer because of off-target product performance and the measured deviation of the product from its intended performance.

Quality: Customers define quality, customers want products and services that, throughout their life, meet their needs and expectations at a cost that represents value.

R/1000: Repairs per 1000 vehicles at various times in service, calculated from warranty/repair data.

Real-World Usage Profiles (RWUP): A quantification of customer usage, frequency, and load conditions that are relevant to a key failure mode. RWUP quantify the stresses that the product must be designed and tested to withstand.

ReDCAS: Reliability Data Collection and Analysis System. This is a software package available to perform Bayesian methods.

Redundancy: The existence of more than one means for accomplishing a given function. Each means of accomplishing the function need not necessarily be identical.

Reliability block diagram: This is a method to break down a system into smaller elements and to show their relationship from a reliability standpoint.

Reliability growth: Machine reliability improvement as a result of identifying and eliminating machinery or equipment failure causes during machine testing and operation.

Reliability, mission: The ability of an item to perform its required functions for the duration of a specified mission profile.

Reliability: Reliability is the probability of a product performing its intended function for a specified life under the operating conditions encountered, in a manner that meets or exceeds customer expectations. Average reliability at a point in time can be obtained using the following equation:

$$R(t) = \frac{N_{\text{Success}}}{N_{\text{TOTAL}}} = \frac{(N_{\text{TOTAL}} - N_{\text{FAILED}})}{N_{\text{TOTAL}}}$$

$$1 - \frac{N_{\text{FAILED}}}{N_{\text{Total}}} = 1 - \text{Unreliability}$$

REP: Robust Engineering Process.
Repairability: Probability that a failed system will be restored to operable condition within a specific repair time.
Robust technology development: An upstream activity that improves the efficiency and effectiveness of ideal function. The engineering activity focuses on developing flexible and reproducible technology before program definition.
Robustness: The ability of a product or process to perform its intended function under a variety of environmental and other uncontrollable conditions throughout the life cycle at the lowest possible cost. Thus, a robust design is insensitive to "noise."
ROCOF: Rate of change of failure, or rate of change of occurrence of failure.
Scheduled (planned) downtime: The elapsed time that the machine is down for scheduled maintenance or turned off for other reasons.
SDS: System Design Specifications.
Serviceability: The ease with which a system can be repaired. Serviceability is a characteristic of system design, primarily considering accessibility.
Servicing: The performance of any act to keep an item in operating condition (i.e., lubrication, fueling, oiling, cleaning, etc.) but not including other preventive maintenance of parts or corrective maintenance.
SIT: Structured Inventive Thinking (TRIZ).
SPC: Statistical Process Control. A statistical technique, using control charts to analyze process output, which serves as a basis for appropriate actions to achieve and maintain the process in a state of statistical control and improve the capability of the process. A specification that defines standards and evaluation criteria for acceptable specified conditions.

Storage life: The length of time an item can be stored under specified conditions and still meet specified requirements.

Stress–strength interference: Identification and quantification of failures probability resulting from the probability of stress exceeding product strength.

Structured Inventive Thinking (SIT): A method for applying TRIZ in a shorter time, with less reliance on external databases.

Subsystem: A combination of sets, groups, etc., that performs an operational function within a system and is a major subdivision of the system, for example, data-processing subsystem, guidance subsystem.

Success tree analysis: Analysis that focuses on what must happen at the top-level event to be a success.

Suspended data: Unfailed units that are still functioning at removal or end of test.

System Design Specification (SDS): A specific term that refers to an engineering specification that defines standards and evaluation criteria for acceptable system performance.

System: General—A composite of equipment and skills, and techniques capable of performing or supporting an operational role, or both. A complete system includes all equipment, related facilities, material, software, services, and personnel required for its operation and support to the degree that it can be considered self-sufficient in its intended operational environment.

Systems engineering: This is a process to transform customers' needs into effective designs. The process enables product engineers to optimize designs within and across systems.

Testability: A design characteristic allowing the following to be determined with a given confidence and in specified time: location of any faults, whether an item is inoperable, operable but degraded, and/or operable.

TGR: Things Gone Right.

TGW: Things Gone Wrong.

Theory of Inventive Problem Solving (TRIZ): This is a method for developing creative solutions to technical problems. It is also known as SIT.

Time To Repair (TTR): Total clock time from the occurrence of failure of a component or system to the time when the component or system is restored to service (i.e., capable of producing good parts or performing operations within acceptable limits). Typical elements of repair time are diagnostic time, troubleshooting time, waiting time for spare parts, replacement/fixing of broken parts, testing time, and restoring.

Time, administrative: That element of delay time that is not included in supply delay time.

Time, delay: That element of downtime during which no maintenance is being accomplished on the item because of either supply or administrative delay.

Time, supply delay: That element of delay time during which a needed replacement item is being obtained.

TIS: Time In Service.

TNI: Trouble Not Identified.

Tolerance design: Determination of the tolerances that should be tightened or loosened and by how much in order to reduce the response variability of the overall system to the desired level. Tolerance design increases product/manufacturing cost and should be done when parameter design has not sufficiently reduced variation.

Total downtime: The elapsed time during which a machine is not capable of operating to specifications. Total downtime = Scheduled downtime + Unscheduled downtime.

TPM: Total Productive Maintenance.

TRIZ: Theory of Inventive Problem Solving (TRIZ is a Russian acronym). Also known as SIT.

Unscheduled (unplanned) downtime: The elapsed time that the machine is incapable of operating to specifications because of unanticipated breakdowns.

Uptime: Total time that a machine is online (powered up) and capable of producing parts. Uptime = Gross operating time + Nonoperating time.

Useful life definition: The total operating time from final manufacturing debugging and wear out. It is also called normal failure period as the failures are random.

Vehicle design: A specific term that refers to an engineering specification.

Wear: Removal of material from the surfaces of the components as a result of their movement.

Wearout: A failure mode characterized by a hazard rate that increases with age; that is, old units are more likely to fail than new units. The product life-cycle phase that begins after the design's expected life.

Weibull analysis: Procedure for finding the Weibull distribution that best describes a sample of unit lifetimes in order to estimate reliability.

Yield: The fraction of products meeting quality standards produced by machinery or equipment.

Chapter 1: Selected Bibliography

Conner, G. (2001). *Lean manufacturing.* Society of Manufacturing Engineers, Dearborn, MI.

Flinchbaugh, J. and A. Carlino (2006). *The hitchhiker's Guide to lean: Lessons from the road.* Society of Manufacturing Engineers, Dearborn, MI.

Nicholas, J. and A. Soni (2006). *The portal to lean production.* Auerbach Publications, Boca Raton, FL.

Rother, M. and J. Shook (February 2000). *Training to see: A value stream mapping workshop.* The Lean Enterprise Institute, Brookline, MA.

Stamatis, D. (2004). *Integrating ISO 9001:2000 with ISO/TS 16949 and AS 9100.* ASQ Quality Press, Milwaukee, WI.

Womack, J. and D. Jones (June 1999). *Learning to see: Value stream mapping to add value and eliminate muda.* Version 1.2. The Lean Enterprise Institute, Brookline, MA.

Womack, J. and D. Jones (1996). *Lean thinking.* Simon and Schuster, New York.

Chapter 2: References

Goldratt, E. M. (May 1992). *The goal: A process of ongoing improvement.* 2nd rev. ed. North River Press, Great Barrington, MA.

Chapter 2: Selected Bibliography

Mouradian, G. (2000). *Handbook of QS-9000: Tooling and equipment certification.* Society of Automotive Engineers, Inc., Warrendale, PA.

Productivity Press (1999). *OEE for operators.* Productivity Press, New York.

Stamatis, D. (1998). *Implementing the TE Supplement to QS-9000: The tooling and equipment supplier's handbook.* Quality Resources, New York.

Chapter 3: Selected Bibliography

Bajaria, H. J., R. P. Copp. (1991). *Statistical problem solving,* Multiface Publishing Company. Garden City, MI.

Brassard, M. and D. Ritter. (1994). The *memory jogger II.* GOAL/QPC.Salem, NH.

Breyfogle, F. W. III. *Statistical methods for testing, development and manufacturing.* J. Wiley and Sons, Inc. NY.

Imai, M. (1986). *Kaizen: The key to Japan's competitive success*. Random House Business Division. NY.
Ishikawa, K. (1982). *Guide to quality control*. Asian Productivity Organization. Tokyo, Japan; Kraus International Publications. NY.
Mouradian, G. (2002). The quality revolution. University Press of America. NY.
Som, R. K. (1996). *Practical sampling techniques*. 2nd ed. Rev. and extended. Marcel Dakker, Inc. NY.
Stamatis, D. H. (2003). *Six sigma and beyond: Statistical process* control. St. Lucie Press. Boca Raton, FL.
Stamatis, D. H. (1997). *TQM engineering handbook*. Marcel Dekker, Inc. NY.
Wilburn, A. J. (1984). *Practical statistical sampling for auditors*. Marcel Dekker, Inc. NY.

Chapter 4: Selected Bibliography

Blanchard, B. S. (1986). *Logistics engineering and management*. 3rd ed. Prentice Hall, Englewood Cliffs, NJ.
Kececioglu, D. (1991). *Reliability engineering handbook*. Vol. 1–2. Prentice Hall, Englewood Cliffs, NJ.
Nelson, W. (1983). *How to analyze reliability data*. Vol. 6. American Society for Quality Control. Statistics Division, Milwaukee, WI.

Chapter 5: References

Ford Motor Company (2008). *G8D*. Ford Motor Company. Dearborn, MI.
Ford Motor Company (September 1987). *Team oriented problem solving*. Ford Motor Company. Power Train, Dearborn, MI.
Stamatis, D. (2003). *Failure mode and effect analysis: FMEA from theory to execution*. 2nd ed. Revised and expanded. ASQ Quality Press, Milwaukee, WI.
www.sv.vt.edu/classes/MSE2094_NoteBook/97ClassProj/num/widas/history.html.

Chapter 5: Selected Bibliography

Cook, R. D., D. S. Malkus, M. E. Plesha, and R. J. Witt (2002). *Concepts and applications of finite element analysis*. 4th ed. John Wiley & Sons, New York.
Productivity Press (1996). *Quick changeover for operators*. Productivity Press, New York.
Shingo, S. (1985). *A revolution in manufacturing: The SMED system*. Productivity Press, New York.
Shingo, S. (1986). *Zero quality control. Source inspection and the poka-yoke system*. Productivity Press, Cambridge, MA.

Shirose, Kunio (1995). *P-M analysis: An advanced step in TPM implementation.* Productivity Press, New York.

Chapter 6: Selected Bibliography

Bhote, K. R. (1991). *World class quality: Using design of experiments to make it happen.* AMACOM, New York.
Chowdhury, S. (2002). *Design for six sigma: The revolutionary process for achieving extraordinary profits.* Dearborn Trade Publishing, Chicago, IL.
Dovich, R. (1990). *Reliability statistics.* ASQ Quality Press, Milwaukee, WI.
Schonberger, R. (1986).*World class manufacturing: The lessons of simplicity applied.* The Free Press, New York.
Shina, S. (1991). *Concurrent engineering and design for manufacture of electronics products.* Van Nostrand Reinhold, New York.
Suh, N. (1990). *The principles of design.* Oxford University Press, New York.
Stamatis, D. (2003). *Six sigma and beyond: Design for six sigma.* St Lucie Press, Boca Raton, FL.
Taguchi, G., S. Chowdhury, and Y. Wu (2001). *The Mahalanobis-Taguchi system.* McGraw-Hill, New York.
Wowk, V. (1991). *Machinery vibration: Measurement and Analysis.* McGraw-Hill, New York.

Chapter 7: References

Kececioglu, D. (1991). *Reliability engineering handbook.* Vol. 1–2. Prentice Hall, Englewood Cliffs, NJ.
Stamatis, D. H. (2003). *Six sigma and beyond: Design for six sigma.* St. Lucie Press, Boca Raton, FL.
Stamatis, D. H. (1998). *Implementing the TE Supplement to QS-9000.* Quality Resources, New York.
Stamatis, D. H. (1997). *TQM engineering handbook.* Marcel Dekker, New York.

Chapter 8: References

Altshuller, G. (1997). *40 Principles: TRIZ keys to technical innovation,* Technical Innovation Center, Worcester, MA.
ASME. (1994). *Dimensioning and tolerancing [ASME Y14.5M-1994].* The American Society of Mechanical Engineers, New York.
ASME. (2009). *Demonstrating and tolerancing [ASME Y14.5M-2009].* The American Society of Mechanical Engineers, NY.
Bothe, D. (1997). *Measuring process capability.* McGraw-Hill, New York.

DaimlerChrysler Corporation, Ford Motor Company, and General Motors Corporation (2002). *Measurement Systems Analysis (MSA)*. 3rd ed. AIAG. Southfield, MI.
Griffith, G. K. (1996). *Statistical process control methods for long and short runs*. 2nd ed. Quality Press, Milwaukee, WI.
Hillier, F. S. (January 1969). X-bar and R chart control limits based on a small number of subgroups. *Journal of Quality Technology*. 17–26.
Kececioglu, D. (1991). *Reliability engineering handbook*. Volume 1. Prentice Hall, Englewood Cliffs, NJ.
Kececioglu, D. (1991). *Reliability engineering handbook*. Volume 2. Prentice Hall, Englewood Cliffs, NJ.
Montgomery, D. C. (1996). *Introduction to quality control*. John Wiley & Sons, New York.
Montgomery, D. C. (1991). *Design and analysis of experiments*. 3rd ed. ASQ Quality Press, Milwaukee, WI.
Pyzdek, T. (1992). *Pyzdek's guide to SPC. Vol. 2. Applications and special topics*. Quality Press, Milwaukee, WI.
Rogers, E. M. (2003). *Diffusion of innovations*. 5th ed. Free Press, New York.
SAE. (August 1999). *Reliability and maintainability guideline for manufacturing machinery and equipment*. Society of Automotive Engineers and National Center for Manufacturing Sciences, Warrendale, PA and Ann Arbor, MI.
Stamatis, D. (2003a). *Six sigma and beyond: Design for six sigma*. St. Lucie Press, Boca Raton, FL.
Stamatis, D. (2003b). *Six sigma and beyond: Statistical process control*. St. Lucie Press, Boca Raton, FL.
Stamatis, D. (2003c). *Failure mode and effect analysis: FMEA from theory to execution*. 2nd ed. ASQ Quality Press, Milwaukee, WI.
Stamatis, D. (1997). *TQM engineering handbook*. Marcel Dekker, New York.
Stamatis, D. (1998). *Implementing the TE supplement to QS-9000: The tooling and equipment supplier's handbook*. Quality Resources, New York.
Terninko, J., A. Zusman, and B. Zlotin. (1996). *Step-by-Step TRIZ: Creating Innovative Solution Concepts*, 3rd ed., Responsible Management Inc., Nottingham, NH.

Chapter 8–9: Selected Bibliography

Altshuller, G. S. (1988). *Creativity as an exact science*. Gordon and Breach, New York.
Ansell, J. I. and M. P. Philips. (1994). *Practical methods for reliability data analysis*. Oxford University Press, Cambridge, U.K.
Bar-El, Z. (May 1996). TRIZ methodology. *The Entrepreneur Network Newsletter*.
Braham, J. (October 12, 1995). Inventive Ideas Grow on TRIZ. *Machine Design*. 58.
Crowder, M. J., A. C. Kimber, R. L. Smith, and T. J. Sweeting. (1991). *Statistical analysis of reliability data*. Chapman/Hall/CRC, New York.

DaimlerChrysler Corporation, Ford Motor Company, and General Motors Corporation. (2002). *Potential failure mode and effect analysis (FMEA)*. 3rd ed. AIAG. Southfield, MI.

Kaplan, S. (1996). *An Introduction to TRIZ: The Russian theory of inventive problem solving*. Ideation International Inc., Southfield, MI.

Krivtsov, V. V. (July 7, 2006). *Practical extensions to NHPP application in repairable system reliability analysis*. Ford Motor Company, MD SCII-604, Office 1CJ32, 14555 Rotunda Drive, Dearborn, MI.

Osborne, A. (1953). *Applied imagination*. Scribner, New York.

Patterson, M. L. (1993). *Accelerating innovation: Improving the process of product development*. Van Nostrand Reinhold, New York.

Pugh, S. (1991). *Total Design—Integrated methods for successful product engineering*. Addison-Wesley, Reading, MA.

SAE. (1994-07). *Potential failure mode and effect analysis (in design FMEA) and potential failure mode and effect analysis in manufacturing and assembly processes (process FMEA) reference manual [SAE J1739]*. Society of Automotive Engineers, Warrendale, PA.

Taguchi, G. (1983). *Introduction to quality engineering*. Asian Productivity Organization, Tokyo.

Terninko, J. (1996). *Systematic innovation: Theory of inventive problem solving (TRIZ/TIPS)*. Responsible Management Inc., Nottingham, NH.

Terninko, J. (1995). *Step by step QFD: Customer-driven product design*. Responsible Management Inc., Nottingham, NH.

Terninko, J. (1996). *Introduction to TRIZ: A work book*. Responsible Management Inc., Nottingham, NH.

Terninko, J. (1989). *Robust design: Key points for world class quality*. Responsible Management Inc., Nottingham, NH.

Verduyn, D. M. (Spring 1997). Systematic innovation using TRIZ. *Automotive Excellence*. 13–14.

von Oech, R. (1983). *A whack on the side of the head*. Warner Books, New York.

Zusman, A. and Terninko, J. (June 1996). TRIZ/Ideation methodology for customer-driven innovation. *8th symposium on quality function deployment*, The QFD Institute, Novi, MI.

Chapter 10–14: Selected Bibliography

Stamatis, D. H. (2003). *Failure mode and effect analysis: FMEA from theory to execution*. 2nd ed. Revised and expanded. ASQ Quality Press, Milwaukee, WI.

Chapter 15: Selected Bibliography

Abernethy, R. (2004). *The new Weibull handbook*. 5th ed. Gulf Publishers, North Palm Beach, FL.

Dodson, B. (2006). *The Weibull analysis handbook*. 2nd ed. Quality Press, Milwaukee, WI.

Lawless, J. F. (1978). Confidence interval estimation for the Weibull and extreme value distributions. *Technometrics*. Vol. 20. pp. 355–364.

McKane, S. W., L. A. Escobar, and W. Q. Meeker (2005). Sample size and number of failure requirements for demonstration tests with log-location-scale distributions and failure censoring. *Technometrics*. Vol. 47. pp. 182–190.

Meeker, W. (1984). A comparison of accelerated life test plans for Weibull and lognormal distribution and type I censored data. *Technometrics*. Vol. 26. pp. 157–171.

Murphy, D. N. P., M. Xie, and R. Jiang (2004). *Weibull models*. John Wiley & Sons, New York.

Nelson, W. and W. Meeker. (1978). Theory for optimum accelerated censored life tests for Weibull and extreme value distributions. *Technometrics*. Vol. 20. pp. 171–177.

Pascual, F. G. and G. Montepiedra. (2005). Lognormal and Weibull accelerated life test plans under distribution misspecification, *IEEE Transactions on Reliability*. Vol. 54. pp. 43–52.

Weibull, A. (1961). *Fatigue testing and analysis of results*. Published by published for advisory group for aeronautical research and development, North Atlantic Treaty Organization, by Pergamon Press, 1961. Original from the University of Michigan. Digitized November 30, 2007.

Yang, G. (2005). Accelerated life tests at higher usage rates. *IEEE Transactions on Reliability*. Vol. 54. pp. 53–57.

Index

A

Accelerated testing, 248–250
Accessibility for maintenance, equipment design, 172–173
Affinity diagrams, 146
Allocation of reliability goals, 323
Allocation of resources, 34
AND gate symbol, 340
Applications engineering, 376
Apportioning reliability model, 324
ARIMA control charts, 83
Arrhenius model, 255–256
Attribute charts, calculation work sheet, 83
Attribute data, charts for, 82
Attributes tests, 259
 timing of, 260–278
Attributing counting, check sheet, 57
Automotive industry, 38–42
Availability of machinery/equipment, 280–282
Average cost of repair, 370
Axiomatic designs, 146

B

Benchmarking, 146
Binomial distribution, 263–265
 sequential test plan for, 267–270
Blind testing, 126
Bonus tolerance, 201
Brainstorming, 51–52, 146
 attributing counting, check sheet for, 57
 cause analysis, mistakes in, 64
 check sheets, 52–66
  cause-and-effect diagrams, 61–62
  histogram, 57–59, 158
  horizontal axis, 60
  Pareto diagram, 59–61
  vertical scale, 61
 constructing cause/effect diagram, 64–65
 data organization, 54
 designing check sheet, 54
 example, 57, 63
 measurable data, check sheet for, 55–56
 number of defects by day, check sheet, 53
 number of defects by shift, check list, 53
 part XYZ outside diameter, check sheet, 55
 procedures for, 51–52
 rules for brainstorming, 52
 steps for construction, 65
 types of cause/effect diagrams, 65–66
 types of cause/effect relationships, 63
Business unit manager, 122

C

C charts, 84
Capability analysis, to improve design, 206–208
Capacity, 32–38
 allocation of resources, 34
 effectiveness *vs.* invested effort, 35
 set specifications, 35
Catastrophic failures, 282
Cause analysis, mistakes in, 64

447

Cause-and-effect diagrams, 61–62
Changeover, 110–124
   die shoe thickness, 123
   die sizes, different, working
      with, 122–123
   extended questionnaire for machine
      failures, 111–118
   external setup, 119
   internal setup, 119
   loading large dies, 123
   locating needed tools, 123
   locating next die, 124
   moving dies in, out, 124
   quick changeover, benefits of, 119–120
   ram adjustments, 123–124
   recognizing component wear, tear, 123
Changing product flow from push to pull, 5
Charts, 70–71, 80–87
   ARIMA control charts, 83
   attribute data, 84–85
   for attribute data, 82
   C chart, 84
   control charting production
      process, 85–86
   CuScore charts, 83
   to display data, 70–71
   for individuals, 84
   interpreting control charts, 86–87
   median chart, 84
   multivariate charts, 83
   NP chart, 84
   P chart, 84
   R chart, 84
   S chart, 84
   time-weighted charts, 82
   toolwear charts, 83
   U chart, 85
   variable data, 84
   for variable data, 82
   X-Bar chart, 84
Check sheets, 56–62
   cause-and-effect diagrams, 61–62
   to collect data, 52–66
   histogram, 57–59, 158
      construction of, 58–59
      horizontal axis, 60
   for measurable data, 55–56

Pareto diagram, 59–61
   vertical scale, 61
Circle symbol, 341–342
Clamps, changing, 123
Clarify purpose for collecting data, 48–50
Classical reliability terms, 242–246
   equipment failure rates, 243–244
   equipment maintainability, 244–246
   mean time between failures, 242
      confidence intervals, 242–243
      point estimates, 242–243
Commonality of design, 156
Comparison of MFMEA/FTA, 137
Complete data, 388
Completing tolerance studies, 346
Component application review, 176–177
Component data file, 96
Component supplier failure data, 99
Components of form, 137–143
   actions taken/revised RPN, 143
   classification, 140
   current controls, 141
   date, responsible party, 143
   detection rating, 142
   failure mode, 138
   MFMEA header information, 137–138
   occurrence ratings, 141
   potential causes, 140–141
   potential effects, 138–139
   recommended actions, 142
   revised RPN, 143
   risk priority number, 142
   severity descriptions, 139
   severity ratings, 139–140
   system/subsystem name function, 138
Composite position, 201
Confidence limits, 393–394
Constant-stress testing, 250–251
Constructing cause/effect diagram, 64–65
Contaminants, performance under, 190
Contamination, 150
   environmental factor, 150
Continued operation of equipment,
      costs, 368–369
Continuous improvement activities, 95–96
Continuously review field history data, 374
Contour plot, 394

Control point analysis, 177
Converting, decommissioning equipment, 290–293
Coordinator, 120–121
Correction, example of waste, 9
Corrosive materials, 150
  environmental factor, 150
Cost of repair, 369
Coverage limitations, 219
Critical control points, 194
Current-state map, 11–13
CuScore charts, 83
Customer's equipment design responsibilities, 318–320
Customer's plant, reliability data, 347
Customer's responsibility
  building, installation of equipment, 363
  reliability growth program, 363

D

Data collection, 93–99, 224–225
  component data file, 96
  continuous improvement activities, 95–96
  direct machine monitoring, 96–99
    component supplier failure data, 99
    field history/service reports, 99
    maintenance system data, 98–99
  failure report form, 96
  machinery data file, 96
  manual recording of data, 95–96
  manual recording system, 96
  master data list, 96
  by sampling, 49
  structure of manual recording, 96
Data for first failure in ten, 413
Data organization, 54
Datum reference frame, 201
Decreasing cycle time, 5
Demonstrated overall equipment effectiveness, 43–44
Derating, 185–188
Design concept, 154–155
Design for maintenance requirements, 174
Design for parts interchangeability, standardization, 175–176

Design for removal, replacement of components, 175
Design margins, mechanical stress analysis, 375–376
Design of experiments, 146, 194
  in reliability applications, 209
Design review, 320–321, 382–383
  guidelines, reliability/maintainability, 377
Design simplicity, 155–156
Design with maintenance tools, 175
Designing against fatigue, 157–158
Designing check sheet, 54
Designing equipment
  maintainability, 315–320
  replacement, 317
Designing equipment reliability, 321
Detailed process mapping, 7–8
Determining factory requirements, 299–300
Diametral tolerance zone, 201
Diamond symbol, 340
Die shoe thickness, 123
Die sizes, different, working with, 122–123
Dimensional prove-out test, 350
Dimensional strength degradation, 147
Direct machine monitoring, 96–99
  component supplier failure data, 99
  field history/service reports, 99
  maintenance system data, 98–99
Dock-to-dock form, 33
Documentation, 300, 378
DOE. *See* Design of experiments
Dry-cycle run, 383–385
Dust, performance under, 190

E

Effectiveness *vs.* invested effort, 35
Effects analysis, 374–375
Efficiency, effectiveness, relationship, 35
Eight D process, 106–110
Electrical components for environment, 190–191
Electrical design/safety margins, 185
Electrical failure modes, 189–190
Electrical failures, preventing, 188–190
Electrical noise, 150
  environmental factor, 150

Electrical power changes, performance under, 190
Electrical reliability, 179–191
   benefits of thermal analysis, 181–183
   derating, 185–188
   electrical components for environment, 190–191
   electrical design/safety margins, 185
   electrical failure modes, 189–190
   electrical failures, preventing, 188–190
   electrical power quality, 188
   electrical stress, example, 186–188
   thermal analysis, 183–185
   thermal properties, electrical equipment, 179–181
Electrical stress, example, 186–188
Electrical stress analysis, derating, 375
Electromagnetic fields, 150
   environmental factor, 150
EMC. *See* Equipment maintenance cost
Engineering analysis, 130–131
Ensuring safety in using equipment, 286
Environment, causes of failure, 152–153
Environmental analysis before installing equipment, 150–151
Environmental conditions, 190
Environmental data sheet, 151
Environmental failure prevention, 151
Equipment conversions, evaluating, 311
Equipment corrosion prevention, 160
Equipment design
   allocation model to evaluate, 235–236
   cause of waste, 10
   variables, 165–166
Equipment lifetime distribution, 390–395
   B_X life, 394
   confidence bounds, uncertainty quantification, 394–395
   contour plot, 394
   failure rate, 394
   location parameter, estimating, 393
   mean life, 394
   Pdf plot, 394
   plots, 393–394
   probability of failure, 394
   probability plot, 394
   reliability, 393–394
   reliability *vs.* time plot, 394
   scale parameter, estimating, 391–393
   shape parameter, estimating, 391
   time plot, 394
   warranty time, 394
Equipment maintainability, designing for, 170–172
Equipment maintainability matrix, 377–378
Equipment maintenance cost, 369
Equipment qualification method, 384
Equipment reliability/maintainability, implementation of, 286–293
Equipment reliability/maintainability goals, 303–304
   documenting failure definition, goals for, 304
   environmental considerations, goals for, 304
   equipment usage, goals for, 304
   failure requirements for equipment, goals for, 304
   maintainability requirements, 304
Equipment safety margins, 160–162
Equipment supplier, machine reliability, 226–227
Equipment wear
   maintenance actions, 159
   methods to reduce, 158–159
   preventing, 158
      minimizing, 158
   recognizing, 123
   types of, 159
Error proofing, 127
Establishing system requirements, 287–288, 355–363
   accelerated testing considerations, 359–360
   automating test, 359–360
   acceleration testing for subsystems, 360
   area for test focusing, 356–357
   equipment supplier, 362–363
   existing data, 357
   increased sample size, 357–358
   integrated system testing, 360–361
   real-world conditions, 358–359
      harsh environment, 358
      imperfect operators, 359
      processing parts, 358–359

reliability growth, 362
subsystem testing, 361
test progress sheets, 361–362
testing guidelines, 361
Existing machinery, 105–143
advisor, 121
business unit manager, 122
changeover, 122–124
  changing clamps, 123
  die shoe thickness, 123
  die sizes, 122–123
  loading large dies, 123
  locating needed tools, 123
  locating next die, 124
  moving dies in, out, 124
  ram adjustments, 123–124
  recognizing component wear, tear, 123
changeover to improve process flexibility, 110–124
  extended questionnaire for machine failures, 111–118
  external setup, 119
  internal setup, 119
  quick changeover, benefits of, 119–120
comparison of MFMEA/FTA, 137
components of form, 137–143
  actions taken/revised RPN, 143
  classification, 140
  current controls, 141
  date, responsible party, 143
  detection rating, 142
  failure mode, 138
  MFMEA header information, 137–138
  occurrence ratings, 141
  potential causes, 140–141
  potential effects, 138–139
  recommended actions, 142
  revised RPN, 143
  risk priority number, 142
  severity descriptions, 139
  severity ratings, 139–140
  system/subsystem name function, 138
coordinator, 120–121
8D process, 108–109
effect analysis, 134–143

eight-discipline process, machinery failure resolution, 106–110
failure mode, 134–143
  effect analysis, 131
fault tree analysis, causes of machine failures, 131–134
fault tree analysis steps, 133
fault tree analysis symbols, 134–135
finite element analysis, product refinement, 128–131
  engineering analysis using FEA, 130–131
  fatigue analysis, 131
  heat transfer analysis, 131
  results of finite element analysis, 131
  structural analysis, 130
  vibrational analysis, 130
maintenance superintendent, 122
maintenance supervisor, 121
manufacturing planning specialist, 121
manufacturing tech, 120
mistake proofing, 124–127
  blind testing, 126
  different criteria, 126
  error proofing, 127
  human errors, minimizing, 126–127
  mechanical screens, 126
  opportunities, prioritizing, 127
  principles, 125–126
P-M analysis, 127–128
Pareto analysis, 106
plant manager, 122
production superintendent, 122
8D Generic Report Form, 107
Exponential distribution, 271–273
  fixed sample test, 270–271
  sequential test plans, 273–277
External environment, 147

F

Factory requirements, 149–150
  duty cycle operating patterns, 149
  jobs per hour, 149
  maintenance, 150
  management, 150
  operator attention, 149
  required quality levels, 149

Failure/causal event symbol, 340
Failure data
    feedback process flowchart, 228
    tracking spare part utilization, 178
Failure data modeling, 395–424
    characteristic life, 403–404
    confidence limits, 393–394, 409–424
        data for first failure in ten, 413
        graphical analysis, 408–409
        hazard functions, 397–401
        left-censored data, 388
        new increment method, 410–411
        plots, 393–394
        sudden death data, 412
        sudden death testing, 411–424
        suspended item analysis, 409–410
    example, 399
    fan data, 400
    graphical estimation of Weilbull
        parameters, 404–405
    median rank tables, 407–408
    plotting positions, 406
    probability density function
        model, 395–397
    rate of failure, 397–401
        hazard functions, 397–401
    three-parameter Weilbull, 401–402
    two-parameter Weilbull, 402–403
    Weilbull probability density
        function, 401–409
    Weilbull slope, 405–406
    Weilbull slope *vs.* failure
        mechanism, 404
Failure list for pump, 153
Failure mode, 374–375
    effect analysis, 131
Failure mode and effect
        analysis, 134–143, 324–339
    discarding, 327
    form to analyze equipment
        failures, 327–339
    preparation, 325–326
    timing of completion, 326
    timing of initiation, 326
    updating of EFMEA, 326
Failure rates, 285–286, 394
    vs. time plot, 394

Failure reports, 97
    "bucket of parts" system, 177–178
    corrective action system, 100–103
    form, 96
Failure-truncated test plans, fixed
        sample test, 270–271
Failures of equipment of
        machinery, 282–285
Fatigue
    fracture, designing equipment
        to prevent, 157
    maintenance action against, 158
Fatigue analysis, 131
Fault tolerance design, 316–317
Fault tree analysis, 342–343
    diagrams, 339–343
    machine failure, 131–134
    steps, 133
    symbols, 134–135
FEA. *See* Finite element analysis
Field history
    report process, 99
    service reports, 99
Finite element analysis, 128–131
    engineering analysis, 130–131
    fatigue analysis, 131
    heat transfer analysis, 131
    results of finite element analysis, 131
    structural analysis, 130
    vibrational analysis, 130
First run percentage, 311–313
5S methodology of organizing
        work area, 15–18
    benefits of, 17–18
    organization, 17
    sorting, 16
    standardizing, 17
    straighten out and set in order, 16
    sustaining, 17
    sweeping, 16–17
    systematic cleaning, 16–17
5-Why analysis, 75–77
    ARIMA control charts, 83
    attribute data, 84–85
    charts, 80–87
        for attribute data, 82
        C chart, 84

control charting production
  process, 85–86
CuScore charts, 83
example, 75
for individuals, 84
interpreting control charts, 86–87
kaizen, 79–80
kanban, 78
median chart, 84
multivariate charts, 83
NP chart, 84
P chart, 84
process flow diagrams, 77–78
R chart, 84
S chart, 84
time-weighted charts, 82
toolwear charts, 83
U chart, 85
variable data, 84
for variable data, 82
X-Bar chart, 84
Fixed labor costs, 369
Fixed sample tests, 259, 270–271
FMEA. *See* Failure mode and
  effect analysis
Force-field analysis, 146
Ford's Accelerated Stress Test, 211–213
Ford's Production Accelerated Stress
  Screen, 213–215
FTA. *See* Fault tree analysis
Functional specification/gauging, 201–202

G

Gauge accuracy, 205
Geometric dimensioning, 146
  tolerancing, 200–202

H

HALT, HASS test processes to improve
  machine design, 210–213
Handling, maintenance, 174
Hazard analysis, 194
  critical control points, 210–215
    Ford's Accelerated Stress
      Test, 211–213

Ford's Production Accelerated
  Stress Screen, 213–215
HALT, HASS test processes to
  improve machine design, 210–213
Hazard functions, 397–401
Heat, performance under, 190
Heat transfer analysis, 131
Histograms, 58–59
  to chart data, 57–58
  construction of, 58–59
  use of, 158
Horizontal axis, 60
Humidity, 150
  environmental factor, 150
  performance under, 190
Hypergeometric distribution, 260–262

I

Ideal cycle time, 31
Identifying subsystem tree model, 324
Immersion, 150
  environmental factor, 150
Impact of life-cycle costs, 370–373
  documenting LCC process, 372–373
  LCC, improving, 372
Improved equipment maintainability,
  benefits of, 279–280
Inadequate training, cause of waste, 10
Incapable processes, cause of waste, 10
Ineffective production planning,
  cause of waste, 10
Infant mortality period, failures
  during, 282–283
Input requirements, 147–149
  control factors, 148
  equipment under design, 147–148
  failure states of equipment, 148–149
  ideal function of equipment, 148
Interchangeability,
  standardization, 317–318
Interfacing parts reliability, design for, 316
Interference, 211
Interference testing, 278
Internal environment, 147
Internal setup, 119
Internal temperatures, 147

Inventories beyond absolute minimum, 3
Inventory, example of waste, 9
Inverse power law model, 253–255
Is-is not analysis, data comparison, 71–75
    example, 71
Is-is not matrix, 73
ISO-capacity, evaluating, 305

J

Japanese Institute of Plant Engineers, 2
JIPE. *See* Japanese Institute of
    Plant Engineers

K

Kaizen, to improve equipment
    effectiveness, 79–80
Kanban, to collect, analyze equipment
    repair data, 78
Kano model, 146

L

Layout, cause of waste, 9
LCC. *See* Life-cycle costs
Lean manufacturing, 5–15
    changing product flow from push
        to pull, 5
    current-state map, 11–13
    decreasing cycle time, 5
    detailed process mapping, 7–8
    future-state map, 11–13
    lean organizations, 6–7
    map value stream of process, 10–11
    mapping processes, 7–8
    problems with lean, 15
    process mapping, 13–15
    value stream mapping, 7–10
    waste
        causes of, 9–10
        examples of, 9
Lean organizations, 6–7
Left-censored data, 388
Length dimension, spindle speed,
    scatter diagram, 68
Levels of innovation, 195

Life-cycle costs, 286, 365–367
    model, 374
Life data analysis, 388
    complete data, 388
    interval data, 388
    suspended or right-censored data, 388
Life of equipment, 369–370
Linearity, 205
Loading large dies, 123
Locating needed tools, 123
Locating next die, 124

M

Machine failure mode effects analysis, 146
Machinery data file, 96
Machinery FMEA, 375
Machinery parts, selection, 346
Machines relating to reliability growth, 225
Maintainability in design, benefits, 171
Maintenance, 279–293
    availability of machinery/
        equipment, 280–282
    benefits of improved equipment
        maintainability, 279–280
    building, installing for reliability/
        maintainability, 289–290
    converting, decommissioning
        equipment, 290–293
    developing, designing for reliability/
        maintainability, 288–289
    ensuring safety in using equipment, 286
    equipment reliability/maintainability,
        implementation, 286–293
    establishing system requirements
        for equipment reliability/
        maintainability, 287–288
    failure rates, 285–286
    failures of equipment of
        machinery, 282–285
    infant mortality period, failures
        during, 282–283
    life-cycle costs, 286
    operation, support of reliability/
        maintainability, 290
    useful life period, failures during, 283–284
    wearout period, failures during, 284–285

Maintenance superintendent, 122
Maintenance supervisor, 121
Maintenance system activities, 98
Maintenance system data, 98–99
Maintenance tools, equipment, 317
Manual recording of data, 95–96
Manual recording system, 96
Manufacturing planning specialist, 121
Manufacturing tech, 120
Map value stream of process, 10–11
Mapping processes, 7–8
Master data list, 96
Material conditions, 202
Material selection, 170
Matrix, reliability/maintainability, 378–380
Mean time between events
   example, 92
   measurement, 91–92
Mean time between failures
   measurement, 92–93
   parameters, 306–307
Mean time to repair, replace, 307
Measurement system analysis, 194, 204–206
Mechanical failure modes, 168
Mechanical reliability, 185–178
   accessibility for maintenance, 172–173
   benefits of maintainability in design, 171
   commonality of design, 156
   component application review, 176–177
   control point analysis, 177
   design, maintenance requirements, 174
   design concept, 154–155
   design for interchangeability, standardization, 175–176
   design for removal of components, 175
   design simplicity, 155–156
   design with maintenance tools, 175
   designing against fatigue, 157–158
   duty cycle operating patterns, 149
   environment, causes of failure, 152–153
   environmental analysis, 150–151
   environmental data sheet, 151
   environmental failures, 151
   equipment corrosion, preventing, maintaining against, 160
   equipment design variables, 146–178
   equipment failures, 151
   equipment maintainability, designing, 170–172
   equipment safety margins, 160–162
   equipment under design, 147–148
   equipment wear, preventing, minimizing, 158
   factory requirements, 149–150
   failure data, tracking spare part utilization, 178
   failure list for pump, 153
   failure reports, "bucket of parts" system, 177–178
   failure states of equipment, 148–149
   fatigue, 157–158
   handling, maintenance, 174
   ideal function of equipment, 148
   input requirements, 147–149
   jobs per hour, 149
   maintenance, 150
   management, 150
   material selection, 170
   mechanical failure modes, 168
   mechanical reliability, understanding mechanical failures, 156–178
   normal stress/strength relationship, 163–167
   operator attention, 149
   reliability/maintainability validation/ verification techniques, 176
   required quality levels, 149
   review working environment, 176
   stress analysis, 162
   stress/strength process, 167–169
   stress/strength relationship, 163
   stresses and bathtub curve, 162
   types of wear found in equipment, 159
   understanding mechanical failures, 156–178
   visual factory, 172
   wear, 158–159
   Z table, normal distribution, 165–166
Mechanical screens, 126
Mechanical shock, 150
   environmental factor, 150
Mechanical stress analysis, to check design margins, 375–376
Median chart, 84
Methodologies, 193–221

MFMEA. *See* Machine failure mode
    effects analysis
Mistake proofing to prevent
    or mitigate errors, 124–127
  human errors, minimizing, 126–127
    blind testing, 126
    different criteria, 126
    error proofing, 127
    mechanical screens, 126
  mistake proofing, 126
  mistake-proofing principles, 125–126
  prioritizing, 127
Monitoring time, 27–32
  cost of quality, 28–29
  equipment availability, 29–30
    example, 30
  equipment downtime, 27–28
  equipment performance, 31–32
  equipment setup time, 30
  equipment uptime, 27
  ideal cycle time, 28, 31
  identify production losses, 30–31
  planned downtime, 31
  process line considerations, 29
  process line quality issues, 29
  quantifying total defects, 28–29
  total available time, 31
  unplanned downtime, 31
Motions of employees, 3
Moving dies in, out, 124
MPS. *See* Manufacturing planning specialist
MSA. *See* Measurement system analysis
MTBE. *See* Mean time between events
MTBF. *See* Mean time between failures
Multivariate charts, 83

N

Negative correlation, scatter diagrams, 70
New increment method, 410–411
New machinery, items of concern
    for, 373–385
No correlation, scatter diagrams, 69
Nonrecurring costs for acquiring
    equipment, 367–370
  continued operation of equipment,
    costs, 368–369

equipment maintenance cost, 369
scheduled equipment maintenance
    costs, 369
unscheduled equipment maintenance
    costs, 369–370
Normal distribution, 273
  Z table, 165–166
Normal stress/strength relationship, 163–167
NP chart, 84
Number of assists, 309–311
Number of defects by day, check sheet, 53
Number of defects by shift, check list, 53

O

Objectives for reliability testing, 241
OEE. *See* Overall equipment effectiveness
Operation, support of equipment reliability/
    maintainability, 290
OR gate symbol, 340
Overall equipment effectiveness, 21–45
  automotive industry, 38–42
  capacity, 32–38
    allocation of resources, 34
    effectiveness *vs.* invested effort, 35
    set specifications, 35
  demonstrated overall equipment
    effectiveness, 43–44
  example, 24
  metrics of, 22–28
    availability, 25
    loading portion of TEEP metric, 25
    overall equipment effectiveness, 24
    performance, 26
    quality, 26
    total effective equipment
        performance, 24–25
  monitoring time, 27–32
    cost of quality, 28–29
    equipment availability, 29–30
    equipment downtime, 27–28
    equipment performance, 31–32
    equipment setup time, 30
    equipment uptime, 27
    ideal cycle time, 28, 31
    identify production losses, 30–31
    planned downtime, 31

process line considerations, 29
process line quality issues, 29
quantifying total defects, 28–29
total available time, 31
unplanned downtime, 31
reducing required overall equipment effectiveness, 44–45
selection process, 42–44
Overprocessing
example of waste, 9
of parts, 3–4
Overproduction
ahead of demand, 3
example of waste, 9

P

P chart, 84
P-M analysis, 127–128
Parallel model to evaluate equipment design, 238–239
Parallel reliability models, 321–322
Pareto analysis, 106
Pareto diagrams, 60, 106
to chart data, 59
creation, 60–61
use, 59–60
Partial median rank table, 249
Performance, 26
example, 26
Performance data feedback
plan, 228–231
benefits of, 231
responsibilities matrix, 229–230
Performance measures, cause of waste, 10
Piece-to-piece variation, 147
Planned downtime, 31
Plant manager, 122
Plots, 393–394
Point estimate for reliability, 90
Poisson distribution, 265–266
Poka-yoke, 4
Poor maintenance, cause of waste, 10
Poor work methods, cause of waste, 10
Poor workplace organization, cause of waste, 10
Positive correlation, scatter diagrams, 68

Positive relationship, scatter diagram, 69
Possible negative correlation, scatter diagrams, 69–70
Possible positive correlation, scatter diagrams, 68–69
Preliminary process capability study, 350
Pressure or vacuum, 150
environmental factor, 150
Preventive maintenance schedule, 369
Probability of failure given time, 394
Probability plot, 394
Process control, 48
Process flow diagrams to visualize processes, 77–78
example, 78
Process mapping, 7, 13–15
Product design, cause of waste, 10
Production of defective parts, 4
Production superintendent, 122
Production time, 305–306
Progressive-stress testing, 252–253
Projected tolerance zone, 202
Pugh concept, 146
Pugh diagrams to improve machine design, 199–200
Pugh selection matrix, 200

Q

Qualitative tools, 203
Quality function deployment, 146
Quantitative tools, 203

R

R chart, 84
Ram adjustments, 123–124
Rate of failure, hazard functions, 397–401
Rationale for data collection, 50
Reading scatter diagrams, 67–68
Recognizing component wear, tear, 123
Reducing required overall equipment effectiveness, 44–45
Reliability approach, 89–103
data collection to monitor equipment performance, 93–99
component data file, 96

continuous improvement
  activities, 95–96
  direct machine monitoring, 96–99
  failure report form, 96
  field history/service reports, 99
  machinery data file, 96
  maintenance system data, 98–99
  manual recording of data, 95–96
  manual recording system, 96
  master data list, 96
  structure of manual recording, 96
failure reporting, 100–103
mean time between events
  example, 92
  measurement, 91–92
mean time between failures,
  measurement, 92–93
reliability point measurement, 90–91
  example, 91
Reliability demonstration tests, 258–260
Reliability given time, 393–394
Reliability growth, 223–278
  attributes tests, 259
    timing of, 260–278
  benefits of performance data
    feedback plan, 231
  binomial distribution, 263–265
    sequential test plan for, 267–270
  classical reliability terms, 242–246
    confidence intervals, 242–243
    equipment failure rates, 243–244
    equipment maintainability, 244–246
    mean time between failures, 242
    point estimates, 242–243
  customer's responsibility, 227
  data collection, 224–225
  equipment design, allocation
    model, 235–236
  equipment supplier, machine
    reliability, 226–227
  exponential distribution, 271–273
    sequential test plans, 273–277
  failure data feedback process
    flowchart, 228
  failure-truncated test plans, fixed
    sample test, 270–271
  fixed-sample tests, 259

hypergeometric distribution, 260–262
interference testing, 278
machines relating to reliability
  growth, 225
normal distribution, 273
objectives for reliability testing, 241
parallel model to evaluate equipment
  design, 238–239
performance data feedback plan, 228–231
performance data feedback
  responsibilities matrix, 229–230
Poisson distribution, 265–266
program, 227–234
reliability demonstration test, 258–260
  characteristics, 257–278
reliability/maintainability feedback
  process, 227–228
reliability/maintainability
  information systems, 231
reliability tests
  rationale for, 239–241
  timing of, 240–241
reports, 231–234
results of sequential failures, 276
sample sizes at point stress, 277
sequential test plans, 273
sequential tests, 259–260
series model to evaluate equipment
  design, 237–238
slope interpretation, 226
success testing, 266
test methods, 260
tests for defining failures, 246–256
  accelerated testing, 248–250
  Arrhenius model, 255–256
  constant-stress testing, 250–251
  inverse power law model, 253–255
  partial median rank table, 249
  progressive-stress testing, 252–253
  step-stress testing, 251–252
  sudden death testing, 246–248
TGR/TGW report, 233–234
variables tests, 259
Weibull distribution, 273
Reliability/maintainability, 295–313
  activities list, 381–382
  analyze performance, 301–303

checklist, 380–382
current equipment baseline, 300
determining factory requirements, 299–300
environmental specifications, 300–301
equipment reliability/maintainability
    goals, 303–304
    documenting failure definition,
        goals for, 304
    environmental considerations,
        goals for, 304
    equipment usage, goals for, 304
    failure requirements for equipment,
        goals for, 304
    maintainability requirements, 304
equipment specifications, 299–303
improving equipment
        performance, 305–313
    equipment conversions, evaluating, 311
    first run percentage, 311–313
    ISO-capacity, evaluating, 305
    mean time between failures
        parameters, 306–307
    mean time to repair, replace, 307
    number of assists, 309–311
    production time, 305–306
    scheduled downtime, 307–309
    starved time, 309
information systems, 231
information to determine fit, 300
performance requirements, 300
specification matrix, 297–298
specs for, 295–298
testing, 377
validation/verification techniques, 176
Reliability point measurement, 90–91
    example, 91
Reliability qualification testing, 346–347,
        351–355
    MTBF value, 351–353
    partial chi-squared table, 352
    RAMGT software, 355
    RQT assumptions, 354
    RQT test plan, 354–355
    validating MTBF parameters, 353–354
Reliability tests
    rationale for, 239–241
    timing of, 240–241

Reliability *vs.* time plot, 394
Repeatability, 205
Replacement, 317
Reporting, reliability/maintainability, 374
Reports on things gone right/things
        gone wrong, 231–234
Reproducibility, 205
Results assessment, 377
Results of finite element analysis, 131
Results of sequential failures, 276
Review working environment, 176
Right-censored data, 388
R&M. *See* Reliability/maintainability
Roadblocks, testing, eliminating, 347–348
    establish confidence through
        design, 348
    reliability testing, 348
    simultaneous engineering, 348
    test time, reducing, 348
    testing everything, 348
Robot arm, 162
ROCOF analysis
    to improve machine design, 215–221
        benefits of using, 218
        cautions for using with warranty
            data, 218–221
Root cause/failure analysis of
        equipment, 347
RQT. *See* Reliability qualification testing
Rules for brainstorming, 52
Runoff assessment techniques, 383

S

S chart, 84
Sample sizes at point stress, 277
Sampling, 50
    defined, 49–50
Scatter diagram
    data, 66–70
        length dimension, spindle speed, 68
        negative correlation, scatter diagrams, 70
        no correlation, scatter diagrams, 69
        positive correlation, scatter diagrams, 68
        positive relationship, 69
        possible negative correlation,
            scatter diagrams, 69–70

possible positive correlation, scatter diagrams, 68–69
reading scatter diagrams, 67–68
Scheduled downtime, 307–309
Scheduled equipment maintenance costs, 369
Scheduling, cause of waste, 10
Seiketsu, 17
Seiri, 16
Seiso, 16–17
Seiton, 16
Selection process, 42–44
Sequential tests, 259–260
  plans, 273
Series model to evaluate equipment design, 237–238
Series reliability models, 322–324
Setup time, cause of waste, 9
Shitsuke, 17
Shock, performance under, 190
Short-run statistical process control, 194, 202–204
  clarifying misunderstandings of SPC, 203
  using SPC for short-run production cycles, 204
Slope interpretation, 226
Specification matrix, 297–298, 379–380
Specs for reliability/maintainability, 295–298
Spindle speed, length dimension, scatter diagram, 68
Starved time, 309
Statistical control, 219
Statistical distributions, equipment lifetimes, 388–389
Step-stress testing, 251–252
Steps for construction, 65
Stress analysis, 162, 346
Stress levels, effects on mechanical failure rates, 163
Stress/strength process, 167–169
Stress/strength relationship, 163
Stresses, bathtub curve, 162
Structural analysis, 130
Structure of manual recording, 96
Sudden death data, 412
Sudden death testing, 246–248, 411–424

Supplier equipment design responsibilities, 320
Supplier quality, cause of waste, 10
Supplier's facility, reliability data, 347
Suspended data, 388
Suspended item analysis, 409–410
Systematic cleaning, 16–17

T

Takt time, 19
TEEP. *See* Total effective equipment performance
Temperature, 150
  environmental factor, 150
Test for overall equipment effectiveness, 350–351
Testing programs, 349–355
  24-hour dry cycle run, 349
Tests for defining failures, 246–256
  accelerated testing, 248–250
  Arrhenius model, 255–256
  constant-stress testing, 250–251
  inverse power law model, 253–255
  partial median rank table, 249
  progressive-stress testing, 252–253
  step-stress testing, 251–252
  sudden death testing, 246–248
TGR/TGW report, 233–234
8D Generic Report Form, 107
Theory of inventive problem solving, 194
Thermal analysis, 183–185, 376
  benefits of, 181–183
Thermal properties of electrical equipment, 179–181
Time-weighted charts, 82
Toolwear charts, 83
Total available time, 31
Total effective equipment performance, 24–25
  example, 24–25
  metric, leading portion, 25
    example, 25
Total preventive maintenance, 1–20
  aim of productive maintenance, 2
  5S methodology, 15–18
    benefits of, 17–18

in organization, 17
sorting, 16
standardizing, 17
straighten out, set in order, 16
sustaining, 17
sweeping, 16–17
systematic cleaning, 16–17
goals of, 3–5
history of, 2
inventories beyond absolute
minimum, 3
Japanese Institute of Plant Engineers, 2
lean manufacturing, 5–15
changing product flow from
push to pull, 5
current-state map, 11–13
decreasing cycle time, 5
detailed process mapping, 7–8
future-state map, 11–13
lean organizations, 6–7
map value stream of process, 10–11
mapping processes, 7–8
problems with lean, 15
process mapping, 13–15
value stream mapping, 7–10
waste, 9–10
motions of employees, 3
overall equipment effectiveness, 1
overprocessing of parts, 3–4
overproduction ahead of demand, 3
poka-yoke, 4
preventive maintenance, 2
production of defective parts, 4
takt time, 19
unnecessary transport of materials, 3
visual factory requires
standardization, 19–20
visual factory to make decisions
quickly, 18–19
waiting for next process step, 3
Traditional simple approach, 47–87
ARIMA control charts, 83
charts, 70–71, 80–87
for attribute data, 82
C chart, 84
control charting production
process, 85–86

CuScore charts, 83
for individuals, 84
interpreting control charts, 86–87
median chart, 84
multivariate charts, 83
NP chart, 84
P chart, 84
R chart, 84
S chart, 84
time-weighted charts, 82
toolwear charts, 83
U chart, 85
for variable data, 82
X-Bar chart, 84
data about overall equipment
effectiveness, 47–50
analysis, 48
clarify purpose for collecting
data, 48–50
data collection by sampling, 49
example, 48–49
process control, 48
rationale for data collection, 50
regulation, 48
sampling, 49–50
tools for evaluating data, 51–87
attribute data, 84–85
brainstorming to gather ideas, 51–52
categories, 66
cause analysis, mistakes in, 64
cause-and-effect diagrams, 61–62
check sheets, 52–66
constructing cause/effect
diagram, 64–65
data organization, 54
designing check sheet, 54
example, 57, 63
5-Why analysis, 75–77
histograms, 57–59, l58
horizontal axis, 60
is-is not analysis, 71–75
is-is not matrix, 73
kaizen, 79–80
kanban, 78
number of defects by shift,
check list, 53
Pareto diagram, 59–61

process flow diagrams to visualize
processes, 77–78
rules for brainstorming, 52
scatter diagrams, 66–70
steps for construction, 65
types of cause/effect diagrams, 65–66
types of cause/effect relationships, 63
variable data, 84
vertical scale, 61
Transportation, example of waste, 9
Triple 5-Why analysis worksheet, 76
TRIZ method, 146, theory of inventive problem solving
  core tool use, TRIZ methodology, 195–196
  design innovation, steps in using, 199
  engineering parameters, 197
  innovation level, 195
  steps, 196
  theory of inventive problem solving, 194–199
  TRIZ method in design situation, 196–199
Types of cause/effect relationships, 63
Types of wear found in equipment, 159

U

U chart, 85
Ultraviolet radiation, 150
  environmental factor, 150
Universal tag, 103
Unnecessary transport of materials, 3
Unplanned downtime, 31
Unscheduled breakdowns, 370
Unscheduled equipment maintenance costs, 369–370
Useful life period, failures during, 283–284
Utility services, 150
  environmental factor, 150

V

Value stream mapping, 7–10
Variable data, charts for, 82
Variables
  no correlation between, scatter diagram, 70
  possible positive correlation between, scatter diagram, 69
  tests, 259
Vertical scale, 61
Vibration, 150
  environmental factor, 150
  measurements test, 349–350
  performance under, 190
Vibrational analysis, 130
Virtual condition, 202
Visual factory, 172
  to make decisions quickly, 18–19
  standardization, 19–20

W

Warranty time, 394
Waste
  causes of, 9–10
  examples of, 9
Wear
  maintenance actions, 159
  methods to reduce, 158–159
  minimizing, 158
  preventing, 158
  recognizing, 123
  types of, 159
Wear-out failures, 282
Wearout period, failures during, 284–285
Weibull distribution, 273, 387–424
  equipment lifetime distribution, 390–395
    B_X life, 394
    confidence bounds to quantify uncertainty, 394–395
    contour plot, 394
    failure rate, 394
    failure rate *vs.* time plot, 394
    location parameter, estimating, 393
    mean life, 394
    Pdf plot, 394
    plots, 393–394
    probability of failure given time, 394
    probability plot, 394
    reliability given time, 393–394
    reliability *vs.* time plot, 394
    scale parameter, estimating, 391–393

shape parameter, estimating, 391
warranty time, 394
failure data modeling, 395–424
    characteristic life, 403–404
    confidence limits, 388, 393–394, 397–401, 409–424
    confidence limits for graphical analysis, 408–409
    example, 399
    fan data, 400
    graphical estimation of Weilbull parameters, 404–405
    median rank tables, 407–408
    plotting positions, 406
    probability density function model, 395–397
    rate of failure, 397–401
    three-parameter Weilbull, 401–402
    two-parameter Weilbull, 402–403
    Weilbull probability density function, 401–409
    Weilbull slope, 405–406
    Weilbull slope *vs.* failure mechanism, 404
life data analysis, 388
    complete data, 388
    interval data, 388
    suspended or right-censored data, 388
    statistical distributions, equipment lifetimes, 388–389

X

X-and-y-axes scatter diagram, 67
X-Bar chart, 84

Z

Z table, 165–166
Zero tolerance, 202
Zone control, 87

About the Author

D. H. Stamatis, Ph.D., AS QC-Fellow, CQE, CMfgE, MSSBB, ISO 9000 Lead Assessor (graduate), is the president of Contemporary Consultants Co. in Southgate, Michigan.

He is a specialist in management consulting, organizational development, and quality science. He has taught project management, operations management, logistics, mathematical modeling, and statistics for both graduate and undergraduate levels at Central Michigan University, the University of Michigan, and Florida Institute of Technology.

With over 30 years of experience in management, quality training, and consulting, Dr. Stamatis has served numerous private sector industries in fields including, but not limited to, steel, automotive, general manufacturing, tooling, electronics, plastics, food, pharmaceutical, chemical, printing, healthcare, and medical devices as well as the U.S. Navy and Department of Defense.

He has consulted for such companies as Ford Motor Co., Federal Mogul, GKN, Siemens, Bosch, SunMicrosystems, Hewlett-Packard, GM-Hydromatic, Motorola, IBM, Dell, Texas Instruments Sandoz, Dawn Foods, Dow Corning Wright, BP Petroleum, Bronx North Central Hospital, Mill Print, St. Claire Hospital, Tokheim, Jabill, Koyoto, SONY, ICM/Krebsoge, Progressive Insurance, B. F. Goodrich, and ORMET, to name just a few.

Dr. Stamatis has created, presented, and implemented quality programs with a focus on total quality management, statistical process control (both normal and short run), design of experiments (both classical and Taguchi), Six Sigma (DMAIC and DFSS), Lean manufacturing/service quality function deployment, failure mode and effects analysis, value engineering, supplier certification, audits, reliability and maintainability, cost of quality, quality planning, ISO 9000, QS-9000, ISO/TS 16949, and TE 9000 series.

Also, he has created, presented, and implemented programs on project management, strategic planning, teams, self-directed teams, facilitating, leadership, benchmarking, and customer service.

He is a Certified Quality Engineer through the American Society of Quality Control, Certified Manufacturing Engineer through the Society of Manufacturing Engineers, Certified Master Black Belt through IABLS, Inc., and a graduate of BSI's ISO 9000 Lead Assessor Training Program.

Dr. Stamatis has written over 70 articles, presented many speeches, and has participated in both national and international conferences on quality. He is a contributing author on several books and the sole author of 30 books. His consulting extends across the United States, South East Asia, Japan, China, India, and Europe. In addition, he has performed over 100 automotive-related audits, 25 preassessment ISO 9000 audits, and has helped several companies attain certification, including Rockwell International–Switching Division (ISO 9001), Transamerica Leasing (ISO 9002), and Detroit Electro Plate (QS-9000).

Dr. Stamatis received his B.S. and B.A. degrees in marketing from Wayne State University, his master's degree from Central Michigan University, and his Ph.D. degree from Wayne State University in instructional technology and business/statistics.

He is an active member of the Detroit Engineering Society, the American Society for Training and Development, an executive member of the American Marketing Association, a member of the American Research Association, and a fellow of the American Society for Quality Control.